MATHEMATICAL
INEQUALITIES

MATHEMATICAL INEQUALITIES

Pietro Cerone
Sever S. Dragomir

CRC Press
Taylor & Francis Group
Boca Raton London New York

CRC Press is an imprint of the
Taylor & Francis Group, an **informa** business
A CHAPMAN & HALL BOOK

CRC Press
Taylor & Francis Group
6000 Broken Sound Parkway NW, Suite 300
Boca Raton, FL 33487-2742

First issued in paperback 2019

ISBN-13: 978-0-367-38327-5

Library of Congress Cataloging-in-Publication Data

Cerone, Pietro.
  Mathematical inequalities : a perspective / Pietro Cerone, Sever S. Dragomir.
      p. cm.
  Includes bibliographical references and index.

  1. Inequalities (Mathematics) I. Dragomir, Sever Silvestru. II. Title.

  QA295.C37 2010
  515'.26--dc22

2010043385

Contents

Preface

An investigation of the Mathematical Reviews database (MathSciNet) reveals that many new mathematical inequalities are discovered every year, some for their intrinsic interest while many others flow from results obtained in various branches of mathematics. A more thorough investigation demonstrates that the growth is exponential since the output is doubling every 10 years or so. The study of inequalities thus reflects a general tendency in modern and contemporary mathematics which may be explained by its intimate connection with the many and various aspects of the discipline and with its applications in science.

There are now numerous applications of inequalities in a wide variety of fields, from mathematical physics and biology to information theory and economics. It is also clear that the impact in applications will grow even more spectacularly in the future due to the wide accessibility of the new results to various scientists who use the Internet as a primary source of information.

In a tableau dominated by numerous monographs devoted to various types of mathematical inequalities, this book endeavours to give the reader a different perspective of the field, which is personalised by the preferences of the authors and the research work they have conducted over the last 10 years. The emphasis here is not only in presenting a number of classical and recent results for both experts and general scientists, but also in presenting a large number of new connections and intimate relationships between various inequalities that have not previously been provided by other authors. This is the main underlying emphasis in providing the many remarks and comments following each section devoted to specific inequalities. They can be used by the reader as inspiration for starting their own research in inequalities or as a basis for finding interesting applications in other fields.

The monograph is partitioned into three parts. The first part consists of but one chapter dealing with some of the most important inequalities for real or complex numbers and sequences in analysis. These include: the Abel inequality; the Cauchy-Bunyakovsky-Schwarz (CBS) inequality; the De Bruijn inequality; the Čebyšev inequality for synchronous sequences; an inequality of Biernacki, Pidek, and Ryll-Nardzewski and its weighted form due to Andrica-Badea refining the Grüss inequality; and the Daykin-Eliezer-Carlitz inequality, which together with Wagner's inequality provide interesting and different generalisations for the CBS inequality. Classical reverses for the CBS inequality due to Pólya-Szegö and Cassels are also addressed. Some fresh insight on the celebrated Hölder and Minkowski inequality for sequences of real numbers and various inequalities for convex functions including the Jensen, Slater, and Petrović inequalities are described. All the results considered are complemented by numerous innovative and evocative remarks and comments, which reveal a network of connections between the above inequalities and create a natural background for research in the related fields or in applications in probability theory and statistics; information theory; coding and guessing; population dynamics; and so forth.

The second part consists of Chapters 2 through 4, which are devoted to integral inequalities. These have a prominent place in the construction of the book due to the fact that integral inequalities play a fundamental role in contemporary mathematical inequalities theory and their applications. A substantial portion of this section is shared between the Ostrowski, Grüss, trapezoidal, and Hermite-Hadamard type inequalities. This is motivated by an unprecedented growth in the last decade related to these inequalities which have grown many times and in different directions including the Riemann-Stieltjes integral, multivariate functions, n-time differentiable mappings, and so forth, in which the authors themselves have played a leading role. The Karamata inequality, Young's inequality, and Steffensen's inequality are considered in different settings and degrees of generality. The relationships between Ostrowski, trapezoidal, and Čebyšev functionals are explored. Many comments and hints for further investigation and applications are added to all the expounded results. Complete references are also given.

Finally, the third section consisting of Chapters 5 and 6, is devoted to a number of fundamental inequalities holding in Hilbert and Banach spaces. Recent results related to the Schwarz, Bessel, Boas-Bellman, Bombieri, Kurepa, Precupanu, Dunkl-Williams, Grüss, and Buzano inequalities are surveyed. Reverses of the Schwarz and the triangle inequalities in both the discrete and continuous forms in Hilbert and Banach spaces are discussed. Generalisations of the Hermite-Hadamard inequalities for isotonic linear functionals or isotonic sublinear functionals are also given. Numerous connections and comments are made to enable the reader to ask further questions leading to new discoveries.

This book is written in a language and with sufficient detail that will enable not only an expert in the area to understand and appreciate, but also make it accessible to undergraduate and graduate students. Any person with an interest in mathematical analysis in general should find this book within their comprehension, although to fully appreciate all the topics covered, knowledge of calculus, elementary real analysis and elementary functional analysis is required.

The authors would like to express their sincere gratitude to Eder Kikianty for her detailed reading and comments that helped to improve the work.

The Authors
Melbourne, Australia

Chapter 1

Discrete Inequalities

Some of the most important inequalities for real or complex numbers and sequences in analysis are presented in this chapter. These include the Abel inequality; the Cauchy-Bunyakovsky-Schwarz (CBS) inequality; De Bruijn's inequality; Čebyšev's inequality for synchronous sequences; the Biernacki, Pidek, and Ryll-Nardzewski inequality and its weighted form due to Andrica-Badea refining the Grüss inequality; and the Daykin-Eliezer-Carlitz inequality, which together with Wagner's inequality provide interesting and different generalisations for the CBS inequality. Classical reverses for the CBS inequality due to Pólya-Szegö and Cassels are also provided. Some fresh insight on the celebrated Hölder and Minkowski inequalities for sequences of real numbers are presented, and various inequalities for convex functions, including those of Jensen, Slater, and Petrović, are investigated.

All the results considered are complemented by numerous innovative and evocative remarks and comments which reveal a network of connections among the above inequalities and create a natural background for further research in related fields or in their application.

1.1 An Elementary Inequality for Two Numbers

The following inequality is known in the literature as the arithmetic–geometric–harmonic mean inequality.

For any positive numbers a and b

$$\frac{a+b}{2} \geq \sqrt{ab} \geq \frac{2}{\frac{1}{a} + \frac{1}{b}}. \tag{1.1}$$

Equality holds if and only if $a = b$ [6, p. 48].

PROOF   Since $a, b > 0$, we may write $a = x^2$, $b = y^2$ with $x, y > 0$ to get

$$\frac{x^2 + y^2}{2} \geq xy$$

which is clearly equivalent to

$$(x - y)^2 \geq 0 \tag{1.2}$$

which is true for any real value of x and y. This proves the first part of (1.1).

Now, write the first inequality in (1.1) for $\frac{1}{a}$ and $\frac{1}{b}$ to get

$$\frac{\frac{1}{a} + \frac{1}{b}}{2} \geq \sqrt{\frac{1}{ab}} = \frac{1}{\sqrt{ab}}$$

which is equivalent to the second part of (1.1).

Now, observe that equality holds in (1.2) if and only if (iff) $x = y$ and so it holds in the first part of (1.1) if and only if $a = b$. The same applies for the second part of (1.1).

□

Comments

(a) For a geometric proof of the first part of (1.1), see Beckenbach and Bellman [6, pp. 50–51].

(b) The inequality (1.2) is a fundamental inequality for real numbers. It can be improved as follows. For

$$\operatorname{sgn}(x) := \begin{cases} 1 & \text{if } x > 0 \\ 0 & \text{if } x = 0 \\ -1 & \text{if } x < 0 \end{cases},$$

we have

$$x^2 - 2xy + y^2 \geq \left| x^2 \operatorname{sgn}(x) + y^2 \operatorname{sgn}(y) - xy \left[\operatorname{sgn}(x) + \operatorname{sgn}(y) \right] \right| \geq 0 \tag{1.3}$$

for any $x, y \in \mathbb{R}$. Indeed, by the continuity property of the modulus, namely

$$|z - t| \geq ||z| - |t||$$

for all $z, t \in \mathbb{R}$, then we can state that

$$x^2 - 2xy + y^2 = (x - y)^2 = |x - y| \, |x - y| \geq |x - y| \, ||x| - |y|| \tag{1.4}$$
$$= |(x - y)(|x| - |y|)| = |x \, |x| + y \, |y| - x \, |y| - |x| \, y|.$$

It is well known that

$$|x| = x \operatorname{sgn}(x) \quad \text{for all } x \in \mathbb{R}.$$

Writing this in inequality (1.4) proves (1.3). Equality holds in both parts of (1.3) simultaneously iff $x = y$.

(c) A similar result to that in (b) holds for complex numbers z, w, namely,

$$|z|^2 - 2\operatorname{Re}(zw) + |w|^2 \geq |z \, |z| + w \, |w| - z \, |w| - |z| \, w| \geq 0. \tag{1.5}$$

Indeed,

$$|z - w|^2 = (z - w)\overline{(z - w)} = (z - w)(\bar{z} - \bar{w}) = z\bar{z} + w\bar{w} - z\bar{w} - \bar{z}w$$
$$= |z|^2 + |w|^2 - 2\operatorname{Re}(zw)$$

and so, as above,

$$|z - w|^2 \geq ||z| - |w|||z - w| = |z||z| + w|w| - z|w| - |z|w| \geq 0.$$

Equality is achieved in both inequalities in (1.5) simultaneously iff $z = w$.

1.2 An Elementary Inequality for Three Numbers

For any three positive numbers a, b, c

$$\frac{a + b + c}{3} \geq \sqrt[3]{abc} \geq \frac{3}{\frac{1}{a} + \frac{1}{b} + \frac{1}{c}}. \tag{1.6}$$

Equality results if and only if $a = b = c$ [6, p. 53].

PROOF Since $a, b, c > 0$, we may set $a = x^3$, $b = y^3$, and $c = z^3$ with $x, y, z > 0$. We claim that

$$x^3 + y^3 + z^3 - 3xyz \geq 0, \tag{1.7}$$

which is equivalent to the first part of (1.6). To prove the claim, we assert that

$$x^3 + y^3 + z^3 - 3xyz$$
$$= (x + y + z)(x^2 + y^2 + z^2 - xy - xz - yz), \quad x, y, z \in \mathbb{R} \tag{1.8}$$

which can be readily verified by multiplication of the right-hand side.

Since $x, y, z > 0$, $x + y + z > 0$; and in order to demonstrate (1.7), it is sufficient to show that the second factor is nonnegative, that is, that

$$x^2 + y^2 + z^2 - xy - xz - yz \geq 0. \tag{1.9}$$

We know that

$$x^2 + y^2 \geq 2xy, \quad y^2 + z^2 \geq 2yz, \quad x^2 + z^2 \geq 2xz$$

for all $x, y, z \in \mathbb{R}$. Adding these inequalities and dividing by 2 produces (1.9). Equality holds in (1.9) iff $x = y = z$. This implies that equality holds in the first part of (1.6) iff $a = b = c$.

To prove the second part of (1.6), we apply the first part for $\frac{1}{a}$, $\frac{1}{b}$, and $\frac{1}{c}$ to get

$$\frac{\frac{1}{a} + \frac{1}{b} + \frac{1}{c}}{3} \geq \sqrt[3]{\frac{1}{a} \cdot \frac{1}{b} \cdot \frac{1}{c}} = \frac{1}{\sqrt[3]{abc}}.$$

□

Comments

(a) The inequality (1.9), which is in itself a classical result, can be improved as follows:

$$x^2 + y^2 + z^2 - xy - xz - yz$$

$$\geq \left| x^2 \operatorname{sgn}(x) + y^2 \operatorname{sgn}(y) + z^2 \operatorname{sgn}(z) - xy \left(\frac{\operatorname{sgn}(x) + \operatorname{sgn}(y)}{2} \right) \right.$$

$$\left. - xz \left(\frac{\operatorname{sgn}(x) + \operatorname{sgn}(z)}{2} \right) - yz \left(\frac{\operatorname{sgn}(y) + \operatorname{sgn}(z)}{2} \right) \right| \geq 0 \quad (1.10)$$

for all $x, y, z \in \mathbb{R}$.

The proof is obtained from result (1.3) by adding the following inequalities:

$$x^2 + y^2 - 2xy \geq \left| x^2 \operatorname{sgn}(x) + y^2 \operatorname{sgn}(y) - xy (\operatorname{sgn}(x) + \operatorname{sgn}(y)) \right| \geq 0,$$
$$y^2 + z^2 - 2yz \geq \left| y^2 \operatorname{sgn}(y) + z^2 \operatorname{sgn}(z) - yz (\operatorname{sgn}(y) + \operatorname{sgn}(z)) \right| \geq 0,$$
$$z^2 + x^2 - 2zx \geq \left| z^2 \operatorname{sgn}(y) + x^2 \operatorname{sgn}(z) - zx (\operatorname{sgn}(z) + \operatorname{sgn}(x)) \right| \geq 0,$$

making use of the triangle inequality and dividing by 2.

(b) We have the following refinement of (1.7) as well:

$$x^3 + y^3 + z^3 - 3xyz$$

$$\geq (x + y + z) \left| x^2 \operatorname{sgn}(x) + y^2 \operatorname{sgn}(y) + z^2 \operatorname{sgn}(z) - xy \left(\frac{\operatorname{sgn}(x) + \operatorname{sgn}(y)}{2} \right) \right.$$

$$\left. - xz \left(\frac{\operatorname{sgn}(x) + \operatorname{sgn}(z)}{2} \right) - yz \left(\frac{\operatorname{sgn}(y) + \operatorname{sgn}(z)}{2} \right) \right| \geq 0 \quad (1.11)$$

for all $x, y, z \in \mathbb{R}$ provided that $x + y + z \geq 0$.

1.3   A Weighted Inequality for Two Numbers

Let $\alpha, \beta \geq 0$ so that $\alpha + \beta > 0$ and $a, b > 0$. Then

$$\frac{\alpha a + \beta b}{\alpha + \beta} \geq a^{\frac{\alpha}{\alpha + \beta}} b^{\frac{\beta}{\alpha + \beta}} \geq \frac{\alpha + \beta}{\frac{\alpha}{a} + \frac{\beta}{b}}. \quad (1.12)$$

Equality holds *iff* $a = b$.

PROOF First, let us prove the following inequality [138, p. 30]:

$$\frac{1}{p}x^p + \frac{1}{q}y^q \geq xy \tag{1.13}$$

for all $x, y \geq 0$ and $p, q > 1$ with $\frac{1}{p} + \frac{1}{q} = 1$.

Consider the mapping $f : [0, \infty) \to \mathbb{R}$, $f(x) = \frac{1}{p}x^p - xy + \frac{1}{q}y^q$. We have

$$f'(x) = x^{p-1} - y,$$

so that

$$f'(x) < 0 \text{ if } x \in \left(0, y^{\frac{1}{p-1}}\right), \ f'\left(y^{\frac{1}{p-1}}\right) = 0, \text{ and } f'(x) > 0 \text{ if } x \in \left(y^{\frac{1}{p-1}}, +\infty\right).$$

This implies that f is monotonic decreasing on $\left[0, y^{\frac{1}{p-1}}\right]$ and monotonic increasing on $\left(y^{\frac{1}{p-1}}, +\infty\right)$, which leads to

$$f(x) \geq f\left(y^{\frac{1}{p-1}}\right) = \frac{1}{p}y^{\frac{p}{p-1}} + \frac{1}{q}y^q - y^{\frac{1}{p-1}} \cdot y = 0$$

and thus proving the inequality (1.13).

Equality holds in (1.13) iff $x = y^{\frac{1}{p-1}}$, so that, $x^p = y^q$.

If in (1.13) we choose

$$p = \frac{\alpha+\beta}{\alpha}, \quad q = \frac{\alpha+\beta}{\beta}, \quad x = a^{\frac{\alpha}{\alpha+\beta}}, \text{ and } y = b^{\frac{\beta}{\alpha+\beta}},$$

then we get the first part of (1.12).

The second part of (1.12) follows by the first part applied for $\frac{1}{a}$ and $\frac{1}{b}$, that is,

$$\frac{\alpha \cdot \frac{1}{a} + \beta \cdot \frac{1}{b}}{\alpha+\beta} \geq \left(\frac{1}{a}\right)^{\frac{\alpha}{\alpha+\beta}} \cdot \left(\frac{1}{b}\right)^{\frac{\beta}{\alpha+\beta}} = \frac{1}{a^{\frac{\alpha}{\alpha+\beta}} \cdot b^{\frac{\beta}{\alpha+\beta}}}.$$

The case of equality follows by the fact that $x^p = a$ and $y^q = b$. $\qquad\Box$

Comments

(a) In (1.13) let us choose $x = \frac{a_i}{b_i}$, $y = \frac{a_j}{b_j}$, where $a_i, b_i > 0$ and $i, j \in \{1, \dots, n\}$ to obtain

$$\frac{1}{p}\left(\frac{a_i}{b_i}\right)^p + \frac{1}{q}\left(\frac{a_j}{b_j}\right)^q \geq \frac{a_i a_j}{b_i b_j} \text{ for all } i, j \in \{1, \dots, n\},$$

which is equivalent to

$$\frac{1}{p}a_i^p b_j^q + \frac{1}{q}a_j^q b_i^p \geq a_i b_i^{p-1} a_j b_j^{q-1} \text{ for all } i, j \in \{1, \dots, n\}.$$

Summing over i and j from 1 to n, we get

$$\frac{1}{p}\sum_{i=1}^{n} a_i^p \sum_{j=1}^{n} b_j^q + \frac{1}{q}\sum_{j=1}^{n} a_j^q \sum_{i=1}^{n} b_i^p \geq \sum_{i=1}^{n} a_i b_i^{p-1} \sum_{j=1}^{n} a_j b_j^{q-1},$$

which gives the inequality [66, p. 7]

$$\frac{1}{p}\sum_{i=1}^{n} a_i^p \sum_{i=1}^{n} b_i^q + \frac{1}{q}\sum_{i=1}^{n} a_i^q \sum_{i=1}^{n} b_i^p \geq \sum_{i=1}^{n} a_i b_i^{p-1} \sum_{i=1}^{n} a_i b_i^{q-1}. \tag{1.14}$$

(b) In (1.13) let us choose $x = \frac{a_i}{a_j}$, $y = \frac{b_i}{b_j}$ to get

$$\frac{1}{p}\left(\frac{a_i}{a_j}\right)^p + \frac{1}{q}\left(\frac{b_i}{b_j}\right)^q \geq \frac{a_i b_i}{a_j b_j} \quad \text{for all } i, j \in \{1, \ldots, n\},$$

which is equivalent to

$$\frac{1}{p}a_i^p b_j^q + \frac{1}{q}b_i^q a_j^p \geq a_i b_i a_j^{p-1} b_j^{q-1} \quad \text{for all } i, j \in \{1, \ldots, n\}.$$

By summing over i and j from 1 to n, we get

$$\frac{1}{p}\sum_{i=1}^{n} a_i^p \sum_{j=1}^{n} b_j^q + \frac{1}{q}\sum_{i=1}^{n} b_i^q \sum_{j=1}^{n} a_j^p \geq \sum_{i=1}^{n} a_i b_i \sum_{j=1}^{n} a_j^{p-1} b_j^{q-1}$$

which is equivalent to [66, p. 8]

$$\sum_{i=1}^{n} a_i^p \sum_{i=1}^{n} b_i^q \geq \sum_{i=1}^{n} a_i b_i \sum_{i=1}^{n} a_i^{p-1} b_i^{q-1}. \tag{1.15}$$

(c) If we choose $p = q = 2$ in (1.14) or (1.15), we get the Cauchy-Bunyakovsky-Schwarz (CBS) inequality [6, p. 66]:

$$\sum_{i=1}^{n} a_i^2 \sum_{i=1}^{n} b_i^2 \geq \left(\sum_{i=1}^{n} a_i b_i\right)^2.$$

The equality case holds in this inequality if and only if $a_i = r b_i$, $i \in \{1, \ldots, n\}$ with $r \in \mathbb{R}$. We discuss this inequality further in Section 1.7 where we also provide an alternative proof.

1.4   The Abel Inequality

Let $a_1, \ldots, a_n$ and $b_1, \ldots, b_n$ with $b_1 \geq \cdots \geq b_n \geq 0$, be two sequences of real numbers, and

$$s_k = a_1 + \cdots + a_k \quad (k = 1, \ldots, n).$$

If $m = \min\limits_{1 \le k \le n} s_k$ and $M = \max\limits_{1 \le k \le n} s_k$, *then*

$$mb_1 \le a_1 b_1 + \cdots + a_n b_n \le M b_1. \tag{1.16}$$

PROOF   We use Abel's well-known identity:

$$\sum_{k=1}^{n} a_k b_k = s_1 b_1 + (s_2 - s_1) b_2 + \cdots + (s_n - s_{n-1}) b_n$$

$$= s_1 (b_1 - b_2) + \cdots + s_{n-1} (b_{n-1} - b_n) + s_n b_n.$$

Since

$$m (b_1 - b_2) \le s_1 (b_1 - b_2) \le M (b_1 - b_2)$$

$$\cdots \cdots \cdots \cdots$$

$$m (b_{n-1} - b_n) \le s_{n-1} (b_{n-1} - b_n) \le M (b_{n-1} - b_n)$$
$$m b_n \le s_n b_n \le M b_n$$

we get inequality (1.16) by adding these inequalities. □

Comments

(a) Let $a = (a_1, \ldots, a_n)$ and $p = (p_1, \ldots, p_n)$ be n-tuples of real numbers such that $a_1 \le \cdots \le a_n$ and $\sum_{k=i}^{n} p_k \ge 0$ for $i = 2, \ldots, n$. Then [105]

$$\sum_{i=1}^{n} p_i a_i \ge a_1 P_n + \left| \sum_{i=1}^{n} p_i |a_i| - |a_1| P_n \right|, \tag{1.17}$$

where $P_n = \sum_{i=1}^{n} p_i$.

First, let us rewrite Abel's identity as follows:

$$\sum_{i=1}^{n} p_i a_i = a_1 \sum_{i=1}^{n} p_i + \sum_{i=2}^{n} \left(\sum_{k=i}^{n} p_k \right) \Delta a_{i-1} \tag{1.18}$$

$$= a_n \sum_{i=1}^{n} p_i - \sum_{i=1}^{n-1} \left(\sum_{k=i}^{n} p_k \right) \Delta a_i,$$

where $\Delta a_i = a_{i+1} - a_i$, $i = \overline{1, n-1}$.

Since a is nondecreasing, we have

$$\Delta a_{i-1} = a_i - a_{i-1} = |a_i - a_{i-1}| \ge ||a_i| - |a_{i-1}|| = |\Delta |a_{i-1}|| \ge 0$$

for all $i = 2, \ldots, n$ and

$$0 \le \sum_{k=i}^{n} p_k = \left| \sum_{k=i}^{n} p_k \right| \quad \text{for all } i = 2, \ldots, n.$$

Then, by the first identity in (1.18), we obtain

$$\sum_{i=1}^{n} p_i a_i - a_1 P_n = \sum_{i=2}^{n} \left(\sum_{k=i}^{n} p_k \right) \Delta a_{i-1} = \sum_{i=2}^{n} \left| \sum_{k=i}^{n} p_k \right| |\Delta a_{i-1}|$$

$$\ge \sum_{i=2}^{n} \left| \sum_{k=i}^{n} p_k \right| |\Delta |a_{i-1}|| = \sum_{i=2}^{n} \left(\sum_{k=i}^{n} p_k \right) \Delta |a_{i-1}|$$

$$\ge \left| \sum_{i=2}^{n} \left(\sum_{k=i}^{n} p_k \right) \Delta |a_{i-1}| \right|.$$

By Abel's identity for $|a| := (|a_1|, \ldots, |a_n|)$, we have also that

$$\sum_{i=1}^{n} p_i |a_i| - |a_1| \sum_{i=1}^{n} p_i = \sum_{i=2}^{n} \left(\sum_{k=i}^{n} p_k \right) \Delta |a_{i-1}|.$$

Thus

$$\sum_{i=1}^{n} p_i a_i - a_1 \sum_{i=1}^{n} p_i \ge \left| \sum_{i=1}^{n} p_i |a_i| - |a_1| P_n \right| \ge 0,$$

which proves (1.17).

(b) Using the second identity in (1.18), we can prove the following similar inequality [105]:

$$a_n P_n - \sum_{i=1}^{n} p_i a_i \ge \left| |a_n| P_n - \sum_{i=1}^{n} p_i |a_i| \right| \ge 0 \qquad (1.19)$$

provided that $a_1 \le \cdots \le a_n$ and $\sum_{k=1}^{i} p_k \ge 0$ $(i = 1, \ldots, n-1)$.

(c) By **(a)** and **(b)** Equations (1.16) and (1.18), we can state that [105]

$$a_n P_n - \left| |a_n| P_n - \sum_{i=1}^{n} p_i |a_i| \right|$$

$$\ge \sum_{i=1}^{n} p_i a_i \ge a_1 P_n + \left| \sum_{i=1}^{n} p_i |a_i| - |a_1| P_n \right|, \qquad (1.20)$$

provided that $a_1 \le \cdots \le a_n$ and $P_n \ge P_i \ge 0$ for all $i = 1, \ldots, n-1$, where $P_i = \sum_{k=1}^{i} p_k$.

1.5 The Biernacki, Pidek, and Ryll-Nardzewski (BPR) Inequality

Let a and b be n-tuples such that $m \leq a_i \leq M$ and $k \leq b_i \leq K$ for $i = 1, 2, \ldots, n$. Then,

$$\left| \frac{1}{n} \sum_{i=1}^{n} a_i b_i - \frac{1}{n} \sum_{i=1}^{n} a_i \cdot \frac{1}{n} \sum_{i=1}^{n} b_i \right| \leq (M - m)(K - k) C(n), \qquad (1.21)$$

where

$$C(n) = \frac{1}{n} \left[\frac{n}{2} \right] \left(1 - \frac{1}{n} \left[\frac{n}{2} \right] \right) \leq \frac{1}{4}, \qquad (1.22)$$

with $[\cdot]$ being the greatest integer function. Equality occurs when n is even.

PROOF We start with the well-known identity

$$n \sum_{i=1}^{n} a_i b_i - \sum_{i=1}^{n} a_i \cdot \sum_{i=1}^{n} b_i = \frac{1}{2} \sum_{i=1}^{n} \sum_{j=1}^{n} (a_i - a_j)(b_i - b_j), \qquad (1.23)$$

which may be shown to hold by straightforward manipulation. Applying the Cauchy-Bunyakovsky-Schwarz inequality (1.34) for double sums, we have

$$\left| \sum_{i=1}^{n} \sum_{j=1}^{n} (a_i - a_j)(b_i - b_j) \right|$$
$$\leq \left[\sum_{i=1}^{n} \sum_{j=1}^{n} (a_i - a_j)^2 \right]^{\frac{1}{2}} \left[\sum_{i=1}^{n} \sum_{j=1}^{n} (b_i - b_j)^2 \right]^{\frac{1}{2}}. \qquad (1.24)$$

Observe from (1.23) that

$$n \sum_{i=1}^{n} a_i^2 - \left(\sum_{i=1}^{n} a_i \right)^2 = \frac{1}{2} \sum_{i=1}^{n} \sum_{j=1}^{n} (a_i - a_j)^2 := D_n(a) \qquad (1.25)$$

and similarly for the b.

We note that $D_n(\cdot)$ is a convex function in the variable $a = (a_1, \ldots, a_n) \in$

$[m, M]^n$. Indeed, if $\alpha \in [0,1]$ and $x, y \in \mathbb{R}^n$, then

$$D_n\left(\alpha x + (1-\alpha)y\right) = \frac{1}{2}\sum_{i=1}^{n}\sum_{j=1}^{n}\left(\alpha x_i + (1-\alpha)y_i - \alpha x_j - (1-\alpha)y_j\right)^2$$

$$= \frac{1}{2}\sum_{i=1}^{n}\sum_{j=1}^{n}\left(\alpha(x_i - x_j) + (1-\alpha)(y_i - y_j)\right)^2$$

$$\leq \frac{1}{2}\sum_{i=1}^{n}\sum_{j=1}^{n}\left[\alpha(x_i - x_j)^2 + (1-\alpha)(y_i - y_j)^2\right]$$

$$= \alpha D_n(x) + (1-\alpha)D_n(y).$$

It is a well-known result that a convex function attains its maximum on the boundary. Let there be β of the a_i that equals m and $n - \beta$ of the a_i that equals M. Then from (1.25), we get

$$0 \leq D_n^*(\beta) = n\left[\beta m^2 + (n-\beta)M^2\right] - \left(\beta m + (n-\beta)M\right)^2.$$

We remark that $D_n^*(\beta)$ is a quadratic function of β with $D_n^*(n) = D_n^*(0) = 0$. Hence, the maximum is attained at $\beta = \left[\frac{n}{2}\right]$ and so

$$D_n^*\left(\left[\frac{n}{2}\right]\right) = \left[\frac{n}{2}\right]\left(n - \left[\frac{n}{2}\right]\right)(M - m)^2. \tag{1.26}$$

We apply similar steps for b. By using (1.23) and (1.24) we obtain the desired inequality (1.21) and (1.22). Equality in (1.22) is easily demonstrated. It remains to be shown that the inequality holds for n odd. Let $n = 2N - 1$, then

$$C(2N-1) = \frac{1}{2N-1}\left[N - \frac{1}{2}\right]\left(1 - \frac{[N - \frac{1}{2}]}{2N-1}\right)$$

$$= \frac{N}{2N-1}\left(1 - \frac{N}{2N-1}\right)$$

$$= \frac{1}{2 - \frac{1}{N}}\left(1 - \frac{1}{2 - \frac{1}{N}}\right) < \frac{1}{2}\left(1 - \frac{1}{2}\right) = \frac{1}{4}.$$

This completes the proof.   □

Comments
(a) The weighted version of (1.21) was proved by Andrica and Badea [2]. Namely, they have shown that:

$$\left|P_n\sum_{i=1}^{n}p_i a_i b_i - \sum_{i=1}^{n}p_i a_i \cdot \sum_{i=1}^{n}p_i b_i\right|$$

$$\leq (M - m)(K - k)\sum_{i \in S}p_i\left(P_n - \sum_{i \in S}p_i\right), \tag{1.27}$$

where S is a subset of $\{1, 2, \ldots, n\}$ which minimises

$$\left| \sum_{i \in S} p_i - \frac{P_n}{2} \right|,$$

and $P_n = \sum_{i=1}^{n} p_i$ with p_i positive.

Taking $p_i = \frac{1}{n}$ recaptures (1.21) from (1.27). This type of inequality is further discussed in Sections 1.8 and 1.10.

(b) If we take $b_i = \frac{1}{a_i}$, then for $0 < m \le a_i \le M$ we have $\frac{1}{M} \le \frac{1}{a_i} \le \frac{1}{m}$ and

$$\frac{1}{n} \sum_{i=1}^{n} a_i \cdot \frac{1}{n} \sum_{i=1}^{n} \frac{1}{a_i} \le C(n) \frac{(M-m)^2}{mM} + 1 \tag{1.28}$$

$$= \frac{\left(\left[\frac{n}{2}\right] M + \left[\frac{n+1}{2}\right] m\right) \left(\left[\frac{n+1}{2}\right] M + \left[\frac{n}{2}\right] m\right)}{mM}.$$

1.6 Čebyšev's Inequality for Synchronous Sequences

Let $a = (a_1, \ldots, a_n)$, $b = (b_1, \ldots, b_n) \in \mathbb{R}^n$. If a, b are synchronous (asynchronous), by which is meant,

$$(a_i - a_j)(b_i - b_j) \ge (\le) 0 \quad \text{for all} \quad i, j \in \{1, \ldots, n\}, \tag{S}$$

then for all $p_i > 0$ we have the inequality

$$\sum_{i=1}^{n} p_i \sum_{i=1}^{n} p_i a_i b_i \ge (\le) \sum_{i=1}^{n} p_i a_i \sum_{i=1}^{n} p_i b_i. \tag{1.29}$$

Equality holds in (1.29) iff $(a_i - a_j)(b_i - b_j) = 0$ for all $i, j \in \{1, \ldots, n\}$.

PROOF First, let us prove Korkine's identity [141]:

$$\sum_{i=1}^{n} p_i \sum_{i=1}^{n} p_i a_i b_i - \sum_{i=1}^{n} p_i a_i \sum_{i=1}^{n} p_i b_i = \sum_{1 \le i < j \le n} p_i p_j (a_i - a_j)(b_i - b_j). \tag{1.30}$$

By symmetry over i and j, we have that

$$\sum_{i,j=1}^{n} p_i p_j (a_i - a_j)(b_i - b_j) = 2 \sum_{1 \le i < j \le n} p_i p_j (a_i - a_j)(b_i - b_j).$$

On the other hand, we also have

$$\sum_{i,j=1}^{n} p_i p_j (a_i - a_j)(b_i - b_j) = \sum_{i,j=1}^{n} p_i p_j (a_i b_i + a_j b_j - a_i b_j - a_j b_i)$$

$$= \sum_{i=1}^{n} p_i a_i b_i \sum_{j=1}^{n} p_j + \sum_{i=1}^{n} p_i \sum_{j=1}^{n} p_j a_j b_j$$

$$- \sum_{i=1}^{n} p_i a_i \sum_{j=1}^{n} p_j b_j - \sum_{i=1}^{n} p_i b_i \sum_{j=1}^{n} p_j a_j$$

$$= 2 \left[\sum_{i=1}^{n} p_i \sum_{i=1}^{n} p_i a_i b_i - \sum_{i=1}^{n} p_i a_i \sum_{i=1}^{n} p_i b_i \right]$$

which proves the identity (1.30).

The inequality (1.29) is now obvious by (1.30) and using (S).

The case of equality is also obvious. $\quad\square$

Comments

For $a, b \in \mathbb{R}^n$ and $p \in \mathbb{R}_+^n$, define

$$C(p; a, b) := P_n \sum_{i=1}^{n} p_i a_i b_i - \sum_{i=1}^{n} p_i a_i \sum_{i=1}^{n} p_i b_i,$$

where $P_n := \sum_{i=1}^{n} p_i$.

If by $|x|$ we denote the vector $(|x_1|, \ldots, |x_n|)$, where $x = (x_1, \ldots, x_n)$, then we have the following refinement of Čebyšev's inequality [107]:

$$C(p; a, b) \geq \max\{C(p; |a|, b), C(p; a, |b|), C(p; |a|, |b|)\} \geq 0. \qquad (1.31)$$

By the continuity property of the modulus, we have

$$(a_i - a_j)(b_i - b_j) = |(a_i - a_j)(b_i - b_j)| \geq |(|a_i| - |a_j|)(|b_i| - |b_j|)| \qquad (1.32)$$

for all $i, j \in \{1, \ldots, n\}$.

If we multiply (1.32) by $p_i p_j \geq 0$, then by summing over i and j from 1 to n we obtain

$$\sum_{i,j=1}^{n} p_i p_j (a_i - a_j)(b_i - b_j) \geq \left| \sum_{i,j=1}^{n} p_i p_j (|a_i| - |a_j|)(|b_i| - |b_j|) \right| \geq 0. \qquad (1.33)$$

Note that

$$\sum_{i,j=1}^{n} p_i p_j (a_i - a_j)(b_i - b_j) = 2C(p; a, b)$$

and

$$\sum_{i,j=1}^{n} p_i p_j \left(|a_i| - |a_j|\right)\left(|b_i| - |b_j|\right) = 2C\left(p; |a|, |b|\right).$$

Therefore, (1.33) gives us

$$C\left(p; a, b\right) \geq \left|C\left(p; |a|, |b|\right)\right|.$$

Similar arguments apply for the first and second parts of (1.31).

1.7 The Cauchy-Bunyakovsky-Schwarz (CBS) Inequality for Real Numbers

Let $a = (a_1, \ldots, a_n)$, $b = (b_1, \ldots, b_n) \in \mathbb{R}^n$. Then

$$\sum_{i=1}^{n} a_i^2 \sum_{i=1}^{n} b_i^2 \geq \left(\sum_{i=1}^{n} a_i b_i\right)^2. \tag{1.34}$$

Equality holds in (1.34) iff a and b are proportional; that is, if there exists a real constant r such that $a_i = r b_i$ for all $i \in \{1, \ldots, n\}$.

PROOF   First, let us prove Lagrange's identity [6, p. 66]:

$$\sum_{i=1}^{n} a_i^2 \sum_{i=1}^{n} b_i^2 - \left(\sum_{i=1}^{n} a_i b_i\right)^2 = \sum_{1 \leq i < j \leq n} (a_i b_j - a_j b_i)^2. \tag{1.35}$$

By symmetry over i and j, we have

$$\sum_{i,j=1}^{n} (a_i b_j - a_j b_i)^2 = 2 \sum_{1 \leq i < j \leq n} (a_i b_j - a_j b_i)^2.$$

On the other hand, we also have

$$\sum_{i,j=1}^{n} (a_i b_j - a_j b_i)^2 = \sum_{i,j=1}^{n} (a_i^2 b_j^2 - 2 a_i b_i a_j b_j + a_j^2 b_i^2)$$

$$= \sum_{i=1}^{n} a_i^2 \sum_{j=1}^{n} b_j^2 - 2 \sum_{i=1}^{n} a_i b_i \sum_{j=1}^{n} a_j b_j + \sum_{i=1}^{n} b_i^2 \sum_{j=1}^{n} a_j^2$$

$$= 2 \left[\sum_{i=1}^{n} a_i^2 \sum_{i=1}^{n} b_i^2 - \left(\sum_{i=1}^{n} a_i b_i\right)^2 \right]$$

which proves the identity (1.35). The inequality (1.34) is now obvious by (1.35).

The case of equality holds iff $(a_i b_j - a_j b_i)^2 = 0$ for all $i, j \in \{1, \ldots, n\}$, which is equivalent to $a_i b_j = a_j b_i$ for all $i, j \in \{1, \ldots, n\}$, that is, when a and b are proportional. □

Comments

We can improve (1.34) as follows [95, p. 79]:

$$\sum_{i=1}^{n} a_i^2 \sum_{i=1}^{n} b_i^2 - \left(\sum_{i=1}^{n} a_i b_i \right)^2 \tag{1.36}$$

$$\geq \left| \sum_{i=1}^{n} a_i^2 \operatorname{sgn}(a_i) \sum_{i=1}^{n} b_i^2 \operatorname{sgn}(b_i) - \sum_{i=1}^{n} a_i b_i \operatorname{sgn}(a_i) \sum_{i=1}^{n} a_i b_i \operatorname{sgn}(b_i) \right|$$

$$\geq 0.$$

Equality results in both inequalities simultaneously iff a and b are proportional.

We apply the elementary inequality (see inequality 1.3) for real numbers x, y, namely,

$$x^2 + y^2 - 2xy \geq \left| x^2 \operatorname{sgn}(x) + y^2 \operatorname{sgn}(y) - xy \left(\operatorname{sgn}(x) + \operatorname{sgn}(y) \right) \right| \geq 0 \quad (1.37)$$

to get

$$a_i^2 b_j^2 + a_j^2 b_i^2 - 2 a_i b_i a_j b_j$$
$$\geq \left| a_i^2 b_j^2 \operatorname{sgn}(a_i b_j) + a_j^2 b_i^2 \operatorname{sgn}(a_j b_i) - a_i b_i a_j b_j \left(\operatorname{sgn}(a_i b_j) + \operatorname{sgn}(a_j b_i) \right) \right|$$
$$= \left| a_i^2 \operatorname{sgn}(a_i) b_j^2 \operatorname{sgn}(b_j) + a_j^2 \operatorname{sgn}(a_j) b_i^2 \operatorname{sgn}(b_i) \right.$$
$$\left. - a_i b_i a_j b_j \left(\operatorname{sgn}(a_i) \operatorname{sgn}(b_j) + \operatorname{sgn}(a_j) \operatorname{sgn}(b_i) \right) \right|$$
$$\geq 0$$

for all $i, j \in \{1, \ldots, n\}$.

By summing over i and j from 1 to n, using the generalised triangle inequality, i.e., $\sum_{k=1}^{m} |z_k| \geq |\sum_{k=1}^{m} z_k|$, $z_k \in \mathbb{R}$, $k \in \{1, \ldots, m\}$ and properties of double sums, we get (1.36).

As in (1.37), equality holds in (1.36) in both parts simultaneously iff $a_i b_j = a_j b_i$ for all $i, j \in \{1, \ldots, n\}$, that is, if the vectors a and b are proportional.

For other classical or new results in connection to the CBS inequality, see Mitrinović, Pečarić, and Fink [141] where further references are given.

1.8 The Andrica-Badea Inequality

The following result is due to Andrica and Badea [2, p. 16]:

Let $x = (x_1, \ldots, x_n) \in I^n = [m, M]^n$ be a sequence of real numbers and S be the subset of $\{1, \ldots, n\}$ that minimises the expression

$$\left| \sum_{i \in S} p_i - \frac{1}{2} P_n \right|, \tag{1.38}$$

where $P_n := \sum_{i=1}^n p_i > 0$, $p = (p_1, \ldots, p_n)$ is a sequence of nonnegative real numbers. Then,

$$\max_{x \in I^n} \left[\frac{1}{P_n} \sum_{i=1}^n p_i x_i^2 - \left(\frac{1}{P_n} \sum_{i=1}^n p_i x_i \right)^2 \right]$$

$$= \frac{(M-m)^2}{P_n^2} \sum_{i \in S} p_i \left(P_n - \sum_{i \in S} p_i \right). \tag{1.39}$$

PROOF We follow the proof by Andrica and Badea [2, p. 161].
Define

$$D_n(x, p) := \frac{1}{P_n} \sum_{i=1}^n p_i x_i^2 - \left(\frac{1}{P_n} \sum_{i=1}^n p_i x_i \right)^2$$

$$= \frac{1}{P_n} \sum_{1 \le i < j \le n} p_i p_j (x_i - x_j)^2.$$

Keeping in mind the convexity of the quadratic function, we have

$$D_n(\alpha x + (1-\alpha) y, p)$$

$$= \frac{1}{P_n^2} \sum_{1 \le i < j \le n} p_i p_j [\alpha x_i + (1-\alpha) y_i - \alpha x_j - (1-\alpha) y_j]^2$$

$$= \frac{1}{P_n^2} \sum_{1 \le i < j \le n} p_i p_j [\alpha (x_i - x_j) + (1-\alpha)(y_i - y_j)]^2$$

$$\le \frac{1}{P_n^2} \sum_{1 \le i < j \le n} p_i p_j \left[\alpha (x_i - x_j)^2 + (1-\alpha)(y_i - y_j)^2 \right]$$

$$= \alpha D_n(x, p) + (1-\alpha) D_n(y, p).$$

Hence $D_n(\cdot, p)$ is a convex function on I^n.

By using a well-known theorem (see, for instance, Andrica and Badea [2, p. 124]), we get that the maximum of $D_n(\cdot, p)$ is attained on the boundary of I^n.

Let $(S, \bar{S})$ be the partition of $\{1, \ldots, n\}$ such that the maximum of $D_n(\cdot, p)$ is obtained for $x_0 = (x_1^0, \ldots, x_n^0)$, where $x_i^0 = m$ if $i \in \bar{S}$ and $x_i^0 = M$ if $i \in S$. In this case we have

$$D_n(x_0, p) = \frac{1}{P_n^2} \sum_{1 \le i < j \le n} p_i p_j (x_i - x_j)^2 \tag{1.40}$$

$$= \frac{(M-m)^2}{P_n^2} \sum_{i \in S} p_i \left(P_n - \sum_{i \in S} p_i \right).$$

The expression

$$\sum_{i \in S} p_i \left(P_n - \sum_{i \in S} p_i \right)$$

is a maximum when the set S minimises the expression

$$\left| \sum_{i \in S} p_i - \frac{1}{2} P_n \right|.$$

From (1.40) it follows that $D_n(x, p)$ is also a maximum. This completes the proof. □

Comments

The following counterpart of the (CBS)-inequality holds (see Dragomir [55, p. 118]):

Let $a = (a_1, \ldots, a_n)$ and $b = (b_1, \ldots, b_n)$ be two sequences of real numbers with $a_i \ne 0$ for $i = 1, \ldots, n$ and

$$-\infty < m \le \frac{b_i}{a_i} \le M < \infty \quad \text{for each } i \in \{1, \ldots, n\}. \tag{1.41}$$

Let S be the subset of $\{1, \ldots, n\}$ that minimises the expression

$$\left| \sum_{i \in S} a_i^2 - \frac{1}{2} \sum_{i=1}^n a_i^2 \right|, \tag{1.42}$$

and denote $\bar{S} := \{1, \ldots, n\} \setminus S$. Then we have the inequality

$$0 \le \sum_{i=1}^n a_i^2 \sum_{i=1}^n b_i^2 - \left(\sum_{i=1}^n a_i b_i \right)^2 \tag{1.43}$$

$$\le (M-m)^2 \sum_{i \in S} a_i^2 \sum_{i \in \bar{S}} a_i^2 \le \frac{1}{4} (M-m)^2 \left(\sum_{i=1}^n a_i^2 \right)^2.$$

PROOF   The proof of the second inequality in (1.43) follows by (1.39) on choosing $p_i = a_i^2$, $x_i = \frac{b_i}{a_i}$, $i \in \{1, \ldots, n\}$.

The third inequality is obvious, since if $\alpha\beta \leq \frac{1}{4}(\alpha + \beta)^2$ for any $\alpha, \beta \in \mathbb{R}$, then

$$\sum_{i \in S} a_i^2 \sum_{i \in \bar{S}} a_i^2 = \sum_{i \in S} a_i^2 \left(\sum_{j=1}^{n} a_j^2 - \sum_{i \in S} a_i^2 \right)$$

$$\leq \frac{1}{4} \left(\sum_{i \in S} a_i^2 + \sum_{j=1}^{n} a_j^2 - \sum_{i \in S} a_i^2 \right)^2 = \frac{1}{4} \left(\sum_{j=1}^{n} a_j^2 \right)^2.$$

This completes the proof.   □

1.9   A Weighted Grüss Type Inequality

Throughout this book, a nonnegative sequence $p = (p_1, \ldots, p_n)$ is referred to as a probability sequence if $\sum_{i=1}^{n} p_i = 1$.

The following Grüss type inequality has been obtained in Cerone and Dragomir [29]:

Let $a = (a_1, \ldots, a_n)$, $b = (b_1, \ldots, b_n)$ be two sequences of real numbers and assume that there are $\gamma, \Gamma \in \mathbb{R}$ such that

$$-\infty < \gamma \leq a_i \leq \Gamma < \infty \text{for each } i \in \{1, \ldots, n\}. \tag{1.44}$$

Then, for any probability sequence $p = (p_1, \ldots, p_n)$, one has the inequality

$$\left| \sum_{i=1}^{n} p_i a_i b_i - \sum_{i=1}^{n} p_i a_i \sum_{i=1}^{n} p_i b_i \right| \leq \frac{1}{2}(\Gamma - \gamma) \sum_{i=1}^{n} p_i \left| b_i - \sum_{k=1}^{n} p_k b_k \right|. \tag{1.45}$$

The constant $\frac{1}{2}$ is sharp in the sense that it cannot be replaced by a smaller constant.

PROOF   We will give here a simple direct proof based on Sonin's identity [141]. A simple calculation shows that:

$$\sum_{i=1}^{n} p_i a_i b_i - \sum_{i=1}^{n} p_i a_i \sum_{i=1}^{n} p_i b_i = \sum_{i=1}^{n} p_i \left(a_i - \frac{\gamma + \Gamma}{2} \right) \left(b_i - \sum_{k=1}^{n} p_k b_k \right). \tag{1.46}$$

By (1.44) we have

$$\left| a_i - \frac{\gamma + \Gamma}{2} \right| \leq \frac{\Gamma - \gamma}{2} \text{ for all } i \in \{1, \ldots, n\}.$$

Thus, by taking the modulus, we have from (1.46)

$$\left| \sum_{i=1}^{n} p_i a_i b_i - \sum_{i=1}^{n} p_i a_i \sum_{i=1}^{n} p_i b_i \right| \le \sum_{i=1}^{n} p_i \left| a_i - \frac{\gamma + \Gamma}{2} \right| \left| b_i - \sum_{k=1}^{n} p_k b_k \right|$$

$$\le \frac{1}{2} \left(\Gamma - \gamma \right) \sum_{i=1}^{n} p_i \left| b_i - \sum_{k=1}^{n} p_k b_k \right|.$$

To prove the sharpness of the constant $\frac{1}{2}$, let us assume that (1.45) holds with a constant $c > 0$ instead of $\frac{1}{2}$, that is,

$$\left| \sum_{i=1}^{n} p_i a_i b_i - \sum_{i=1}^{n} p_i a_i \sum_{i=1}^{n} p_i b_i \right| \le c \left(\Gamma - \gamma \right) \sum_{i=1}^{n} p_i \left| b_i - \sum_{k=1}^{n} p_k b_k \right|, \qquad (1.47)$$

provided a_i satisfies (1.44).

If we choose $n = 2$ in (1.47) and take into account that

$$\sum_{i=1}^{2} p_i a_i b_i - \sum_{i=1}^{2} p_i a_i \sum_{i=1}^{2} p_i b_i = p_1 p_2 \left(a_1 - a_2 \right) \left(b_1 - b_2 \right),$$

provided $p_1 + p_2 = 1$, $p_1, p_2 \in [0, 1]$, and since

$$\sum_{i=1}^{2} p_i \left| b_i - \sum_{k=1}^{2} p_k b_k \right| = p_1 \left| (p_1 + p_2) b_1 - p_1 b_1 - p_2 b_2 \right|$$

$$+ p_2 \left| (p_1 + p_2) b_2 - p_1 b_1 - p_2 b_2 \right|$$

$$= 2 p_1 p_2 \left| b_1 - b_2 \right|,$$

then we deduce by (1.47)

$$p_1 p_2 \left| a_1 - a_2 \right| \left| b_1 - b_2 \right| \le 2c \left(\Gamma - \gamma \right) \left| b_1 - b_2 \right| p_1 p_2. \qquad (1.48)$$

If we assume that $p_1, p_2 \ne 0$, $b_1 \ne b_2$ and $a_1 = \Gamma$, $a_2 = \gamma$, then by (1.48) we deduce $c \ge \frac{1}{2}$. This proves the sharpness of the constant $\frac{1}{2}$. ◻

Comments

(a) *Assume that $a = (a_1, \ldots, a_n)$ satisfies the assumption (1.44) and p is a probability sequence. Then*

$$0 \le \sum_{i=1}^{n} p_i a_i^2 - \left(\sum_{i=1}^{n} p_i a_i \right)^2 \le \frac{1}{2} \left(\Gamma - \gamma \right) \sum_{i=1}^{n} p_i \left| a_i - \sum_{k=1}^{n} p_k a_k \right|. \qquad (1.49)$$

The constant $\frac{1}{2}$ is best possible in the sense mentioned above.
The following counterpart of the (CBS)-inequality may be stated.

(b) *Assume that* $x = (x_1, \ldots, x_n)$ *and* $y = (y_1, \ldots, y_n)$ *are sequences of real numbers with* $y_i \neq 0$ $(i = 1, \ldots, n)$. *If there exist real numbers* m *and* M *such that*

$$m \leq \frac{x_i}{y_i} \leq M \quad \text{for each } i \in \{1, \ldots, n\}, \tag{1.50}$$

then we have the inequality [55, p. 134]:

$$0 \leq \sum_{i=1}^{n} x_i^2 \sum_{i=1}^{n} y_i^2 - \left(\sum_{i=1}^{n} x_i y_i \right)^2 \tag{1.51}$$

$$\leq \frac{1}{2} (M - m) \sum_{i=1}^{n} |y_i| \left| \sum_{k=1}^{n} y_k \cdot \begin{vmatrix} x_i & y_i \\ x_k & y_k \end{vmatrix} \right|.$$

PROOF   If we choose $p_i = \frac{y_i^2}{\sum_{k=1}^{n} y_k^2}$, $a_i = \frac{x_i}{y_i}$ for $i = 1, \ldots, n$ and $\gamma = m$, $\Gamma = M$ in (1.49), then we deduce

$$\frac{\sum_{i=1}^{n} x_i^2}{\sum_{k=1}^{n} y_k^2} - \left(\frac{1}{\sum_{k=1}^{n} y_k^2} \sum_{i=1}^{n} x_i y_i \right)^2$$

$$\leq \frac{1}{2} (M - m) \frac{1}{\sum_{k=1}^{n} y_k^2} \sum_{i=1}^{n} y_i^2 \left| \frac{x_i}{y_i} - \frac{1}{\sum_{k=1}^{n} y_k^2} \sum_{k=1}^{n} x_k y_k \right|$$

$$= \frac{1}{2} (M - m) \frac{1}{\left(\sum_{k=1}^{n} y_k^2 \right)^2} \sum_{i=1}^{n} |y_i| \left| x_i \sum_{k=1}^{n} y_k^2 - y_i \sum_{k=1}^{n} x_k y_k \right|$$

$$= \frac{1}{2} (M - m) \frac{1}{\left(\sum_{k=1}^{n} y_k^2 \right)^2} \sum_{i=1}^{n} |y_i| \left| \sum_{k=1}^{n} y_k \cdot \begin{vmatrix} x_i & y_i \\ x_k & y_k \end{vmatrix} \right|.$$

This gives the desired inequality (1.51).   ∎

1.10   Andrica-Badea's Refinement of the Grüss Inequality

In 1988, Andrica and Badea [2] established a weighted version of the Grüss inequality:

If $m_1 \leq a_i \leq M_1$, $m_2 \leq b_i \leq M_2$ $(i \in \{1, \ldots, n\})$, *and* S *is the subset of* $\{1, \ldots, n\}$ *which minimises the expression*

$$\left| \sum_{i \in S} p_i - \frac{1}{2} P_n \right|, \tag{1.52}$$

where $P_n := \sum_{i=1}^n p_i > 0$, then

$$\left| P_n \sum_{i=1}^n p_i a_i b_i - \sum_{i=1}^n p_i a_i \cdot \sum_{i=1}^n p_i b_i \right| \tag{1.53}$$

$$\leq (M_1 - m_1)(M_2 - m_2) \sum_{i \in S} p_i \left(P_n - \sum_{i \in S} p_i \right)$$

$$\leq \frac{1}{4} P_n^2 (M_1 - m_1)(M_2 - m_2).$$

Note that the first part of Equation (1.53) has been obtained in Section 1.5 (cf. inequality 1.27).

PROOF By using the result obtained above in Section 1.8, we have from Equation (1.39)

$$\frac{1}{P_n} \sum_{i=1}^n p_i a_i^2 - \left(\frac{1}{P_n} \sum_{i=1}^n p_i a_i \right)^2 \leq \frac{(M_1 - m_1)^2}{P_n^2} \sum_{i \in S} p_i \left(P_n - \sum_{i \in S} p_i \right) \tag{1.54}$$

and

$$\frac{1}{P_n} \sum_{i=1}^n p_i b_i^2 - \left(\frac{1}{P_n} \sum_{i=1}^n p_i b_i \right)^2 \leq \frac{(M_2 - m_2)^2}{P_n^2} \sum_{i \in S} p_i \left(P_n - \sum_{i \in S} p_i \right). \tag{1.55}$$

By utilising the CBS inequality for double sums,

$$\sum_{i,j=1}^n p_{ij} x_{ij}^2 \sum_{i,j=1}^n p_{ij} y_{ij}^2 \geq \left(\sum_{i,j=1}^n p_{ij} x_{ij} y_{ij} \right)^2$$

with the changes

$$p_{ij} = p_i p_j, \quad x_{ij} = a_i - a_j, \quad y_{ij} = b_i - b_j$$

produces

$$\left(\frac{1}{P_n} \sum_{i=1}^n p_i a_i b_i - \frac{1}{P_n} \sum_{i=1}^n p_i a_i \cdot \frac{1}{P_n} \sum_{i=1}^n p_i b_i \right)^2$$

$$\leq \left[\frac{1}{P_n} \sum_{i=1}^n p_i a_i^2 - \left(\frac{1}{P_n} \sum_{i=1}^n p_i a_i \right)^2 \right]$$

$$\times \left[\frac{1}{P_n} \sum_{i=1}^n p_i b_i^2 - \left(\frac{1}{P_n} \sum_{i=1}^n p_i b_i \right)^2 \right]. \tag{1.56}$$

This proves the first part of (1.53).

The second part follows by the elementary inequality

$$ab \leq \frac{1}{4}(a+b)^2, \quad a, b \in \mathbb{R}$$

for the choices $a := \sum_{i \in S} p_i$, $b := P_n - \sum_{i \in S} p_i$. ⬚

Comments

(a) In 1914, Schweitzer [159] proved the following result:

If $\mathbf{a} = (a_1, \ldots, a_n)$ *is a sequence of real numbers such that* $0 < m \leq a_i \leq M < \infty$ $(i \in \{1, \ldots, n\})$, *then*

$$\left(\frac{1}{n} \sum_{i=1}^{n} a_i \right) \left(\frac{1}{n} \sum_{i=1}^{n} \frac{1}{a_i} \right) \leq \frac{(M+m)^2}{4mM}. \tag{1.57}$$

(b) In 1972, Lupaş [133] (see also Equation 1.28) proved the following refinement of Schweitzer's result which also gives the best bound when n is odd:

If $\mathbf{a} = (a_1, \ldots, a_n)$ *is a sequence of real numbers such that* $0 < m \leq a_i \leq M < \infty$ $(i \in \{1, \ldots, n\})$, *one has*

$$\left(\frac{1}{n} \sum_{i=1}^{n} a_i \right) \left(\frac{1}{n} \sum_{i=1}^{n} \frac{1}{a_i} \right) \leq \frac{\left(\left[\frac{n}{2} \right] M + \left[\frac{n+1}{2} \right] m \right) \left(\left[\frac{n+1}{2} \right] M + \left[\frac{n}{2} \right] m \right)}{Mm}, \tag{1.58}$$

where $[\cdot]$ *is the greatest integer function.*

(c) The following weighted version of Schweitzer's result may be stated:

If $0 < m \leq a_i \leq M < \infty$, $i \in \{1, \ldots, n\}$ *and* S *is a subset of* $\{1, \ldots, n\}$ *that minimises the expression*

$$\left| \sum_{i \in S} p_i - \frac{P_n}{2} \right|,$$

then we have the inequality

$$\left(\sum_{i=1}^{n} p_i a_i \right) \left(\sum_{i=1}^{n} \frac{p_i}{a_i} \right) \leq P_n^2 + \frac{(M-m)^2}{Mm} \sum_{i \in S} p_i \left(P_n - \sum_{i \in S} p_i \right) \tag{1.59}$$

$$\leq \frac{(M+m)^2}{4Mm} P_n^2.$$

PROOF We follow the proof of Andrica and Badea [2]. We obtain from (1.53) with $b_i = \frac{1}{a_i}$, $m_1 = m$, $M_1 = M$, $m_2 = \frac{1}{M}$, $M_2 = \frac{1}{m}$, the following result:

$$\left| P_n^2 - \sum_{i=1}^{n} p_i a_i \sum_{i=1}^{n} p_i \frac{1}{a_i} \right| \leq (M-m) \left(\frac{1}{m} - \frac{1}{M} \right) \sum_{i \in S} p_i \left(P_n - \sum_{i \in S} p_i \right),$$

that leads, in a simple manner, to (1.59).   ▯

(d) We may now prove the following counterpart for the weighted (CBS)-inequality that improves the additive version of Cassels' inequality:

Let $a = (a_1, \ldots, a_n)$, $b = (b_1, \ldots, b_n)$ be two sequences of positive real numbers with the property that

$$0 < m \le \frac{b_i}{a_i} \le M < \infty \quad \text{for each } i \in \{1, \ldots, n\} \tag{1.60}$$

and $p = (p_1, \ldots, p_n)$ a sequence of nonnegative real numbers such that $P_n := \sum_{i=1}^n p_i > 0$. If S is a subset of $\{1, \ldots, n\}$ that minimises the expression

$$\left| \sum_{i \in S} p_i a_i b_i - \frac{1}{2} \sum_{i=1}^n p_i a_i b_i \right|, \tag{1.61}$$

then we have the inequality

$$\sum_{i=1}^n p_i a_i^2 \sum_{i=1}^n p_i b_i^2 - \left(\sum_{i=1}^n p_i a_i b_i \right)^2 \tag{1.62}$$

$$\le \frac{(M-m)^2}{Mm} \sum_{i \in S} p_i a_i b_i \left(\sum_{i=1}^n p_i a_i b_i - \sum_{i \in S} p_i a_i b_i \right)$$

$$\le \frac{(M-m)^2}{4Mm} \left(\sum_{i=1}^n p_i a_i b_i \right)^2.$$

PROOF   Applying (1.53) for $a_i = x_i$, $b_i = \frac{1}{x_i}$, and $p_i = q_i x_i$ we may deduce the inequality

$$\sum_{i=1}^n q_i x_i^2 \sum_{i=1}^n q_i - \left(\sum_{i=1}^n q_i x_i \right)^2$$

$$\le \frac{(M-m)^2}{Mm} \sum_{i \in S} q_i x_i \left(\sum_{i=1}^n q_i x_i - \sum_{i \in S} q_i x_i \right), \tag{1.63}$$

provided $q_i \ge 0$, $\sum_{i=1}^n q_i > 0$, $0 < m \le x_i \le M < \infty$, for $i \in \{1, \ldots, n\}$ and S is a subset of $\{1, \ldots, n\}$ that minimises the expression

$$\left| \sum_{i \in S} q_i x_i - \frac{1}{2} \sum_{i=1}^n q_i x_i \right|. \tag{1.64}$$

Further, if in (1.63) we choose $q_i = p_i a_i^2$, $x_i = \frac{b_i}{a_i} \in [m, M]$ for $i \in \{1, \ldots, n\}$, we deduce the desired result in (1.62).   ▯

(e) The following result provides a refinement of Cassels' inequality:

With the above assumptions we have the inequality

$$1 \leq \frac{\sum_{i=1}^{n} p_i a_i^2 \sum_{i=1}^{n} p_i b_i^2}{\left(\sum_{i=1}^{n} p_i a_i b_i\right)^2} \tag{1.65}$$

$$\leq 1 + \frac{(M-m)^2}{Mm} \cdot \frac{\sum_{i \in S} p_i a_i b_i}{\sum_{i=1}^{n} p_i a_i b_i} \left(1 - \frac{\sum_{i \in S} p_i a_i b_i}{\sum_{i=1}^{n} p_i a_i b_i}\right)$$

$$\leq \frac{(M+m)^2}{4Mm}.$$

In particular, we have the unweighted version of Cassels' inequality as follows:

Assume that a and b satisfy (1.60). If S is a subset of $\{1, \ldots, n\}$ that minimises the expression

$$\left|\sum_{i \in S} a_i b_i - \frac{1}{2} \sum_{i=1}^{n} a_i b_i\right| \tag{1.66}$$

then one has the inequality

$$1 \leq \frac{\sum_{i=1}^{n} a_i^2 \sum_{i=1}^{n} b_i^2}{\left(\sum_{i=1}^{n} a_i b_i\right)^2} \tag{1.67}$$

$$\leq 1 + \frac{(M-m)^2}{Mm} \cdot \frac{\sum_{i \in S} a_i b_i}{\sum_{i=1}^{n} a_i b_i} \left(1 - \frac{\sum_{i \in S} a_i b_i}{\sum_{i=1}^{n} a_i b_i}\right)$$

$$\leq \frac{(M+m)^2}{4Mm}.$$

In particular, we may obtain the following refinement of the Pólya-Szegö inequality:

Assume that

$$0 < \alpha \leq a_i \leq A < \infty, \quad 0 < \beta \leq b_i \leq B < \infty \quad \text{for} \quad i \in \{1, \ldots, n\}. \tag{1.68}$$

If S is a subset of $\{1, \ldots, n\}$ that minimises the expression (1.66), then one has the inequality

$$1 \leq \frac{\sum_{i=1}^{n} a_i^2 \sum_{i=1}^{n} b_i^2}{\left(\sum_{i=1}^{n} a_i b_i\right)^2} \tag{1.69}$$

$$\leq 1 + \frac{(AB - \alpha\beta)^2}{\alpha\beta AB} \cdot \frac{\sum_{i \in S} a_i b_i}{\sum_{i=1}^{n} a_i b_i} \left(1 - \frac{\sum_{i \in S} a_i b_i}{\sum_{i=1}^{n} a_i b_i}\right)$$

$$\leq \frac{(AB + \alpha\beta)^2}{4\alpha\beta AB}.$$

1.11   Čebyšev Type Inequalities

For $p = (p_1, \ldots, p_n)$, $a = (a_1, \ldots, a_n)$, and $b = (b_1, \ldots, b_n)$ n-tuples of real numbers, consider the Čebyšev functional

$$C_n(p; a, b) := P_n \sum_{i=1}^n p_i a_i b_i - \sum_{i=1}^n p_i a_i \cdot \sum_{i=1}^n p_i b_i, \qquad (1.70)$$

where $P_n := \sum_{i=1}^n p_i$.

In 1882–1883, Čebyšev [15, 16] proved that if a and b are monotonic in the same (opposite) sense and p is nonnegative, then

$$C_n(p; a, b) \geq (\leq) 0. \qquad (1.71)$$

The inequality (1.71) was mentioned by Hardy, Littlewood, and Pólya in their book *Inequalities* [121] in 1934 in a more general setting of synchronous sequences, that is, if a and b are synchronous (asynchronous), meaning that

$$(a_i - a_j)(b_i - b_j) \geq (\leq) 0 \text{ for each } i, j \in \{1, \ldots, n\}, \qquad (1.72)$$

then (1.71) holds.

A relaxation of the synchronicity condition was provided by Biernacki in 1951 [9], who showed that for a nonnegative p, if a and b are monotonic in mean in the same sense, that is, if for $P_k := \sum_{i=1}^k p_i$,

$$\frac{1}{P_k} \sum_{i=1}^k p_i a_i \leq (\geq) \frac{1}{P_{k+1}} \sum_{i=1}^{k+1} p_i a_i, k \in \{1, \ldots, n-1\} \text{ and} \qquad (1.73)$$

$$\frac{1}{P_k} \sum_{i=1}^k p_i b_i \leq (\geq) \frac{1}{P_{k+1}} \sum_{i=1}^{k+1} p_i b_i, k \in \{1, \ldots, n-1\},$$

then (1.71) holds true for the " $\geq$ " sign. If they are monotonic in mean in the opposite sense, then (1.71) holds true for the " $\leq$ " sign.

For general real weights p, Mitrinović and Pečarić [140] have shown that the inequality (1.71) holds if

$$0 \leq P_k \leq P_n \text{ for } k \in \{1, \ldots, n-1\}, \qquad (1.74)$$

and a, b are monotonic in the same (opposite) sense.

The following identity is well known in the literature as *Sonin's identity* (see Sonin [161] and Mitrinović, Pečarić, and Fink [141, p. 246]):

$$C_n(p; a, b) = \sum_{i=1}^n p_i \left(P_n a_i - \sum_{j=1}^n p_j a_j \right)(b_i - \beta), \qquad (1.75)$$

for any real number β.

Another well-known identity in terms of double sums is the following one known in the literature as *Korkine's identity* (see Korkine [129] and Mitrinović, Pečarić, and Fink [141, p. 242]):

$$C_n\left(p;a,b\right) = \sum_{i=1}^{n}\sum_{j=1}^{n}p_ip_j\left(a_i - a_j\right)\left(b_i - b_j\right). \tag{1.76}$$

The following new identity provides a different insight (see Dragomir [55]):

Let $p = (p_1,\ldots,p_n)$, $a = (a_1,\ldots,a_n)$, and $b = (b_1,\ldots,b_n)$ be n-tuples of real numbers. If we define

$$P_i := \sum_{k=1}^{i}p_k, \bar{P}_i := P_n - P_i, \quad i \in \{1,\ldots,n-1\},$$

$$A_i\left(p\right) := \sum_{k=1}^{i}p_ka_k, \bar{A}_i\left(p\right) := A_n\left(p\right) - A_i\left(p\right), \quad i \in \{1,\ldots,n-1\},$$

then we have the identity

$$C_n\left(p;a,b\right) \tag{1.77}$$

$$= \sum_{i=1}^{n-1}\det\left(\begin{array}{cc} P_i & P_n \\ A_i\left(p\right) & A_n\left(p\right) \end{array}\right)\cdot\Delta b_i$$

$$= P_n\sum_{i=1}^{n-1}P_i\left(\frac{A_n\left(p\right)}{P_n} - \frac{A_i\left(p\right)}{P_i}\right)\cdot\Delta b_i \quad (\text{if } P_n, P_i \neq 0, i \in \{1,\ldots,n-1\})$$

$$= \sum_{i=1}^{n-1}P_i\bar{P}_i\left(\frac{\bar{A}_i\left(p\right)}{\bar{P}_i} - \frac{A_i\left(p\right)}{P_i}\right)\cdot\Delta b_i \quad (\text{if } P_i, \bar{P}_i \neq 0, i \in \{1,\ldots,n-1\})$$

where $\Delta b_i := b_{i+1} - b_i$ $(i \in \{1,\ldots,n-1\})$ is the forward difference.

PROOF We use the following well-known summation by parts formula:

$$\sum_{l=p}^{q-1}d_l\Delta v_l = d_lv_l\big|_p^q - \sum_{l=p}^{q-1}v_{l+1}\Delta d_l, \tag{1.78}$$

where d_l, v_l are real numbers, and $l = p,\ldots,q-1$ $(q > p; p,q$ are natural numbers$)$.

If in (1.78) we choose $p = 1, q = n, d_i = P_iA_n\left(p\right) - P_nA_i\left(p\right)$ and $v_i =$

b_i $(i \in \{1, \ldots, n-1\})$, then we get

$$\sum_{i=1}^{n-1} (P_i A_n (p) - P_n A_i (p)) \cdot \Delta b_i$$

$$= [P_i A_n (p) - P_n A_i (p)] \cdot b_i|_1^n - \sum_{i=1}^{n-1} \Delta (P_i A_n (p) - P_n A_i (p)) \cdot b_{i+1}$$

$$= [P_n A_n (p) - P_n A_n (p)] \cdot b_n - [P_1 A_n (p) - P_n A_1 (p)] \cdot b_1$$

$$\quad - \sum_{i=1}^{n-1} [P_{i+1} A_n (p) - P_n A_{i+1} (p) - P_i A_n (p) + P_n A_i (p)] \cdot b_{i+1}$$

$$= P_n p_1 a_1 b_1 - p_1 b_1 A_n (p) - \sum_{i=1}^{n-1} (p_{i+1} A_n (p) - P_n p_{i+1} a_{i+1}) \cdot b_{i+1}$$

$$= P_n p_1 a_1 b_1 - p_1 b_1 A_n (p) - A_n (p) \sum_{i=1}^{n-1} p_{i+1} b_{i+1} + P_n \sum_{i=1}^{n-1} p_{i+1} a_{i+1} b_{i+1}$$

$$= P_n \sum_{i=1}^{n} p_i a_i b_i - \sum_{i=1}^{n} p_i a_i \cdot \sum_{i=1}^{n} p_i b_i$$

$$= C_n (p; a, b),$$

which produces the first identity in (1.77).

The second and third identities are obvious; and we omit the details. □

Before proving the second result, we need the following interesting identity (see Dragomir [55]):

Let $p = (p_1, \ldots, p_n)$ and $a = (a_1, \ldots, a_n)$ be n-tuples of real numbers. Then we have the result

$$\det \begin{pmatrix} P_i & P_n \\ A_i (p) & A_n (p) \end{pmatrix} = \sum_{j=1}^{n-1} P_{\min\{i,j\}} \bar{P}_{\max\{i,j\}} \cdot \Delta a_j, \qquad (1.79)$$

for each $i \in \{1, \ldots, n-1\}$.

PROOF   Define, for $i \in \{1, \ldots, n-1\}$,

$$K (i) := \sum_{j=1}^{n-1} P_{\min\{i,j\}} \bar{P}_{\max\{i,j\}} \cdot \Delta a_j.$$

We have

$$K(i) = \sum_{j=1}^{i} P_{\min\{i,j\}} \bar{P}_{\max\{i,j\}} \cdot \Delta a_j + \sum_{j=i+1}^{n-1} P_{\min\{i,j\}} \bar{P}_{\max\{i,j\}} \cdot \Delta a_j \quad (1.80)$$

$$= \sum_{j=1}^{i} P_j \bar{P}_i \cdot \Delta a_j + \sum_{j=i+1}^{n-1} P_i \bar{P}_j \cdot \Delta a_j$$

$$= \bar{P}_i \sum_{j=1}^{i} P_j \cdot \Delta a_j + P_i \sum_{j=i+1}^{n-1} \bar{P}_j \cdot \Delta a_j.$$

Using the summation by parts formula, we have

$$\sum_{j=1}^{i} P_j \cdot \Delta a_j = P_j \cdot a_j \Big|_{1}^{i+1} - \sum_{j=1}^{i} (P_{j+1} - P_j) \cdot a_{j+1} \quad (1.81)$$

$$= P_{i+1} a_{i+1} - p_1 a_1 - \sum_{j=1}^{i} p_{j+1} \cdot a_{j+1}$$

$$= P_{i+1} a_{i+1} - \sum_{j=1}^{i+1} p_j \cdot a_j$$

and

$$\sum_{j=i+1}^{n-1} \bar{P}_j \cdot \Delta a_j = \bar{P}_j \cdot a_j \Big|_{i+1}^{n} - \sum_{j=i+1}^{n-1} (\bar{P}_{j+1} - \bar{P}_j) \cdot a_{j+1} \quad (1.82)$$

$$= \bar{P}_n a_n - \bar{P}_{i+1} a_{i+1} - \sum_{j=i+1}^{n-1} (P_n - P_{j+1} - P_n + P_j) \cdot a_{j+1}$$

$$= -\bar{P}_{i+1} a_{i+1} + \sum_{j=i+1}^{n-1} p_{j+1} \cdot a_{j+1}.$$

By employing (1.81) and (1.82) we have from (1.80)

$$K(i) = \bar{P}_i \left(P_{i+1} a_{i+1} - \sum_{j=1}^{i+1} p_j \cdot a_j \right) + P_i \left(\sum_{j=i+1}^{n-1} p_{j+1} \cdot a_{j+1} - \bar{P}_{i+1} a_{i+1} \right)$$

$$= \bar{P}_i P_{i+1} a_{i+1} - P_i \bar{P}_{i+1} a_{i+1} - \bar{P}_i \sum_{j=1}^{i+1} p_j \cdot a_j + P_i \sum_{j=i+1}^{n-1} p_{j+1} \cdot a_{j+1}$$

$$= a_{i+1} \left((P_n - P_i) P_{i+1} - P_i (P_n - P_{i+1}) \right)$$

$$+ P_i \sum_{j=i+1}^{n-1} p_{j+1} \cdot a_{j+1} - \bar{P}_i \sum_{j=1}^{i+1} p_j \cdot a_j$$

$$= P_n p_{i+1} a_{i+1} + P_i \sum_{j=i+1}^{n-1} p_{j+1} \cdot a_{j+1} - \bar{P}_i \sum_{j=1}^{i+1} p_j \cdot a_j$$

$$= \left(P_i + \bar{P}_i \right) p_{i+1} a_{i+1} + P_i \sum_{j=i+1}^{n-1} p_{j+1} \cdot a_{j+1} - \bar{P}_i \sum_{j=1}^{i+1} p_j \cdot a_j$$

$$= P_i \sum_{j=i+1}^{n} p_j \cdot a_j - \bar{P}_i \sum_{j=1}^{i} p_j \cdot a_j = P_i \bar{A}_i \left(p \right) - \bar{P}_i A_i \left(p \right)$$

$$= \det \begin{pmatrix} P_i & P_n \\ A_i \left(p \right) & A_n \left(p \right) \end{pmatrix}$$

which proves the identity. ☐

We are able now to state and prove the following identity for the Čebyšev functional in terms of the forward differences:

With the above assumptions, we have the equality

$$C_n \left(p; a, b \right) = \sum_{i=1}^{n-1} \sum_{j=1}^{n-1} P_{\min\{i,j\}} \bar{P}_{\max\{i,j\}} \cdot \Delta a_j \cdot \Delta b_j. \qquad (1.83)$$

The proof is obvious by the above identities, (1.77) and (1.79), and we omit the details.

The identity (1.83) was stated without a proof in Mitrinović and Pečarić [140]. It also may be found in Mitrinović, Pečarić, and Fink [141, p. 281], again without a proof.

Comments

We may point out the following result concerning the positivity of the Čebyšev functional [55]:

Let $p = (p_1, \ldots, p_n), a = (a_1, \ldots, a_n)$, and $b = (b_1, \ldots, b_n)$ be n-tuples of real numbers. If b is monotonic nondecreasing and either

(i) $\det \begin{pmatrix} P_i & P_n \\ A_i \left(p \right) & A_n \left(p \right) \end{pmatrix} \geq 0$ *for each $i \in \{1, \ldots, n-1\}$;*

or

(ii) $P_i > 0$ *for any $i \in \{1, \ldots, n\}$ and*

$$\frac{A_n \left(p \right)}{P_n} \geq \frac{A_i \left(p \right)}{P_i} \text{ for each } i \in \{1, \ldots, n-1\};$$

or

(iii) $0 < P_i < P_n$ *for every* $i \in \{1, \ldots, n-1\}$ *and*

$$\frac{\bar{A}_i(p)}{\bar{P}_i} \geq \frac{A_i(p)}{P_i} \text{ for each } i \in \{1, \ldots, n-1\};$$

then

$$C_n(p; a, b) \geq 0. \tag{1.84}$$

If b *is monotonic nonincreasing and either* (i) *or* (ii) *or* (iii) *from above holds, then the inequality* (1.84) *is reversed.*

The proof of inequality (1.84) follows from the identities incorporated in (1.77); and we omit the details.

By using the second identity, (1.83), we may state the following result as well [55]:

Let $p = (p_1, \ldots, p_n)$, $a = (a_1, \ldots, a_n)$, *and* $b = (b_1, \ldots, b_n)$ *be* n-*tuples of real numbers. If* a *and* b *are monotonic in the same sense and*

$$P_{\min\{i,j\}} \bar{P}_{\max\{i,j\}} \geq 0 \text{ for each } i, j \in \{1, \ldots, n-1\}, \tag{1.85}$$

then (1.84) *holds. If* a *and* b *are monotonic in the opposite sense and the condition* (1.85) *is valid, then the reverse inequality in* (1.84) *is true.*

1.12 De Bruijn's Inequality

If $a = (a_1, \ldots, a_n)$ *is an* n-*tuple of real numbers and* $z = (z_1, \ldots, z_n)$ *is an* n-*tuple of complex numbers, then*

$$\left| \sum_{k=1}^{n} a_k z_k \right|^2 \leq \frac{1}{2} \sum_{k=1}^{n} a_k^2 \left[\sum_{k=1}^{n} |z_k|^2 + \left| \sum_{k=1}^{n} z_k^2 \right| \right]. \tag{1.86}$$

Equality holds in (1.86) *if and only if for* $k \in \{1, \ldots, n\}$, $a_k = \text{Re}(\lambda z_k)$, *where* λ *is a complex number such that* $\lambda^2 \sum_{k=1}^{n} z_k^2$ *is a nonnegative real number.*

PROOF We follow the proof in Mitrinović, Pečarić, and Fink [141, pp. 89–90].

By a simultaneous rotation of all the $z_k's$ about the origin, we get

$$\sum_{k=1}^{n} a_k z_k \geq 0.$$

This rotation does not affect the moduli

$$\left|\sum_{k=1}^{n} a_k z_k\right|, \qquad \left|\sum_{k=1}^{n} z_k^2\right| \quad \text{and} \quad |z_k| \quad \text{for} \quad k \in \{1, \ldots, n\}.$$

Hence, it is sufficient to prove the inequality (1.86) for the case where $\sum_{k=1}^{n} a_k z_k \geq 0$.

If we consider that $z_k := x_k + i y_k$, $k \in \{1, \ldots, n\}$, then, by the Cauchy-Bunyakovsky-Schwarz inequality for real numbers, we have

$$\left|\sum_{k=1}^{n} a_k z_k\right|^2 = \left(\sum_{k=1}^{n} a_k z_k\right)^2 = \left(\sum_{k=1}^{n} a_k x_k\right)^2 \leq \sum_{k=1}^{n} a_k^2 \sum_{k=1}^{n} x_k^2. \qquad (1.87)$$

Since

$$2x_k^2 = |z_k|^2 + \operatorname{Re}\left(z_k^2\right)$$

for any $k \in \{1, \ldots, n\}$, and so by (1.87), we obtain that

$$\left|\sum_{k=1}^{n} a_k z_k\right|^2 \leq \frac{1}{2} \sum_{k=1}^{n} a_k^2 \left[\sum_{k=1}^{n} |z_k|^2 + \sum_{k=1}^{n} \operatorname{Re}\left(z_k^2\right)\right]. \qquad (1.88)$$

Since

$$\sum_{k=1}^{n} \operatorname{Re}\left(z_k^2\right) = \operatorname{Re}\left(\sum_{k=1}^{n} z_k^2\right) \leq \left|\sum_{k=1}^{n} z_k^2\right|, \qquad (1.89)$$

we deduce the desired inequality (1.86) from (1.89). □

Comment

If one chooses $a_k = 1$, $k \in \{1, \ldots, n\}$, then

$$\left|\frac{1}{n} \sum_{k=1}^{n} z_k\right| \leq \frac{1}{2}\left[\left|\frac{1}{n} \sum_{k=1}^{n} z_k^2\right| + \frac{1}{n} \sum_{k=1}^{n} |z_k|^2\right] \qquad (1.90)$$

for any complex n-tuple $z = (z_1, \ldots, z_n)$.

1.13   Daykin-Eliezer-Carlitz's Inequality

Let $a = (a_1, \ldots, a_n)$ and $b = (b_1, \ldots, b_n)$ be two sequences of positive numbers $f, g : [0, \infty) \times [0, \infty) \to (0, \infty)$. The inequality

$$\left(\sum_{i=1}^{n} a_i b_i\right)^2 \leq \sum_{i=1}^{n} f\left(a_i, b_i\right) \sum_{i=1}^{n} g\left(a_i, b_i\right) \leq \sum_{i=1}^{n} a_i^2 \sum_{i=1}^{n} b_i^2 \qquad (1.91)$$

holds if and only if

$$f(a,b) g(a,b) = a^2 b^2, \tag{1.92}$$

$$f(ka, kb) = k^2 f(a,b), \tag{1.93}$$

and

$$\frac{bf(a,1)}{af(b,1)} + \frac{af(b,1)}{bf(a,1)} \le \frac{a}{b} + \frac{b}{a} \tag{1.94}$$

for any $a, b, k > 0$.

PROOF   We will follow the proof in Mitrinović, Pečarić, and Fink [141, pp. 88–89].

Necessity: Indeed, for $n = 1$, the inequality (1.91) becomes

$$(ab)^2 \le f(a,b) g(a,b) \le a^2 b^2, \quad a, b > 0$$

which gives (1.92).

For $n = 2$ in (1.91), using (1.92), we get

$$2a_1 b_1 a_2 b_2 \le f(a_1, b_1) g(a_2, b_2) + f(a_2, b_2) g(a_1, b_1) \tag{1.95}$$

$$\le a_1^2 b_2^2 + a_2^2 b_1^2$$

for any $a_i, b_i > 0$, $i \in \{1,2\}$.

By eliminating g in (1.95), we get

$$2 \le \frac{f(a_1, b_1)}{f(a_2, b_2)} \cdot \frac{a_2 b_2}{a_1 b_1} + \frac{f(a_2, b_2)}{f(a_1, b_1)} \cdot \frac{a_1 b_1}{a_2 b_2} \le \frac{a_1 b_2}{a_2 b_1} + \frac{a_2 b_1}{a_1 b_2} \tag{1.96}$$

for any $a_i, b_i > 0$, $i \in \{1,2\}$.

By substituting in (1.96) a, b for a_1, b_1 and ka, kb for a_2, b_2 $(k > 0)$, we get

$$2 \le \frac{f(a,b)}{f(ka,kb)} k^2 + \frac{f(ka,kb)}{f(a,b)} k^{-2} \le 2.$$

This is valid only if $k^2 f(a,b) / f(ka, kb) = 1$, which is the condition (1.93).

Using (1.96) for $a_1 = a$, $b_1 = 1$, $a_2 = b$, and $b_2 = 1$, we have

$$2 \le \frac{\frac{f(a,1)}{a}}{\frac{f(b,1)}{b}} + \frac{\frac{f(b,1)}{b}}{\frac{f(a,1)}{a}} \le \frac{a}{b} + \frac{b}{a}. \tag{1.97}$$

The first inequality in (1.97) is always satisfied while the second inequality is equivalent to (1.94).

Sufficiency: Suppose that (1.92) holds. Then inequality (1.91) can be written in the form

$$2 \sum_{1 \le i < j \le n} a_i b_i a_j b_j \le \sum_{1 \le i < j \le n} [f(a_i, b_i) g(a_j, b_j) + f(a_j, b_j) g(a_i, b_i)]$$

$$\le \sum_{1 \le i < j \le n} (a_i^2 b_j^2 + a_j^2 b_i^2).$$

Therefore it is enough to prove that

$$2a_i b_i a_j b_j \leq f(a_i, b_i) g(a_j, b_j) + f(a_j, b_j) g(a_i, b_i) \qquad (1.98)$$
$$\leq a_i^2 b_j^2 + a_j^2 b_i^2.$$

Suppose that (1.94) holds true. Then (1.97) is valid; and by putting $a = \frac{a_i}{b_i}$, $b = \frac{a_i}{b_i}$ in (1.96) and using (1.93), we get

$$2 \leq \frac{f(a_i, b_i)}{f(a_j, b_j)} \cdot \frac{a_j b_j}{a_i b_i} + \frac{f(a_j, b_j)}{f(a_i, b_i)} \cdot \frac{a_i b_i}{a_j b_j} \leq \frac{a_i b_j}{a_j b_i} + \frac{a_j b_i}{a_i b_j}.$$

By multiplying the last inequality by $a_i b_i a_j b_j > 0$, $i, j \in \{1, \ldots, n\}$ and using (1.92), we obtain (1.98). $\qquad \square$

In the original version of the result, the condition (1.94) is given as [44]

$$f(b, 1) \leq f(a, 1) \quad \text{and} \quad \frac{f(a, 1)}{a^2} \leq \frac{f(b, 1)}{b^2}$$

for $a \geq b > 0$. In this book, we use the version considered in Mitrinović, Pečarić, and Fink [141, p. 87].

Comments

(a) The choice

$$f(x, y) = x^2 + y^2, \qquad g(x, y) = \frac{x^2 y^2}{x^2 + y^2}$$

provides *Milne's inequality*:

$$\left(\sum_{i=1}^{n} a_i b_i \right)^2 \leq \sum_{i=1}^{n} (a_i^2 + b_i^2) \sum_{i=1}^{n} \frac{a_i^2 b_i^2}{a_i^2 + b_i^2} \leq \sum_{i=1}^{n} a_i^2 \sum_{i=1}^{n} b_i^2. \qquad (1.99)$$

(b) The choice

$$f(x, y) = x^{1+\alpha} y^{1-\alpha}, \qquad g(x, y) = x^{1-\alpha} y^{1+\alpha}, \qquad \alpha \in [0, 1]$$

provides *Callebaut's inequality*:

$$\left(\sum_{i=1}^{n} a_i b_i \right)^2 \leq \sum_{i=1}^{n} a_i^{1+\alpha} b_i^{1-\alpha} \sum_{i=1}^{n} a_i^{1-\alpha} b_i^{1+\alpha} \leq \sum_{i=1}^{n} a_i^2 \sum_{i=1}^{n} b_i^2. \qquad (1.100)$$

1.14 Wagner's Inequality

Let $a = (a_1, \ldots, a_n)$ and $b = (b_1, \ldots, b_n)$ be sequences of real numbers. If $0 \leq x \leq 1$, then

$$\left(\sum_{k=1}^{n} a_k b_k + x \sum_{1 \leq i \neq j \leq n} a_i b_j \right)^2$$

$$\leq \left[\sum_{k=1}^{n} a_k^2 + 2x \sum_{1 \leq i < j \leq n} a_i a_j \right] \left[\sum_{k=1}^{n} b_k^2 + 2x \sum_{1 \leq i < j \leq n} b_i b_j \right]. \quad (1.101)$$

PROOF We follow the proof in Mitrinović, Pečarić, and Fink [141, p. 85]. For any $x \in [0, 1]$, consider the following quadratic polynomial in y:

$$P(y) := (1 - x) \sum_{k=1}^{n} (a_k y - b_k)^2 + x \left[\sum_{k=1}^{n} (a_k y - b_k) \right]^2$$

$$= (1 - x) \left[y^2 \sum_{k=1}^{n} a_k^2 - 2y \sum_{k=1}^{n} a_k b_k + \sum_{k=1}^{n} b_k^2 \right]$$

$$+ x \left[y^2 \left(\sum_{k=1}^{n} a_k \right)^2 - 2y \left(\sum_{k=1}^{n} a_k \right) \left(\sum_{k=1}^{n} b_k \right) + \left(\sum_{k=1}^{n} b_k \right)^2 \right]$$

$$= \left[(1 - x) \sum_{k=1}^{n} a_k^2 + x \left(\sum_{k=1}^{n} a_k \right)^2 \right] y^2$$

$$- 2y \left[(1 - x) \sum_{k=1}^{n} a_k b_k + x \sum_{k=1}^{n} a_k \sum_{k=1}^{n} b_k \right]$$

$$+ (1 - x) \sum_{k=1}^{n} b_k^2 + x \left(\sum_{k=1}^{n} b_k \right)^2$$

$$= \left\{ \sum_{k=1}^{n} a_k^2 + x \left[\left(\sum_{k=1}^{n} a_k \right)^2 - \sum_{k=1}^{n} a_k^2 \right] \right\} y^2$$

$$- 2y \left[\sum_{k=1}^{n} a_k b_k + x \left(\sum_{k=1}^{n} a_k \sum_{k=1}^{n} b_k - \sum_{k=1}^{n} a_k b_k \right) \right]$$

$$+ \sum_{k=1}^{n} b_k^2 + x \left[\left(\sum_{k=1}^{n} b_k \right)^2 - \sum_{k=1}^{n} b_k^2 \right].$$

Since

$$\left(\sum_{k=1}^{n} a_k\right)^2 - \sum_{k=1}^{n} a_k^2 = 2 \sum_{1 \le i < j \le n} a_i a_j,$$

$$\sum_{k=1}^{n} a_k \sum_{k=1}^{n} b_k - \sum_{k=1}^{n} a_k b_k = \sum_{1 \le i \ne j \le n} a_i b_j,$$

and

$$\left(\sum_{k=1}^{n} b_k\right)^2 - \sum_{k=1}^{n} b_k^2 = 2 \sum_{1 \le i < j \le n} b_i b_j,$$

we have

$$P(y) = \left(\sum_{k=1}^{n} a_k^2 + 2x \sum_{1 \le i < j \le n} a_i a_j\right) y^2$$

$$- 2\left(\sum_{k=1}^{n} a_k b_k + x \sum_{1 \le i \ne j \le n} a_i b_j\right) y + \sum_{k=1}^{n} b_k^2 + 2x \sum_{1 \le i < j \le n} b_i b_j.$$

By the definition of P, we have that $P(y) \ge 0$ for any $y \in \mathbb{R}$. It follows that the discriminant $\Delta \le 0$, that is,

$$\frac{1}{4}\Delta = \left(\sum_{k=1}^{n} a_k b_k + x \sum_{1 \le i \ne j \le n} a_i b_j\right)^2$$

$$- \left(\sum_{k=1}^{n} a_k^2 + 2x \sum_{1 \le i < j \le n} a_i a_j\right)\left(\sum_{k=1}^{n} b_k^2 + 2x \sum_{1 \le i < j \le n} b_i b_j\right).$$

This completes the proof. □

If $x = 0$, then from (1.101) one may recapture the Cauchy-Bunyakovsky-Schwarz inequality, namely,

$$\left(\sum_{k=1}^{n} a_k b_k\right)^2 \le \sum_{k=1}^{n} a_k^2 \sum_{k=1}^{n} b_k^2. \tag{1.102}$$

Comments

If $a = (a_1, \ldots, a_n)$, $b = (b_1, \ldots, b_n)$ are complex n-tuples and $x \in [0, 1]$, then one may get the following version of Wagner's inequality obtained by

Dragomir [55, pp. 25–27],

$$\left[\sum_{k=1}^{n} \operatorname{Re}\left(a_k \overline{b_k}\right) + x \sum_{1 \le i \ne j \le n} \operatorname{Re}\left(a_i \overline{b_j}\right)\right]^2$$

$$\le \left[\sum_{k=1}^{n} |a_k|^2 + 2x \sum_{1 \le i < j \le n} \operatorname{Re}\left(a_i \overline{a_j}\right)\right]$$

$$\times \left[\sum_{k=1}^{n} |b_k|^2 + 2x \sum_{1 \le i < j \le n} \operatorname{Re}\left(b_i \overline{b_j}\right)\right], \quad (1.103)$$

where $x \in [0,1]$ and $\bar{z}$ denotes the complex conjugate of z.

1.15   The Pólya-Szegö Inequality

Let $a = (a_1, \ldots, a_n)$ and $b = (b_1, \ldots, b_n)$ be two sequences of positive numbers. If

$$0 < \alpha \le a_i \le A < \infty, \qquad 0 < \beta \le b_i \le B < \infty \qquad (1.104)$$

for each $i \in \{1 \ldots, n\}$, then

$$\frac{\sum_{i=1}^{n} a_i^2 \cdot \sum_{i=1}^{n} b_i^2}{\left(\sum_{i=1}^{n} a_i b_i\right)^2} \le \frac{(\alpha\beta + AB)^2}{4\alpha\beta AB}. \qquad (1.105)$$

The equality holds in (1.105) if and only if

$$p = n \cdot \frac{A}{\alpha} \Big/ \left(\frac{A}{\alpha} + \frac{B}{\beta}\right) \quad \text{and} \quad q = n \cdot \frac{B}{\beta} \Big/ \left(\frac{A}{\alpha} + \frac{B}{\beta}\right)$$

are integers and if p of the numbers $a_1, \ldots, a_n$ are equal to α and q of these numbers are equal to A, and if the corresponding numbers b_i are equal to B and β, respectively.

PROOF   We follow the original proof of Pólya and Szegö [153, pp. 71–71, 253–255].

We may, without loss of generality, suppose that $a_1 \ge \cdots \ge a_n$. Then, to maximise the left-hand side of (1.105) we must have that the critical b_i's be reversely ordered (for if $b_k > b_m$ with $k < m$, then we can interchange b_k and b_m such that $b_k^2 + b_m^2 = b_m^2 + b_k^2$ and $a_k b_k + a_m b_m \ge a_k b_m + a_m b_k$) so that $b_1 \le \cdots \le b_n$.

Define the nonnegative numbers u_i and v_i for $i = 1, \ldots, n-1$ and $n > 2$ such that

$$a_i^2 = u_i a_1^2 + v_i a_n^2 \quad \text{and} \quad b_i^2 = u_i b_1^2 + v_i b_n^2.$$

Since $a_i b_i > u_i a_1 b_1 + v_i a_n b_n$, the left-hand side of (1.105) may be written as

$$\frac{\sum_{i=1}^n a_i^2 \cdot \sum_{i=1}^n b_i^2}{\left(\sum_{i=1}^n a_i b_i\right)^2} \le \frac{\left(U a_1^2 + V a_n^2\right)\left(U b_1^2 + V b_n^2\right)}{\left(U a_1 b_1 + V a_n b_n\right)^2},$$

where $U = \sum_{i=1}^n u_i$ and $V = \sum_{i=1}^n v_i$.

This reduces the problem to that with $n = 2$, which is solvable by elementary methods, leading to

$$\frac{\sum_{i=1}^n a_i^2 \cdot \sum_{i=1}^n b_i^2}{\left(\sum_{i=1}^n a_i b_i\right)^2} \le \frac{(a_1 b_1 + a_n b_n)^2}{4 (a_1 a_n b_1 b_n)},$$

where, since the a_i's and b_i's here are reversely ordered, we have

$$a_1 = \max_{i=\overline{1,n}} \{a_i\}, \quad a_n = \min_{i=\overline{1,n}} \{a_i\}, \quad b_1 = \min_{i=\overline{1,n}} \{b_i\}, \quad b_n = \max_{i=\overline{1,n}} \{b_i\}.$$

If we now assume, as in (1.104), that

$$0 < \alpha \le a_i \le A, \qquad 0 < \beta \le b_i \le B, \qquad i = 1, \ldots, n,$$

then

$$\frac{(a_1 b_1 + a_n b_n)^2}{4 a_1 a_n b_1 b_n} \le \frac{(\alpha\beta + AB)^2}{4\alpha\beta AB}$$

because

$$\frac{(k-1)^2}{4k} \le \frac{(\alpha+1)^2}{4\alpha}$$

for $k \le \alpha$; and the inequality (1.105) is proved. $\quad\square$

Comments

With the assumption (1.104), we may state the following additive versions of the Pólya-Szegö inequality [55, p. 95]:

$$0 \le \left(\sum_{i=1}^n a_i^2 \cdot \sum_{i=1}^n b_i^2\right)^{\frac{1}{2}} - \sum_{i=1}^n a_i b_i \le \frac{1}{2} \cdot \frac{\left(\sqrt{AB} - \sqrt{\alpha\beta}\right)^2}{\sqrt{\alpha\beta AB}} \sum_{i=1}^n a_i b_i \quad (1.106)$$

and

$$0 \le \sum_{i=1}^n a_i^2 \cdot \sum_{i=1}^n b_i^2 - \left(\sum_{i=1}^n a_i b_i\right)^2 \le \frac{1}{4} \cdot \frac{(AB - \alpha\beta)^2}{\alpha\beta AB} \left(\sum_{i=1}^n a_i b_i\right)^2. \quad (1.107)$$

The constants $\frac{1}{2}$ and $\frac{1}{4}$ in (1.106) and (1.107) are best possible in the sense that they cannot be replaced by smaller quantities.

For related results, see the recent monograph of Dragomir [55].

1.16   The Cassels Inequality

Let $a = (a_1, \ldots, a_n)$ and $b = (b_1, \ldots, b_n)$ be *sequences of positive real numbers and* $w = (w_1, \ldots, w_n)$ *a sequence of nonnegative real numbers. Suppose that*

$$m = \min_{i=\overline{1,n}} \left\{ \frac{a_i}{b_i} \right\} \quad \text{and} \quad M = \max_{i=\overline{1,n}} \left\{ \frac{a_i}{b_i} \right\}. \tag{1.108}$$

Then one has the inequality

$$\frac{\sum_{i=1}^{n} w_i a_i^2 \sum_{i=1}^{n} w_i b_i^2}{\left(\sum_{i=1}^{n} w_i a_i b_i \right)^2} \leq \frac{(m+M)^2}{4mM}. \tag{1.109}$$

Equality holds in (1.109) *when* $w_1 = \frac{1}{a_1 b_1}$, $w_n = \frac{1}{a_n b_n}$, $w_2 = \cdots = w_{n-1} = 0$, $m = \frac{a_n}{b_1}$, *and* $M = \frac{a_1}{b_n}$.

PROOF   (1) We will first give Cassels' proof from 1951 which is of interest in itself. We follow the proof from Watson, Hlpargu, and Styan [165].

We begin with the assertion that

$$\frac{(1+kw)\left(1+k^{-1}w\right)}{(1+w)^2} \leq \frac{(1+k)\left(1+k^{-1}\right)}{4}, \quad k > 0, w \geq 0 \tag{1.110}$$

which, being an equivalent form of (1.109) for $n = 2$, shows that it holds for $n = 2$.

To prove that the maximum of (1.109) is obtained when we have more than two w_i's being nonzero, we note that if, for example, $w_1, w_2, w_3 \neq 0$ leads to an extremum M of $\frac{XY}{4}$, then we would have the linear equations

$$a_n^2 X + b_n^2 Y - 2M a_n b_n Z = 0, \quad k = 1, 2, 3.$$

Nontrivial solutions exist if and only if the three vectors $\left[a_n^2, b_n^2, a_n b_n \right]$ are linearly dependent. However, this will be so only if for some $i \neq j$ $(i, j = 1, 2, 3)$ $a_i = \gamma a_j$, $b_i = \gamma b_j$. Furthermore, if that were true, we could, for example, drop the a_i, b_i terms and thus deal with the same problem with no fewer variables. If only one $w_i \neq 0$, then $M = 1$, the lower bound. Therefore, we need only examine all pairs $w_i \neq 0$, $w_j \neq 0$. The result (1.109) then quickly follows.

(2) To give the second proof of (1.109), we use the *baricentric method* of Fréchet and Watson (see, for instance Watson, Hlpargu, and Styan [165]). We substitute $w_i = \frac{u_i}{b_i^2}$ in the left-hand side of (1.109), which may be expressed as the ratio

$$\frac{M}{D^2}$$

where

$$N = \sum_{i=1}^{n} \left(\frac{a_i}{b_i}\right)^2 u_i \quad \text{and} \quad D = \sum_{i=1}^{n} \left(\frac{a_i}{b_i}\right) u_i,$$

by assuming without loss of generality that $\sum_{i=1}^{n} a_i = 1$. But the point with co-ordinates (D, N) must lie within the convex closure of the n-points $\left(\frac{a_i}{b_i}, \frac{a_i^2}{b_i^2}\right)$. Therefore, the volume of $\frac{N}{D^2}$ at points on the parabola is one unit. If $m = \min_{i=\overline{1,n}} \left\{\frac{a_i}{b_i}\right\}$ and $M = \max_{i=\overline{1,n}} \left\{\frac{a_i}{b_i}\right\}$, then the minimum must lie on the chord joining the point (m, m^2) and (M, M^2). Some easy calculus then leads to (1.109). ∎

The following unweighted version of Cassels' inequality holds.
If a and b satisfy the assumptions in (1.109), then

$$\frac{\sum_{i=1}^{n} a_i^2 \sum_{i=1}^{n} b_i^2}{\left(\sum_{i=1}^{n} a_i b_i\right)^2} \leq \frac{(m+M)^2}{4mM}. \tag{1.111}$$

Comments
The following additive versions of Cassels' inequality may be stated [55, pp. 92–93]:

$$0 \leq \left(\sum_{i=1}^{n} w_i a_i^2 \sum_{i=1}^{n} w_i b_i^2\right)^{\frac{1}{2}} - \sum_{i=1}^{n} w_i a_i b_i \tag{1.112}$$

$$\leq \frac{\left(\sqrt{M} - \sqrt{m}\right)^2}{2\sqrt{mM}} \sum_{i=1}^{n} w_i a_i b_i$$

and

$$0 \leq \sum_{i=1}^{n} w_i a_i^2 \sum_{i=1}^{n} w_i b_i^2 - \left(\sum_{i=1}^{n} w_i a_i b_i\right)^2 \tag{1.113}$$

$$\leq \frac{(M-m)^2}{4mM} \left(\sum_{i=1}^{n} w_i a_i b_i\right)^2,$$

provided that a and b satisfy (1.108).
With the same assumptions on a and b, we have the unweighted inequalities

$$0 \leq \left(\sum_{i=1}^{n} a_i^2 \sum_{i=1}^{n} b_i^2\right)^{\frac{1}{2}} - \sum_{i=1}^{n} a_i b_i \leq \frac{\left(\sqrt{M} - \sqrt{m}\right)^2}{2\sqrt{mM}} \sum_{i=1}^{n} a_i b_i \tag{1.114}$$

and

$$0 \le \sum_{i=1}^{n} a_i^2 \sum_{i=1}^{n} b_i^2 - \left(\sum_{i=1}^{n} a_i b_i \right)^2 \le \frac{(M-m)^2}{4mM} \left(\sum_{i=1}^{n} a_i b_i \right)^2. \qquad (1.115)$$

1.17 Hölder's Inequality for Sequences of Real Numbers

Let $a = (a_1, \ldots, a_n)$, $b = (b_1, \ldots, b_n) \in \mathbb{R}_+^n$, and $p > 1$, $\frac{1}{p} + \frac{1}{q} = 1$. Then

$$\sum_{i=1}^{n} a_i b_i \le \left(\sum_{i=1}^{n} a_i^p \right)^{\frac{1}{p}} \left(\sum_{i=1}^{n} b_i^q \right)^{\frac{1}{q}}. \qquad (1.116)$$

The sign of equality holds if and only if the vectors a^p and b^q are proportional, that is, $a_i^p = k b_i^q$ for all $i \in \{1, \ldots, n\}$.

PROOF We know that for $x, y \ge 0$, the inequality (see Equation 1.13)

$$x^{\frac{1}{p}} y^{\frac{1}{q}} \le \frac{x}{p} + \frac{y}{q} \qquad (1.117)$$

holds, provided that $p > 1$ and $\frac{1}{p} + \frac{1}{q} = 1$; and equality holds in (1.117) iff $x = y$.

In (1.117) set

$$x = \frac{a_i^p}{A}, \quad A = \sum_{i=1}^{n} a_i^p$$

$$y = \frac{b_i^q}{B}, \quad B = \sum_{i=1}^{n} b_i^q.$$

Adding over $i = 1, 2, \ldots, n$ produces

$$\frac{\sum_{i=1}^{n} a_i b_i}{A^{\frac{1}{p}} B^{\frac{1}{q}}} \le \frac{1}{p} \sum_{i=1}^{n} \frac{a_i^p}{A} + \frac{1}{q} \sum_{i=1}^{n} \frac{b_i^q}{B} = \frac{1}{p} + \frac{1}{q} = 1.$$

Equality holds in (1.116) iff $\frac{a_i^p}{A} = \frac{b_i^q}{B}$, that is, if the vectors a^p and b^p are proportional. ☐

The weighted version of (1.116) is:

$$\sum_{i=1}^{n} m_i a_i b_i \le \left(\sum_{i=1}^{n} m_i a_i^p \right)^{\frac{1}{p}} \left(\sum_{i=1}^{n} m_i b_i^q \right)^{\frac{1}{q}}, \qquad (1.118)$$

provided $m_i \geq 0$ for $i = 1, \ldots, n$.

Comments

(a) We have the elementary inequality

$$(x + y)^{\frac{1}{p}} (z + w)^{\frac{1}{q}} \geq x^{\frac{1}{p}} z^{\frac{1}{q}} + y^{\frac{1}{p}} w^{\frac{1}{q}} \qquad (1.119)$$

for all $x, y, z, w \geq 0$ and $p, q > 1$ with $\frac{1}{p} + \frac{1}{q} = 1$.

The proof follows from Hölder's inequality (1.116) by choosing $n = 2$, $a_1 = x^{\frac{1}{p}}$, $a_2 = y^{\frac{1}{p}}$, and $b_1 = z^{\frac{1}{q}}$, $b_2 = w^{\frac{1}{q}}$.

(b) For $a, b \in \mathbb{R}^n$ and $x \in \mathbb{R}_+^n$, define $H(x, a, b)$ by

$$H(x, a, b) := \left(\sum_{i=1}^{n} x_i |a_i|^p \right)^{\frac{1}{p}} \left(\sum_{i=1}^{n} x_i |b_i|^q \right)^{\frac{1}{q}} - \left| \sum_{i=1}^{n} x_i a_i b_i \right|,$$

where p, q are as above.

We have the following *superadditivity property* [84]:

$$H(x + y, a, b) \geq H(x, a, b) + H(y, a, b) \geq 0 \qquad (1.120)$$

for all $x, y \in \mathbb{R}_+^n$.

It can be shown by employing the inequality (1.119) that

$$H(x + y, a, b)$$

$$= \left(\sum_{i=1}^{n} x_i |a_i|^p + \sum_{i=1}^{n} y_i |a_i|^p \right)^{\frac{1}{p}} \left(\sum_{i=1}^{n} x_i |b_i|^q + \sum_{i=1}^{n} y_i |b_i|^q \right)^{\frac{1}{q}}$$

$$- \left| \sum_{i=1}^{n} x_i a_i b_i + \sum_{i=1}^{n} y_i a_i b_i \right|$$

$$\geq \left(\sum_{i=1}^{n} x_i |a_i|^p \right)^{\frac{1}{p}} \left(\sum_{i=1}^{n} x_i |b_i|^q \right)^{\frac{1}{q}} + \left(\sum_{i=1}^{n} y_i |a_i|^p \right)^{\frac{1}{p}} \left(\sum_{i=1}^{n} y_i |b_i|^q \right)^{\frac{1}{q}}$$

$$- \left| \sum_{i=1}^{n} x_i a_i b_i \right| - \left| \sum_{i=1}^{n} y_i a_i b_i \right| = H(x, a, b) + H(y, a, b).$$

Also, we have the following *monotonicity property* [84]:

$$H(x, a, b) \geq H(y, a, b) \geq 0 \qquad (1.121)$$

provided $x, y \in \mathbb{R}_+^n$, $x \geq y$, that is, $x_i \geq y_i$ for all $i \in \{1, \ldots, n\}$.

By employing the superadditivity property (1.120), it can be shown that

$$H(x, a, b) = H((x - y) + y, a, b) \geq H(x - y, a, b) + H(y, a, b)$$
$$\geq H(y, a, b).$$

1.18 The Minkowski Inequality for Sequences of Real Numbers

Let $a = (a_1, \ldots, a_n)$, $b = (b_1, \ldots, b_n) \in \mathbb{R}^n$, and $p > 1$. Then

$$\left[\sum_{i=1}^{n} (a_i + b_i)^p \right]^{\frac{1}{p}} \leq \left(\sum_{i=1}^{n} a_i^p \right)^{\frac{1}{p}} + \left(\sum_{i=1}^{n} b_i^p \right)^{\frac{1}{p}}. \qquad (1.122)$$

Equality holds if a and b are proportional.

PROOF   Write

$$\sum_{i=1}^{n} (a_i + b_i)^p = \sum_{i=1}^{n} a_i (a_i + b_i)^{p-1} + \sum_{i=1}^{n} b_i (a_i + b_i)^{p-1}. \qquad (1.123)$$

Applying Hölder's inequality (1.116), we obtain

$$\sum_{i=1}^{n} a_i (a_i + b_i)^{p-1} \leq \left(\sum_{i=1}^{n} a_i^p \right)^{\frac{1}{p}} \left(\sum_{i=1}^{n} (a_i + b_i)^{(p-1)q} \right)^{\frac{1}{q}}$$

$$= \left(\sum_{i=1}^{n} a_i^p \right)^{\frac{1}{p}} \left(\sum_{i=1}^{n} (a_i + b_i)^p \right)^{\frac{1}{q}}$$

and similarly,

$$\sum_{i=1}^{n} b_i (a_i + b_i)^{p-1} \leq \left(\sum_{i=1}^{n} b_i^p \right)^{\frac{1}{p}} \left(\sum_{i=1}^{n} (a_i + b_i)^p \right)^{\frac{1}{q}},$$

for p and q satisfying $\frac{1}{p} + \frac{1}{q} = 1$.

Consequently, (1.123) becomes

$$\sum_{i=1}^{n} (a_i + b_i)^p \leq \left[\sum_{i=1}^{n} (a_i + b_i)^p \right]^{\frac{1}{q}} \left[\left(\sum_{i=1}^{n} a_i^p \right)^{\frac{1}{p}} + \left(\sum_{i=1}^{n} b_i^p \right)^{\frac{1}{p}} \right].$$

Dividing by $[\sum_{i=1}^{n} (a_i + b_i)^p]^{\frac{1}{p}}$ and taking into account that $\frac{1}{p} = 1 - \frac{1}{q}$, we get (1.122).

Equality holds if and only if the vectors a^p and b^p are both proportional to $(a + b)^p$; or, equivalently, iff a and b are proportional. □

The weighted version of (1.122) is

$$\left[\sum_{i=1}^{n} m_i (a_i + b_i)^p \right]^{\frac{1}{p}} \leq \left(\sum_{i=1}^{n} m_i a_i^p \right)^{\frac{1}{p}} + \left(\sum_{i=1}^{n} m_i b_i^p \right)^{\frac{1}{p}}, \qquad (1.124)$$

provided that $m_i \geq 0$ for $i = 1, \ldots, n$.

Comments

(a) We have the elementary inequality

$$\left[(a+b)^{\frac{1}{p}} + (c+d)^{\frac{1}{p}} \right]^p \geq \left(a^{\frac{1}{p}} + c^{\frac{1}{p}} \right)^p + \left(b^{\frac{1}{p}} + d^{\frac{1}{p}} \right)^p \qquad (1.125)$$

for all $a, b, c, d \geq 0$ and $p \geq 1$, which follows by applying Minkowski's inequality for $n = 2$, $a_1 = a^{\frac{1}{p}}$, $a_2 = b^{\frac{1}{p}}$, $b_1 = c^{\frac{1}{p}}$, and $b_2 = d^{\frac{1}{p}}$.

(b) For $a, b \in \mathbb{R}^n$ and $x \in \mathbb{R}^n_+$, define $M(x, a, b)$ by

$$M(x, a, b) := \left[\left(\sum_{i=1}^n x_i |a_i|^p \right)^{\frac{1}{p}} \left(\sum_{i=1}^n x_i |b_i|^p \right)^{\frac{1}{p}} \right]^p - \sum_{i=1}^n x_i |a_i + b_i|^p$$

where $p \geq 1$.

The mapping $M(\cdot, a, b)$ is superadditive on $\mathbb{R}^n$ [84].

Indeed, for all $x, y \in \mathbb{R}^n_+$, we have, by (1.125), that

$$M(x+y, a, b)$$

$$= \left[\left(\sum_{i=1}^n x_i |a_i|^p + \sum_{i=1}^n y_i |a_i|^p \right)^{\frac{1}{p}} \left(\sum_{i=1}^n x_i |b_i|^p + \sum_{i=1}^n y_i |b_i|^p \right)^{\frac{1}{p}} \right]^p$$

$$- \sum_{i=1}^n x_i |a_i + b_i|^p - \sum_{i=1}^n y_i |a_i + b_i|^p$$

$$\geq \left[\left(\sum_{i=1}^n x_i |a_i|^p \right)^{\frac{1}{p}} + \left(\sum_{i=1}^n x_i |b_i|^p \right)^{\frac{1}{p}} \right]^p$$

$$+ \left[\left(\sum_{i=1}^n y_i |a_i|^p \right)^{\frac{1}{p}} + \left(\sum_{i=1}^n y_i |b_i|^p \right)^{\frac{1}{p}} \right]^p$$

$$- \sum_{i=1}^n x_i |a_i + b_i|^p - \sum_{i=1}^n y_i |a_i + b_i|^p$$

$$= M(x, a, b) + M(y, a, b).$$

The mapping $M(\cdot, a, b)$ is monotonic nondecreasing on $\mathbb{R}^n$ [84]. This proposition follows by the superadditivity of M on its first variable.

1.19   Jensen's Discrete Inequality

Let $f : I \subseteq \mathbb{R} \to \mathbb{R}$ be a convex mapping on the interval I. That is,

$$f(tx + (1-t)y) \leq tf(x) + (1-t)f(y)$$

for all $x, y \in I$ and $t \in [0,1]$. If $x_i \in I$, $p_i \geq 0$ $(i \in \{1,\ldots,n\})$ and $P_n := \sum_{i=1}^{n} p_i > 0$, then

$$f\left(\frac{1}{P_n}\sum_{i=1}^{n} p_i x_i\right) \leq \frac{1}{P_n}\sum_{i=1}^{n} p_i f(x_i). \qquad (1.126)$$

PROOF   The proof follows by mathematical induction.

For $n = 2$, we have to prove

$$f\left(\frac{p_1 x_1 + p_2 x_2}{p_1 + p_2}\right) \leq \frac{p_1 f(x_1) + p_2 f(x_2)}{p_1 + p_2}, \qquad (1.127)$$

where $x_1, x_2 \in I$, $p_1, p_2 \geq 0$ with $p_1 + p_2 > 0$.

Now, observe that (1.127) follows by the definition of convexity for $t = \frac{p_1}{p_1+p_2}$, $x = x_1$, and $y = x_2$.

Assume that (1.126) holds for "n" and let us prove it for "$n+1$", that is, we wish to prove

$$f\left(\frac{1}{P_{n+1}}\sum_{i=1}^{n+1} p_i x_i\right) \leq \frac{1}{P_{n+1}}\sum_{i=1}^{n+1} p_i f(x_i) \qquad (1.128)$$

for $x_i \in I$, $p_i \geq 0$ $(i = 1,\ldots,n+1)$ with $P_{n+1} > 0$.

If $p_1 = \cdots = p_n = 0$, then (1.128) is obvious.

Assume that $P_n > 0$, then,

$$f\left(\frac{1}{P_{n+1}}\sum_{i=1}^{n+1} p_i x_i\right) = f\left(\frac{P_n}{P_{n+1}} \cdot \frac{1}{P_n}\sum_{i=1}^{n} p_i x_i + \frac{p_{n+1}}{P_{n+1}} x_{n+1}\right) \qquad (1.129)$$

$$\leq \frac{P_n}{P_{n+1}} f\left(\frac{1}{P_n}\sum_{i=1}^{n} p_i x_i\right) + \frac{p_{n+1}}{P_{n+1}} f(x_{n+1})$$

by the definition of convexity for $t = \frac{P_n}{P_{n+1}}$, $x = \frac{1}{P_n}\sum_{i=1}^{n} p_i x_i$ and $y = x_{n+1}$.

By using the inductive hypothesis, we get

$$\frac{P_n}{P_{n+1}} f \left(\frac{1}{P_n} \sum_{i=1}^{n} p_i x_i \right) + \frac{p_{n+1}}{P_{n+1}} f(x_{n+1}) \tag{1.130}$$

$$\leq \frac{P_n}{P_{n+1}} \frac{1}{P_n} \sum_{i=1}^{n} p_i f(x_i) + \frac{p_{n+1}}{P_{n+1}} f(x_{n+1})$$

$$= \frac{1}{P_{n+1}} \sum_{i=1}^{n+1} p_i f(x_i).$$

Now, (1.129) and (1.130) give the desired inequality (1.128). ▯

When the function f is concave, the inequality (1.126) is reversed.

Comments

Define the means

$$\begin{cases} A_n(p, x) := \dfrac{1}{P_n} \sum_{i=1}^{n} p_i x_i & \text{(weighted arithmetic mean)} \\[3mm] G_n(p, x) := \left(\displaystyle\prod_{i=1}^{n} x_i^{p_i} \right)^{\frac{1}{P_n}} & \text{(weighted geometric mean)} \\[3mm] H_n(p, x) := \dfrac{P_n}{\sum_{i=1}^{n} \frac{p_i}{x_i}} & \text{(weighted harmonic mean)} \end{cases} \tag{1.131}$$

where $p_i \geq 0$ with $P_n > 0$ and $x_i > 0$ $(i = 1, \ldots, n)$.

Apply Jensen's inequality (1.126) for the convex mapping $f(x) = -\ln x$ on $(0, \infty)$ to get:

$$-\ln \left(\frac{1}{P_n} \sum_{i=1}^{n} p_i x_i \right) \leq -\frac{1}{P_n} \sum_{i=1}^{n} p_i \ln x_i$$

which is equivalent to

$$\ln [A_n(p, x)] \geq \ln [G_n(p, x)]$$

or

$$A_n(p, x) \geq G_n(p, x). \tag{1.132}$$

If we apply (1.132) for $\frac{1}{x_i}$ instead of x_i $(i = 1, \ldots, n)$ we get

$$\frac{1}{H_n(p, x)} \geq \frac{1}{G_n(p, x)} \tag{1.133}$$

and then, by (1.132) and (1.133), we obtain

$$A_n(p, x) \geq G_n(p, x) \geq H_n(p, x) \tag{1.134}$$

which is the well-known *arithmetic mean–geometric mean-harmonic mean inequality*. If $p_i > 0$ $(i = 1, \ldots, n)$, then equality holds in (1.134) iff $x_1 = \cdots = x_n$. Note that inequality (1.1) is a special case of (1.134) when $n = 2$.

1.20 A Converse of Jensen's Inequality for Differentiable Mappings

Let $f : [a, b] \to \mathbb{R}$ be a *convex differentiable mapping on* (a, b). *If* $x_i \in (a, b)$, $p_i \geq 0$ $(i = 1, \ldots, n)$ *with* $P_n > 0$, *then we have the inequality* [96]:

$$0 \leq \frac{1}{P_n} \sum_{i=1}^{n} p_i f(x_i) - f\left(\frac{1}{P_n} \sum_{i=1}^{n} p_i x_i\right) \tag{1.135}$$

$$\leq \frac{1}{P_n} \sum_{i=1}^{n} p_i x_i f'(x_i) - \frac{1}{P_n} \sum_{i=1}^{n} p_i x_i \cdot \frac{1}{P_n} \sum_{i=1}^{n} p_i f'(x_i).$$

PROOF Since f is differentiable convex on (a, b), for all $x, y \in (a, b)$ we have

$$f(x) - f(y) \geq f'(y)(x - y). \tag{1.136}$$

Choose in (1.136) $x = \frac{1}{P_n} \sum_{i=1}^{n} p_i x_i$ and $y = x_j$ $(j = 1, \ldots, n)$ to get

$$f\left(\frac{1}{P_n} \sum_{i=1}^{n} p_i x_i\right) - f(x_j) \geq f'(x_j) \left(\frac{1}{P_n} \sum_{i=1}^{n} p_i x_i - x_j\right) \tag{1.137}$$

for all $j \in \{1, \ldots, n\}$.

By multiplying (1.137) by $p_j \geq 0$ and summing over j from 1 to n, we get

$$P_n f\left(\frac{1}{P_n} \sum_{i=1}^{n} p_i x_i\right) - \sum_{j=1}^{n} p_j f(x_j) \geq \frac{1}{P_n} \sum_{i=1}^{n} p_i x_i \sum_{j=1}^{n} p_j f'(x_j) - \sum_{j=1}^{n} p_j x_j f'(x_j)$$

which is clearly equivalent to (1.135). $\square$

Comments
(a) Apply (1.135) for the convex mapping $f(x) = -\ln x$ to get

$$0 \leq \ln\left(\frac{1}{P_n} \sum_{i=1}^{n} p_i x_i\right) - \frac{1}{P_n} \left(\sum_{i=1}^{n} p_i \ln x_i\right) \tag{1.138}$$

$$\leq \frac{1}{P_n} \sum_{i=1}^{n} p_i x_i \cdot \frac{1}{P_n} \sum_{i=1}^{n} \frac{p_i}{x_i} - \frac{1}{P_n} \sum_{i=1}^{n} p_i$$

$$= \frac{1}{P_n} \sum_{i=1}^{n} p_i x_i \left(\frac{P_n}{\sum_{i=1}^{n} \frac{p_i}{x_i}} \right)^{-1} - 1 = \frac{A_n(p, x)}{H_n(p, x)} - 1,$$

which gives the following counterpart of the arithmetic mean–geometric mean inequality:

$$1 \leq \frac{A_n(p, x)}{G_n(p, x)} \leq \exp\left[\frac{A_n(p, x)}{H_n(p, x)} - 1 \right], \quad x_i > 0, \ p_i \geq 0, \ P_n > 0. \qquad (1.139)$$

If we apply (1.138) for $\frac{1}{x_i}$ instead of x_i $(i = 1, \ldots, n)$, we get

$$1 \leq \frac{G_n(p, x)}{H_n(p, x)} \leq \exp\left[\frac{A_n(p, x)}{H_n(p, x)} - 1 \right]. \qquad (1.140)$$

(b) Assume that p_i $(i = 1, \ldots, n)$ is a probability distribution, that is, $\sum_{i=1}^{n} p_i = 1$ and consider the (Shannon) *entropy mapping* $H_n(p) = -\sum_{i=1}^{n} p_i \ln p_i$. Then we have the inequality [94]:

$$0 \leq \ln n - H_n(p) \leq \sum_{1 \leq i < j \leq n} (p_i - p_j)^2. \qquad (1.141)$$

Indeed, if in (1.138) we choose $x_i = \frac{1}{p_i}$ $(i = 1, \ldots, n)$, we get

$$0 \leq \ln n - H_n(p) \leq n \sum_{i=1}^{n} p_i^2 - 1$$

$$= n \sum_{i=1}^{n} p_i^2 - \left(\sum_{i=1}^{n} p_i \right)^2 = \frac{1}{n} \sum_{i,j=1}^{n} (p_i - p_j)^2$$

$$= \sum_{1 \leq i < j \leq n} (p_i - p_j)^2$$

and the inequality (1.141) is proved.

Equality holds in (1.141) iff $(p_1, \ldots, p_n) = \left(\frac{1}{n}, \ldots, \frac{1}{n} \right)$.

1.21   The Petrović Inequality for Convex Functions

Let f be a convex function on the segment $\bar{I} = [0, a]$. If $x_i \in \bar{I}$ $(i = 1, \ldots, n)$ and $x_1 + \cdots + x_n \in \bar{I}$, then

$$f(x_1) + \cdots + f(x_n) \leq f(x_1 + \cdots + x_n) + (n-1) f(0). \qquad (1.142)$$

PROOF   We follow the proof from Mitrinović [138]. As f is convex on $\bar{I}$, then for all $x, y \in \bar{I}$ and $p, q > 0$ with $p + q > 0$, we have

$$f\left(\frac{px + qy}{p + q}\right) \le \frac{pf(x) + qf(y)}{p + q}. \tag{1.143}$$

By putting in (1.143) $p = x_1$, $q = x_2$, $x = x_1 + x_2$, and $y = 0$ $(x_1 + x_2 > 0)$ we get

$$f(x_1) \le \frac{x_1 f(x_1 + x_2)}{x_1 + x_2} + \frac{x_2}{x_1 + x_2} f(0). \tag{1.144}$$

Interchanging x_1 and x_2, we find that

$$f(x_2) \le \frac{x_2 f(x_1 + x_2)}{x_1 + x_2} + \frac{x_1}{x_1 + x_2} f(0). \tag{1.145}$$

Adding (1.144) and (1.145) we get

$$f(x_1) + f(x_2) \le f(x_1 + x_2) + f(0), \tag{1.146}$$

and so (1.142) is true for $n = 2$.

Suppose that it holds for some n. Then, by (1.146), we have

$$f(x_1 + \cdots + x_n + x_{n+1}) = f((x_1 + \cdots + x_n) + x_{n+1})$$
$$\ge f(x_1 + \cdots + x_n) + f(x_{n+1}) - f(0)$$

and by the inductive hypothesis,

$$f(x_1) + \cdots + f(x_n) + f(x_{n+1}) \le f(x_1 + \cdots + x_n) + 1 + nf(0).$$

This completes the inductive proof. $\square$

Comments

(**a**) If we assume that f is convex on $\bar{I}$ and $f(0) = 0$, then f is superadditive, namely,

$$f(x_1 + x_2) \ge f(x_1) + f(x_2) \quad \text{for all } x_1, x_2 \in \bar{I}. \tag{1.147}$$

(**b**) Let $a, b > 0$ and $p \ge 1$. Then we have the inequality:

$$(a + b)^p \ge a^p + b^p. \tag{1.148}$$

This follows by (1.147) applied for the convex mapping $f : [0, \infty) \to \mathbb{R}$, $f(x) = x^p$ for which we have $f(0) = 0$.

(**c**) The mapping $f : [0, \infty) \to \mathbb{R}$, $f(x) = -\ln(1 + x)$ is convex on $[0, \infty)$ and $f(0) = 0$. Then, by (1.142), we get

$$-\sum_{i=1}^{n} \ln(1 + x_i) \le -\ln\left(1 + \sum_{i=1}^{n} x_i\right)$$

which is equivalent to

$$\ln\left(\prod_{i=1}^{n}(1+x_i)\right) \geq \ln\left(1+\sum_{i=1}^{n}x_i\right),$$

producing the result

$$\prod_{i=1}^{n}(1+x_i) \geq 1+\sum_{i=1}^{n}x_i, \text{ for all } x_i \in [0,\infty),\ i=1,\ldots,n. \qquad (1.149)$$

(**d**) Consider the mapping $f : [0,\infty) \to \mathbb{R}$,

$$f(x) = \begin{cases} x\ln x, & x > 0 \\ 0, & \text{if } x = 0. \end{cases}$$

This mapping is continuous and convex on $[0,\infty)$. Applying the inequality (1.142) we get

$$\sum_{i=1}^{n}x_i\ln x_i \leq \sum_{i=1}^{n}x_i\ln\left(\sum_{i=1}^{n}x_i\right)$$

which is equivalent to:

$$\prod_{i=1}^{n}x_i^{x_i} \leq \left(\sum_{i=1}^{n}x_i\right)^{\sum_{i=1}^{n}x_i} \qquad \text{for all } x_i > 0,\ i=(1,\ldots,n). \qquad (1.150)$$

1.22   Bounds for the Jensen Functional in Terms of the Second Derivative

Let $f : [a,b] \to \mathbb{R}$ *be a twice differentiable mapping on* (a,b) *with the property that* $m \leq f''(x) \leq M$ *for all* $x \in (a,b)$. *If* $x_i \in (a,b)$, $p_i \geq 0$ *with* $P_n := \sum_{i=1}^{n}p_i > 0$, *then we have the inequality*

$$\frac{m}{P_n^2}\sum_{1\leq i<j\leq n}p_ip_j(x_i-x_j)^2 \leq \frac{1}{P_n}\sum_{i=1}^{n}p_if(x_i) - f\left(\frac{1}{P_n}\sum_{i=1}^{n}p_ix_i\right) \qquad (1.151)$$

$$\leq \frac{M}{P_n^2}\sum_{1\leq i<j\leq n}p_ip_j(x_i-x_j)^2.$$

PROOF   We follow the proof by Dragomir [80].
The mapping $g : [a,b] \to \mathbb{R}$, $g(x) = f(x) - \frac{1}{2}mx^2$ is twice differentiable on (a,b) and $g''(x) = f''(x) - m \geq 0$, which shows that g is convex on $[a,b]$.

Applying Jensen's inequality (1.126) for the mapping g we can state

$$f\left(\frac{1}{P_n}\sum_{i=1}^n p_i x_i\right) - \frac{1}{2}m\left(\frac{1}{P_n}\sum_{i=1}^n p_i x_i\right)^2 \le \frac{1}{P_n}\sum_{i=1}^n p_i\left[f(x_i) - \frac{1}{2}mx_i^2\right],$$

which is equivalent to

$$\frac{1}{P_n}\sum_{i=1}^n p_i f(x_i) - f\left(\frac{1}{P_n}\sum_{i=1}^n p_i x_i\right)$$

$$\ge \frac{1}{2}m\left[\frac{1}{P_n}\sum_{i=1}^n p_i x_i^2 - \left(\frac{1}{P_n}\sum_{i=1}^n p_i x_i\right)^2\right].$$

However, a simple calculation shows that

$$\frac{1}{P_n}\sum_{i=1}^n p_i x_i^2 - \left(\frac{1}{P_n}\sum_{i=1}^n p_i x_i\right)^2 = \frac{1}{P_n^2}\sum_{i,j=1}^n p_i p_j (x_i - x_j)^2$$

$$= \frac{2}{P_n^2}\sum_{1\le i<j\le n} p_i p_j (x_i - x_j)^2,$$

which proves the first inequality in (1.151).

The second inequality follows likewise for the convex mapping $h : [a, b] \to \mathbb{R}$, $h(x) = \frac{1}{2}Mx^2 - f(x)$; and we omit the details. □

Comments
Consider the mapping $f : [a, b] \subset (0, \infty) \to \mathbb{R}$, $f(x) = -\ln x$. Then $f''(x) = \frac{1}{x^2}$ and $m = \inf_{x\in[a,b]} f''(x) = \frac{1}{b^2}$, $M = \sup_{x\in[a,b]} f''(x) = \frac{1}{a^2}$. Applying the inequality (1.151) for the above mapping, we get for $A_n(p, x)$, $G_n(p, x)$ as given by (1.131) the following inequalities:

$$\frac{1}{b^2 P_n^2}\sum_{1\le i<j\le n} p_i p_j (x_i - x_j)^2 \le \ln A_n(p, x) - \ln G_n(p, x)$$

$$\le \frac{1}{a^2 P_n^2}\sum_{1\le i<j\le n} p_i p_j (x_i - x_j)^2.$$

This is equivalent to

$$\exp\left[\frac{1}{b^2 P_n^2}\sum_{1\le i<j\le n} p_i p_j (x_i - x_j)^2\right]$$

$$\le \frac{A_n(p, x)}{G_n(p, x)} \le \exp\left[\frac{1}{a^2 P_n^2}\sum_{1\le i<j\le n} p_i p_j (x_i - x_j)^2\right], \tag{1.152}$$

where $x_i \in [a, b]$, $p_i \geq 0$ and $P_n > 0$.

Note that

$$\exp\left[\frac{1}{b^2 P_n^2} \sum_{1 \leq i < j \leq n} p_i p_j (x_i - x_j)^2\right] \geq 1$$

so that (1.152) can be seen as both a refinement of the *arithmetic mean–geometric mean* inequality $A_n(p, x) \geq G_n(p, x)$ and a counterpart of it.

From (1.152) we can also obtain a refinement and a counterpart of the *geometric mean–harmonic mean* inequality $G_n(p, x) \geq H_n(p, x)$ as follows.

Choose in (1.152) $\frac{1}{x_i}$ instead of x_i and note that $\frac{1}{x_i} \in \left[\frac{1}{b}, \frac{1}{a}\right]$. Then, by (1.152),

$$\exp\left[\frac{a^2}{P_n^2} \sum_{1 \leq i < j \leq n} p_i p_j \frac{(x_i - x_j)^2}{x_i^2 x_j^2}\right]$$

$$\leq \frac{A_n\left(p, \frac{1}{x}\right)}{G_n\left(p, \frac{1}{x}\right)} \leq \exp\left[\frac{b^2}{P_n^2} \sum_{1 \leq i < j \leq n} p_i p_j \frac{(x_i - x_j)^2}{x_i^2 x_j^2}\right] \qquad (1.153)$$

and as $A_n\left(p, \frac{1}{x}\right) = H_n^{-1}(p, x)$ and $G_n\left(p, \frac{1}{x}\right) = G^{-1}(p, x)$, then by (1.153) one gets

$$\exp\left[\frac{a^2}{P_n^2} \sum_{1 \leq i < j \leq n} p_i p_j \frac{(x_i - x_j)^2}{x_i^2 x_j^2}\right]$$

$$\leq \frac{G_n(p, x)}{H_n(p, x)} \leq \exp\left[\frac{b^2}{P_n^2} \sum_{1 \leq i < j \leq n} p_i p_j \frac{(x_i - x_j)^2}{x_i^2 x_j^2}\right] \qquad (1.154)$$

for all $x_i \in [a, b]$, $p_i \geq 0$ and $P_n > 0$.

1.23   Slater's Inequality for Convex Functions

Suppose that I is an interval of real numbers with interior $\overset{\circ}{I}$ and $f : I \to \mathbb{R}$ is a convex function on I. Then f is continuous on $\overset{\circ}{I}$ and has finite left and right derivatives at each point of $\overset{\circ}{I}$. Moreover, if $x, y \in \overset{\circ}{I}$ and $x < y$, then $D^- f(x) \leq D^+ f(x) \leq D^- f(y) \leq D^+ f(y)$, which implies that both $D^- f$ and $D^+ f$ are nondecreasing functions on $\overset{\circ}{I}$. It is also known that a convex function must be differentiable except for at most countably many points.

For a convex function $f : I \to \mathbb{R}$, the *subdifferential* of f denoted by ∂f is the set of all functions $\varphi : I \to [-\infty, \infty]$ such that $\varphi\left(\overset{\circ}{I}\right) \subset \mathbb{R}$ and

$$f(x) \geq f(a) + (x-a)\varphi(a) \quad \text{for any} \quad x, a \in I. \tag{1.155}$$

It is also well known that if f is convex on I, then ∂f is nonempty, $D^+ f, D^- f \in \partial f$ and if $\varphi \in \partial f$, then

$$D^- f(x) \leq \varphi(x) \leq D^+ f(x) \tag{1.156}$$

for every $x \in \overset{\circ}{I}$. In particular, φ is a nondecreasing function.

If f is differentiable convex on $\overset{\circ}{I}$, then $\partial f = \{f'\}$.

The following inequality is well known in the literature as *Slater's inequality* [160]:

Let $f : I \to \mathbb{R}$ *be a nondecreasing (nonincreasing) convex function*, $x_i \in I$, $p_i \geq 0$ *with* $P_n := \sum_{i=1}^n p_i > 0$ *and* $\sum_{i=1}^n p_i \varphi(x_i) \neq 0$ *where* $\varphi \in \partial f$. *Then one has the inequality:*

$$\frac{1}{P_n} \sum_{i=1}^n p_i f(x_i) \leq f\left(\frac{\sum_{i=1}^n p_i x_i \varphi(x_i)}{\sum_{i=1}^n p_i \varphi(x_i)}\right). \tag{1.157}$$

PROOF First, observe that since, for example, f is nondecreasing, $\varphi(x) \geq 0$ for any $x \in I$. Thus,

$$\frac{\sum_{i=1}^n p_i x_i \varphi(x_i)}{\sum_{i=1}^n p_i \varphi(x_i)} \in I, \tag{1.158}$$

since it is a convex combination of x_i with the positive weights

$$\frac{x_i \varphi(x_i)}{\sum_{i=1}^n p_i \varphi(x_i)}, \quad i = 1, \dots, n.$$

A similar argument applies if f is nonincreasing.

Now, if we use the inequality (1.155), then we deduce

$$f(x) - f(x_i) \geq (x - x_i)\varphi(x_i) \quad \text{for any} \quad x, x_i \in I, \ i = 1, \dots, n. \tag{1.159}$$

Multiplying (1.159) by $p_i \geq 0$ and summing over i from 1 to n, we deduce

$$f(x) - \frac{1}{P_n} \sum_{i=1}^n p_i f(x_i) \geq x \cdot \frac{1}{P_n} \sum_{i=1}^n p_i \varphi(x_i) - \frac{1}{P_n} \sum_{i=1}^n p_i x_i \varphi(x_i) \tag{1.160}$$

for any $x \in I$.

If in (1.160), we choose

$$x = \frac{\sum_{i=1}^n p_i x_i \varphi(x_i)}{\sum_{i=1}^n p_i \varphi(x_i)}$$

(which we have proved belongs to I), then we may deduce the desired inequality (1.157).

If one would like to drop the assumption of monotonicity for the function f, then one can state and prove the following result in a similar manner:

Let $f : I \to \mathbb{R}$ be a convex function, $x_i \in I$, $p_i \geq 0$ with $P_n > 0$ and $\sum_{i=1}^{n} p_i \varphi(x_i) \neq 0$, where $\varphi \in \partial f$. If

$$\frac{\sum_{i=1}^{n} p_i x_i \varphi(x_i)}{\sum_{i=1}^{n} p_i \varphi(x_i)} \in I, \tag{1.161}$$

then the inequality (1.157) holds.

Comments

The following result in connection to the (CBS)-inequality holds [55, p. 208]:

(a) *Assume that $f : \mathbb{R}_+ \to \mathbb{R}$ is a convex function on $\mathbb{R}_+ := [0, \infty)$, $a_i, b_i \geq 0$ with $a_i \neq 0$, $i \in \{1, \ldots, n\}$ and $\sum_{i=1}^{n} a_i^2 \varphi\left(\frac{b_i}{a_i}\right) \neq 0$ where $\varphi \in \partial f$.*

(i) *If f is monotonic nondecreasing (nonincreasing) in $[0, \infty)$ then*

$$\sum_{i=1}^{n} a_i^2 \varphi\left(\frac{b_i}{a_i}\right) \leq \sum_{i=1}^{n} a_i^2 \cdot f\left(\frac{\sum_{i=1}^{n} a_i b_i \varphi\left(\frac{b_i}{a_i}\right)}{\sum_{i=1}^{n} a_i^2 \varphi\left(\frac{b_i}{a_i}\right)}\right). \tag{1.162}$$

(ii) *If*

$$\frac{\sum_{i=1}^{n} a_i b_i \varphi\left(\frac{b_i}{a_i}\right)}{\sum_{i=1}^{n} a_i^2 \varphi\left(\frac{b_i}{a_i}\right)} \geq 0, \tag{1.163}$$

then (1.162) also holds.

(b) *Consider the function $f : [0, \infty) \to \mathbb{R}$, $f(x) = x^p$, $p \geq 1$, then f is convex and monotonic nondecreasing and $f'(x) = px^{p-1}$. Applying (1.162), we may deduce the following inequality:*

$$p\left(\sum_{i=1}^{n} a_i^{3-p} b_i^{p-1}\right)^{p+1} \leq \sum_{i=1}^{n} a_i^2 \left(\sum_{i=1}^{n} a_i^{2-p} b_i^p\right)^p \tag{1.164}$$

for $p \geq 1$, $a_i, b_i \geq 0$, $i = 1, \ldots, n$.

1.24 A Jensen Type Inequality for Double Sums

The following result for convex functions via Jensen's inequality also holds [55, p. 211].

Let $f : \mathbb{R} \to \mathbb{R}$ be a convex (concave) function and $x = (x_1, \ldots, x_n)$, $p = (p_1, \ldots, p_n)$ real sequences with the property that $p_i \geq 0$ $(i = 1, \ldots, n)$ and $\sum_{i=1}^{n} p_i = 1$. Then one has the inequality:

$$f\left[\frac{\sum_{i=1}^{n} p_i x_i^2 - \left(\sum_{i=1}^{n} p_i x_i\right)^2}{\sum_{i=1}^{n} i^2 p_i - \left(\sum_{i=1}^{n} i p_i\right)^2}\right]$$

$$\leq (\geq) \frac{\sum_{1 \leq i < j \leq n} p_i p_j \left[\sum_{k,l=i}^{j-1} f(\Delta x_k \cdot \Delta x_l)\right]}{\sum_{i=1}^{n} i^2 p_i - \left(\sum_{i=1}^{n} i p_i\right)^2}, \quad (1.165)$$

where $\Delta x_k := x_{k+1} - x_k$ $(k = 1, \ldots, n-1)$ is the forward difference.

PROOF   We have, by Jensen's inequality for multiple sums, that

$$f\left[\frac{\sum_{i=1}^{n} p_i x_i^2 - \left(\sum_{i=1}^{n} p_i x_i\right)^2}{\sum_{i=1}^{n} i^2 p_i - \left(\sum_{i=1}^{n} i p_i\right)^2}\right] = f\left[\frac{\sum_{1 \leq i < j \leq n} p_i p_j (x_i - x_j)^2}{\sum_{1 \leq i < j \leq n} p_i p_j (j - i)^2}\right] \quad (1.166)$$

$$= f\left[\frac{\sum_{1 \leq i < j \leq n} p_i p_j (j - i)^2 \frac{(x_j - x_i)^2}{(j-i)^2}}{\sum_{1 \leq i < j \leq n} p_i p_j (j - i)^2}\right]$$

$$\leq \frac{\sum_{1 \leq i < j \leq n} p_i p_j (j - i)^2 f\left(\frac{(x_j - x_i)^2}{(j-i)^2}\right)}{\sum_{1 \leq i < j \leq n} p_i p_j (j - i)^2}$$

$$=: I.$$

On the other hand, for $j > i$ one has

$$x_j - x_i = \sum_{k=i}^{j-1} (x_{k+1} - x_k) = \sum_{k=i}^{j-1} \Delta x_k \quad (1.167)$$

and thus

$$(x_j - x_i)^2 = \left(\sum_{k=i}^{j-1} \Delta x_k\right)^2 = \sum_{k,l=i}^{j-1} \Delta x_k \cdot \Delta x_l.$$

By applying the Jensen inequality for multiple sums once more, we deduce

$$f\left[\frac{(x_j - x_i)^2}{(j-i)^2}\right] = f\left[\frac{\sum_{k,l=i}^{j-1} \Delta x_k \cdot \Delta x_l}{(j-i)^2}\right] \quad (1.168)$$

$$\leq (\geq) \frac{\sum_{k,l=i}^{j-1} f(\Delta x_k \cdot \Delta x_l)}{(j-i)^2}.$$

Thus, by (1.168) we have

$$I \le (\ge) \frac{\sum_{1 \le i < j \le n} p_i p_j (j-i)^2 \frac{\sum_{k,l=i}^{j-1} f(\Delta x_k \cdot \Delta x_l)}{(j-i)^2}}{\sum_{1 \le i < j \le n} p_i p_j (j-i)^2} \tag{1.169}$$

$$= \frac{\sum_{1 \le i < j \le n} p_i p_j \left[\sum_{k,l=i}^{j-1} f(\Delta x_k \cdot \Delta x_l) \right]}{\sum_{i=1}^{n} i^2 p_i - \left(\sum_{i=1}^{n} i p_i \right)^2}.$$

Finally, from (1.166) and (1.169) we obtain the desired inequality (1.165). $\square$

Comments

The following inequality connected with the (CBS)-inequality may be stated [55, p. 212]:

Let $f : \mathbb{R} \to \mathbb{R}$ be a convex (concave) function and $a = (a_1, \dots, a_n)$, $b = (b_1, \dots, b_n)$, $w = (w_1, \dots, w_n)$ sequences of real numbers such that $b_i \ne 0$, $w_i \ge 0$ $(i = 1, \dots, n)$ and not all w_i are zero. Then one has the inequality

$$f \left[\frac{\sum_{i=1}^{n} w_i a_i^2 \sum_{i=1}^{n} w_i b_i^2 - \left(\sum_{i=1}^{n} w_i a_i b_i \right)^2}{\sum_{i=1}^{n} w_i b_i^2 \sum_{i=1}^{n} i^2 w_i b_i^2 - \left(\sum_{i=1}^{n} i w_i b_i \right)^2} \right]$$

$$\le (\ge) \frac{\sum_{1 \le i < j \le n} w_i w_j b_i^2 b_j^2 \left[\sum_{k,l=i}^{j-1} f \left(\Delta \left(\frac{a_k}{b_k} \right) \cdot \Delta \left(\frac{a_l}{b_l} \right) \right) \right]}{\sum_{i=1}^{n} w_i b_i^2 \sum_{i=1}^{n} i^2 w_i b_i^2 - \left(\sum_{i=1}^{n} i w_i b_i \right)^2}. \tag{1.170}$$

The proof follows by (1.165) on choosing $p_i = w_i b_i^2$ and $x_i = \frac{a_i}{b_i}$, $i = 1, \dots, n$. We omit the details.

Chapter 2

Integral Inequalities for Convex Functions

The concept of convexity and its various generalisations and extensions plays an important role in modern analysis.

Numerous Hermite-Hadamard (H.-H.) type integral inequalities are presented for both convex functions as well as for log-convex, quasi-convex, s-convex, and other related classes of functions such as the Godunova-Levin class of functions. Since the logarithmic, identric, and p-logarithmic means can be obtained as integral means of some particular convex functions, we provide here various applications of the H.-H. integral inequalities which complement, improve, and provide reverses of some classical inequalities for means.

Many remarks and comments are given which reveal a rich interconnection between the above inequalities and create a natural platform for further research and applications.

2.1 The Hermite-Hadamard Integral Inequality

Let $f : [a, b] \to \mathbb{R}$ be a convex mapping on $[a, b]$. Then

$$f\left(\frac{a+b}{2}\right) \leq \frac{1}{b-a}\int_a^b f(x)\,dx \leq \frac{f(a)+f(b)}{2}, \qquad (2.1)$$

which is known as the Hermite-Hadamard (H.-H.) integral inequality.

PROOF   By the convexity of f on $[a, b]$, we have

$$f(ta + (1-t)b) \leq tf(a) + (1-t)f(b)$$

for all $t \in [0, 1]$.

Integrating over t on $[0, 1]$, we obtain

$$\int_0^1 f(ta + (1-t)b)\,dt \leq f(a)\int_0^1 t\,dt + f(b)\int_0^1 (1-t)\,dt. \qquad (2.2)$$

Since

$$\int_0^1 t\,dt = \int_0^1 (1-t)\,dt = \frac{1}{2}$$

and, by the change of variable $x = ta + (1-t)\,b$,

$$\int_0^1 f(ta + (1-t)\,b)\,dt = \frac{1}{b-a}\int_a^b f(x)\,dx,$$

we get the second part of (2.1) from (2.2).

By the convexity of f we also have

$$\frac{1}{2}\left[f(ta + (1-t)\,b) + f((1-t)\,a + tb)\right]$$

$$\geq f\left[\frac{ta + (1-t)\,b + (1-t)\,a + tb}{2}\right] = f\left(\frac{a+b}{2}\right).$$

Integrating this inequality over t on $[0,1]$, we get

$$f\left(\frac{a+b}{2}\right) \leq \frac{1}{2}\left[\int_0^1 f(ta + (1-t)\,b)\,dt + \int_0^1 f((1-t)\,a + tb)\,dt\right]$$

$$= \frac{1}{b-a}\int_a^b f(x)\,dx$$

which proves the first part of (2.1). ⬚

Comments

(a) Assume that $g : [a,b] \to \mathbb{R}$ is twice differentiable on (a,b) and $m \leq g''(x) \leq M$ for $x \in (a,b)$. Then we have the inequality [88]

$$\frac{m}{24}(b-a)^2 \leq \frac{1}{b-a}\int_a^b g(x)\,dx - g\left(\frac{a+b}{2}\right) \leq \frac{M}{24}(b-a)^2. \qquad (2.3)$$

Indeed, if $f(x) = g(x) - \frac{m}{2}x^2$, then $f''(x) = g''(x) - m \geq 0$, which shows that f is convex on (a,b). Applying the Hermite-Hadamard inequality (2.1) for f we get

$$g\left(\frac{a+b}{2}\right) - \frac{m}{2}\left(\frac{a+b}{2}\right)^2$$

$$= f\left(\frac{a+b}{2}\right) \leq \frac{1}{b-a}\int_a^b f(x)\,dx = \frac{1}{b-a}\int_a^b \left[g(x) - \frac{m}{2}x^2\right]dx$$

$$= \frac{1}{b-a}\int_a^b g(x)\,dx - \frac{m}{2}\cdot\frac{b^3 - a^3}{3(b-a)}$$

$$= \frac{1}{b-a}\int_a^b g(x)\,dx - \frac{m}{2}\cdot\frac{a^2 + ab + b^2}{3}.$$

This implies that

$$\frac{m}{2} \cdot \frac{a^2 + ab + b^2}{3} - \frac{m}{2}\left(\frac{a+b}{2}\right)^2 \le \frac{1}{b-a}\int_a^b g(x)\,dx - g\left(\frac{a+b}{2}\right),$$

which proves the first inequality in (2.3) upon some simplification.

To prove the second part of (2.3), we apply the same argument for the convex mapping $h(x) = \frac{M}{2}x^2 - g(x)$, $x \in [a, b]$.

(b) If g is as above, then [89]

$$\frac{m}{12}(b-a)^2 \le \frac{g(a)+g(b)}{2} - \frac{1}{b-a}\int_a^b g(x)\,dx \le \frac{M}{12}(b-a)^2. \qquad (2.4)$$

This can be proven by applying the second part of the Hermite-Hadamard inequality (2.1) for the mapping $f(x) = g(x) - \frac{m}{2}x^2$ as follows:

$$\frac{g(a)+g(b)}{2} - \frac{m}{2}\cdot\frac{(a^2+b^2)}{2}$$
$$= \frac{f(a)+f(b)}{2} \ge \frac{1}{b-a}\int_a^b f(x)\,dx = \frac{1}{b-a}\int_a^b \left[g(x) - \frac{m}{2}x^2\right]dx$$
$$= \frac{1}{b-a}\int_a^b g(x)\,dx - \frac{m}{2}\cdot\frac{a^2+ab+b^2}{3}.$$

This is equivalent to

$$\frac{m}{2}\left[\frac{a^2+b^2}{2} - \frac{a^2+ab+b^2}{3}\right] \le \frac{g(a)+g(b)}{2} - \frac{1}{b-a}\int_a^b g(x)\,dx,$$

which proves the first part of (2.4). The second part is established in a similar manner; and we omit the details.

2.2 Hermite-Hadamard Related Inequalities

The following result is a generalisation of the first part of the Hermite-Hadamard inequality (2.1) (see Dragomir and Pearce [103]).

Let $f : I \subseteq \mathbb{R} \to \mathbb{R}$ be a convex function on I and $a, b \in \overset{\circ}{I}$ with $a < b$. Then for all $t \in [a, b]$ and $\lambda \in \left[f'_-(t), f'_+(t)\right]$ one has the inequality:

$$f(t) + \lambda\left(\frac{a+b}{2} - t\right) \le \frac{1}{b-a}\int_a^b f(x)\,dx. \qquad (2.5)$$

PROOF   Let $t \in [a, b]$. It is known [151, Theorem 1.6] that for all $\lambda \in \left[f'_- (t), f'_+ (t) \right]$ one has the inequality:

$$f (x) - f (t) \geq \lambda (x - t) \quad \text{for all } x \in [a, b].$$

Integrating this inequality on $[a, b]$ over x we have

$$\int_a^b f (x) \, dx - (b - a) f (t) \geq \lambda (b - a) \left(\frac{a + b}{2} - t \right)$$

which proves the inequality (2.5).   ☐

For $t = \frac{a+b}{2}$ we get the first part of the Hermite-Hadamard inequality (2.5). Furthermore, we have the following particular cases of (2.5) [103]:

Let f be as above and $0 \leq a < b$.

(i) *If $f'_+ \left(\sqrt{ab} \right) \geq 0$, then*

$$\frac{1}{b - a} \int_a^b f (x) \, dx \geq f \left(\sqrt{ab} \right);$$

(ii) *If $f'_+ \left(\frac{2ab}{a+b} \right) \geq 0$, then*

$$\frac{1}{b - a} \int_a^b f (x) \, dx \geq f \left(\frac{2ab}{a + b} \right);$$

(iii) *If f is differentiable in a and b, then*

$$\frac{1}{b - a} \int_a^b f (x) \, dx \geq \max \left\{ f (a) + f' (a) \frac{b - a}{2}, f (b) + f' (b) \frac{a - b}{2} \right\}$$

and

$$0 \leq \frac{f (a) + f (b)}{2} - \frac{1}{b - a} \int_a^b f (x) \, dx \leq \frac{f' (b) + f' (a)}{2} (b - a);$$

(iv) *If $x_i \in [a, b]$ are points of differentiability for f, $p_i \geq 0$ are such that $P_n := \sum_{i=1}^n p_i > 0$ and*

$$\frac{a + b}{2} \sum_{i=1}^n f' (x_i) p_i \geq \sum_{i=1}^n p_i f' (x_i) x_i,$$

then one has the inequality

$$\frac{1}{b - a} \int_a^b f (x) \, dx \geq \frac{1}{P_n} \sum_{i=1}^n p_i f (x_i).$$

It we assume that f is differentiable on (a, b), we recapture some of the results from Dragomir [62] and Dragomir and Ionescu [97].

The second part of the H.-H. inequality can be extended as follows [62]:

Let f and a, b be as above. Then for all $t \in [a, b]$ we have the inequality:

$$\frac{1}{b - a} \int_a^b f(x) \, dx \leq \frac{f(t)}{2} + \frac{1}{2} \cdot \frac{bf(b) - af(a) - t(f(b) - f(a))}{b - a}. \quad (2.6)$$

PROOF First, we remark that the class of differentiable convex mappings on (a, b) is dense in uniform topology in the class of all convex functions defined on (a, b). Thus, without loss of generality, we may assume that f is differentiable on (a, b). Therefore we can write the inequality:

$$f(t) - f(x) \geq (t - x) f'(x) \text{ for all } t, x \in (a, b).$$

Integrating this inequality over x on $[a, b]$ we get:

$$(b - a) f(t) - \int_a^b f(x) \, dx \geq t(f(b) - f(a)) - \int_a^b x f'(x) \, dx. \quad (2.7)$$

A simple computation shows that

$$\int_a^b x f'(x) \, dx = bf(b) - af(a) - \int_a^b f(x) \, dx,$$

and the inequality (2.7) becomes

$$(b - a) f(t) - t(f(b) - f(a)) + bf(b) - af(a) \geq 2 \int_a^b f(x) \, dx,$$

which is equivalent to (2.6). ⬜

With the above assumptions, and under the condition that $0 \leq a < b$, one has the inequality:

$$\frac{1}{b - a} \int_a^b f(x) \, dx \leq \min \{H_f(a, b), G_f(a, b), A_f(a, b)\} \quad (2.8)$$

where:

$$H_f(a, b) := \frac{1}{2} \left[f\left(\frac{2ab}{a + b}\right) + \frac{bf(b) + af(a)}{b + a} \right],$$

$$G_f(a, b) := \frac{1}{2} \left[f\left(\sqrt{ab}\right) + \frac{\sqrt{b}f(b) + \sqrt{a}f(a)}{\sqrt{b} + \sqrt{a}} \right],$$

$$A_f(a, b) := \frac{1}{2} \left[f\left(\frac{a + b}{2}\right) + \frac{f(b) + f(a)}{2} \right].$$

The inequality (2.8) for $A_f(a,b)$ has been proved by Bullen in 1978 [151, p. 140] and the inequality (2.8) for $G_f(a,b)$ has been proved by Sándor in 1988 [158].

The following generalisation of the above result has been proved by Dragomir and Pearce [106], which gives a refinement for the second part of the H.-H. inequality.

Let $f : I \subseteq \mathbb{R} \to \mathbb{R}$ be a convex function, $a, b \in \mathring{I}$ with $a < b$ and $x_i \in [a, b]$, $p_i \geq 0$ with $P_n > 0$. Then

$$\frac{1}{b-a} \int_a^b f(x)\, dx \tag{2.9}$$

$$\leq \frac{1}{2} \left\{ \frac{1}{P_n} \sum_{i=1}^n p_i f(x_i) + \frac{1}{b-a} \left[(b - x_p) f(b) + (x_p - a) f(a) \right] \right\}$$

$$\leq \frac{1}{2} (f(a) + f(b)),$$

where

$$x_p = \frac{1}{P_n} \sum_{i=1}^n p_i x_i.$$

PROOF   Similar to the preceding proof, it is sufficient to prove (2.9) for convex functions which are differentiable on (a, b). For any $x, y \in (a, b)$, we have

$$f(y) - f(x) \geq f'(x)(y - x) \text{ for all } x, y \in (a, b).$$

Thus, we obtain:

$$f(x_i) - f(x) \geq f'(x)(x_i - x) \text{ for all } i \in \{1, ..., n\}.$$

By integrating on $[a, b]$ over x we have

$$f(x_i) - \frac{1}{b-a} \int_a^b f(x)\, dx$$

$$\geq \frac{x_i}{b-a} (f(b) - f(a)) - \frac{1}{b-a} \int_a^b x f'(x)\, dx$$

$$= \frac{x_i}{b-a} (f(b) - f(a)) - \frac{1}{b-a} (bf(b) - af(a)) + \frac{1}{b-a} \int_a^b f(x)\, dx.$$

Furthermore, multiplying by $p_i \geq 0$ and summing over i from 1 to n, we obtain

$$\frac{1}{P_n} \sum_{i=1}^n p_i f(x_i) - \frac{1}{b-a} \int_a^b f(x)\, dx$$

$$\geq \frac{x_p}{b-a} (f(b) - f(a)) - \frac{1}{b-a} (bf(b) - af(a)) + \frac{1}{b-a} \int_a^b f(x)\, dx.$$

Thus,

$$\frac{2}{b-a} \int_a^b f(x)\, dx \le \frac{1}{P_n} \sum_{i=1}^n p_i f(x_i) + \frac{1}{b-a} \left[(b-x_p)f(b) + (x_p - a)f(a)\right].$$

This proves the first inequality in (2.9). Let

$$\alpha = \frac{b-x_i}{b-a}, \quad \beta = \frac{x_i - a}{b-a}, \quad x_i \in [a,b], \quad i \in \{1,...,n\}.$$

It is clear that $\alpha + \beta = 1$. From the convexity of f we have

$$\frac{b-x_i}{b-a} f(a) + \frac{x_i - a}{b-a} f(b) \ge f(x_i)$$

for all $i \in \{1,...,n\}$. By multiplying with $p_i \ge 0$ and summing over i from 1 to n, we derive the following:

$$\frac{1}{b-a}\left[(b-x_p)f(a) + (x_p - a)f(b)\right] \ge \frac{1}{P_n} \sum_{i=1}^n p_i f(x_i). \qquad (2.10)$$

The inequality (2.10) is well known in the literature as the Lah-Ribarić inequality [141, p. 9]. Using this inequality, we have

$$\frac{1}{P_n} \sum_{i=1}^n p_i f(x_i) + \frac{1}{b-a}\left[(b-x_p)f(b) + (x_p - a)f(a)\right]$$

$$\le \frac{1}{b-a}\left[(b-x_p)f(a) + (x_p - a)f(b) + (b-x_p)f(b) + (x_p - a)f(a)\right]$$

$$= f(a) + f(b).$$

Thus, the inequality (2.9) is now completely proven. □

Observe that, with the above assumptions for f, a, b and if $t \in [a,b]$, then we have

$$\frac{f(t)}{2} + \frac{1}{2} \cdot \frac{bf(b) - af(a) - t(f(b) - f(a))}{b-a} \le \frac{f(a) + f(b)}{2}.$$

PROOF The argument follows by the above result if we choose $x_i = t$, $i \in \{1,...,n\}$. We shall omit the details. □

The inequality (2.9) is also a generalisation of Bullen's result. We recapture this result when $x_i = \frac{a+b}{2}, i = 1, ..., n$.

Before proceeding to further comments, it is worthwhile to recall definitions of some well-known special means. These are

- *The arithmetic mean:* $A = A(a, b) := (a + b)/2, \quad a, b \geq 0$;

- *The geometric mean:* $G = G(a, b) := \sqrt{ab}, \quad a, b \geq 0$;

- *The harmonic mean:*

$$H = H(a, b) := \frac{2}{\frac{1}{a} + \frac{1}{b}}, \qquad a, b > 0;$$

- *The logarithmic mean:*

$$L = L(a, b) := \begin{cases} \frac{b-a}{\ln b - \ln a} & \text{if } a \neq b, \\ a & \text{if } a = b, \end{cases} \qquad a, b > 0.$$

Note that for the convex mapping $f = \frac{1}{t} : (0, \infty) \rightarrow \mathbb{R}$, we have

$$\frac{1}{b-a} \int_a^b f(t)dt = L^{-1}(a, b) \quad \text{for} \quad a \neq b.$$

- *The p-logarithmic mean:*

$$L_p = L_p(a, b) := \begin{cases} \left[\frac{b^{p+1} - a^{p+1}}{(p+1)(b-a)}\right]^{1/p} & \text{if } a \neq b, \\ a & \text{if } a = b, \end{cases} \qquad p \in \mathbb{R}\setminus\{-1, 0\}, \ a, b > 0.$$

For the convex (or concave) mapping $f(t) = t^p, p \in (-\infty, 0) \cup [1, \infty)\setminus\{-1\}$ (or $p \in (0, 1)$), we have

$$\frac{1}{b-a} \int_a^b f(t)dt = L_p^p(a, b) \quad \text{for} \quad a \neq b.$$

- *The identric mean:*

$$I = I(a, b) := \begin{cases} \frac{1}{e}\left(\frac{b^b}{a^a}\right)^{\frac{1}{b-a}} & \text{if } a \neq b, \\ a & \text{if } a = b, \end{cases} \qquad a, b > 0.$$

For the convex mapping $f(t) = -\ln x$, we have

$$\frac{1}{b-a} \int_a^b f(t)dt = \ln I(a, b) \quad \text{if } a \neq b.$$

These means are often used in numerical approximation and in other areas. However, the following simple relationships are known in the literature:

$$H \leq G \leq L \leq I \leq A.$$

It is also known that L_p is monotonically increasing in $p \in \mathbb{R}$ with $L_0 = I$ and $L_{-1} = L$.

Comments

(a) We shall start with the following result [103]:

Let $p \in (-\infty, 0) \cup [1, \infty) \setminus \{-1\}$ and $[a, b] \subset (0, \infty)$. Then one has the inequality:

$$\frac{L_p^p - t^p}{pt^{p-1}} \geq A - t \tag{2.11}$$

for all $t \in [a, b]$.

PROOF If in (2.5) we choose $f : [a, b] \to [0, \infty)$, $f(x) = x^p$ and p as specified above, then we get

$$\frac{1}{b-a} \int_a^b x^p dx \geq t^p + pt^{p-1} \left(\frac{a+b}{2} - t \right)$$

for all $t \in [a, b]$. Since

$$\frac{1}{b-a} \int_a^b x^p dx = L_p^p(a, b) = L_p^p$$

we get the desired inequality (2.11).   ☐

By using the above inequality we deduce the following particular results which are related to various means:

(i) $\dfrac{L_p^p - I^p}{pI^{p-1}} \geq A - I \geq 0, \qquad \dfrac{L_p^p - L^p}{pL^{p-1}} \geq A - L \geq 0;$

(ii) $\dfrac{L_p^p - G^p}{pG^{p-1}} \geq A - G \geq 0, \qquad \dfrac{L_p^p - H^p}{pH^{p-1}} \geq A - H \geq 0;$

(iii) $\dfrac{L_p^p - a^p}{pa^{p-1}} \geq A - a \geq 0, \qquad 0 \leq \dfrac{L_p^p - b^p}{pb^{p-1}} \leq b - A.$

The following result also holds:

(b) Let $0 < a < b$. Then for all $t \in [a, b]$ we have the inequality:

$$\frac{L - t}{L} \leq \frac{A - t}{t}. \tag{2.12}$$

PROOF If in (2.5) we choose $f(x) = \frac{1}{x}$ for $x \in [a, b]$, then we have:

$$\frac{1}{b-a} \int_a^b \frac{dx}{x} \geq \frac{1}{t} - \frac{1}{t^2} \left(\frac{a+b}{2} - t \right),$$

which is equivalent to

$$\frac{1}{L} \geq \frac{1}{t} - \frac{1}{t^2}(A - t),$$

and proves the inequality (2.12).   ▯

Using the above inequality we can state the following interesting inequalities:

$$\frac{L_p - L}{L} \geq \frac{L_p - A}{L_p}, \quad \frac{A - G}{G} \geq \frac{L - G}{L},$$

$$\frac{A - H}{H} \geq \frac{L - H}{L}, \quad \frac{L - a}{L} \geq \frac{A - a}{a}$$

and

$$\frac{b - L}{L} \geq \frac{b - A}{b}.$$

Finally, we have the following additional result [103]:

(c) *Let $0 < a < b$. Then one has the inequality*

$$\ln I - \ln t \leq \frac{A - t}{t} \tag{2.13}$$

for all $t \in [a, b]$.

PROOF   In (2.5), choose $f(x) = -\ln x, x \in [a, b]$ to obtain

$$-\frac{1}{b - a}\int_a^b \ln x \, dx \geq -\ln t - \frac{1}{t}\left(\frac{a + b}{2} - t\right),$$

which is equivalent to

$$-\ln I\,(a, b) = -\ln I \geq -\ln t - \frac{1}{t}(A - t),$$

which proves (2.13).   ▯

Using the inequality (2.13), we get for $p \geq 1$ that

$$\ln L_p - \ln I \geq \frac{L_p - A}{L_p} \geq 0, \quad \ln b - \ln I \geq \frac{b - A}{b}$$

and

$$0 \leq \ln I - \ln L \leq \frac{A - L}{L}, \quad 0 \leq \ln I - \ln G \leq \frac{A - G}{G}$$

and

$$0 \leq \ln I - \ln H \leq \frac{A - H}{H}, \quad 0 \leq \ln I - \ln a \leq \frac{A - a}{a},$$

respectively.

(d) In the following, we shall give some natural applications of (2.6) [103].

Let $p \in (-\infty, 0) \cup [1, \infty) \setminus \{-1\}$ and $[a, b] \subset [0, \infty)$. Then one has the inequality:

$$L_p^p - t^p \leq p \left(L_p^p - t L_{p-1}^{p-1} \right) \tag{2.14}$$

for all $t \in [a, b]$.

PROOF   In (2.6), we choose the convex function $f : [a, b] \to [0, \infty)$, $f(x) = x^p$ to get:

$$\frac{1}{b-a} \int_a^b x^p \, dx \leq \frac{t^p}{2} + \frac{1}{2} \left[\frac{b^{p+1} - a^{p+1}}{b-a} - t \cdot \frac{b^p - a^p}{b-a} \right]$$

for all $t \in [a, b]$. As

$$\frac{1}{b-a} \int_a^b x^p \, dx = L_p^p, \qquad \frac{b^{p+1} - a^{p+1}}{b-a} = (p+1) L_p^p$$

and

$$\frac{b^p - a^p}{b-a} = p L_{p-1}^{p-1},$$

we have

$$L_p^p \leq \frac{t^p}{2} + \frac{1}{2} \left[(p+1) L_p^p - t p L_{p-1}^{p-1} \right]$$
$$= \frac{t^p}{2} + \frac{L_p^p}{2} + \frac{1}{2} p \left(L_p^p - t L_{p-1}^{p-1} \right)$$

from the above inequality, which is equivalent to (2.14).   ⬚

In particular, for $p \geq 1$, we have

$$0 \leq L_p^p - A^p \leq p \left(L_p^p - A L_{p-1}^{p-1} \right), \quad 0 \leq L_p^p - L^p \leq p \left(L_p^p - L L_{p-1}^{p-1} \right)$$

and

$$0 \leq L_p^p - I^p \leq p \left(L_p^p - I L_{p-1}^{p-1} \right), \quad 0 \leq L_p^p - G^p \leq p \left(L_p^p - G L_{p-1}^{p-1} \right).$$

(e) The following proposition also holds [103]:

Let $0 < a < b$. Then for all $t \in [a, b]$ we have the following inequality:

$$\frac{t - L}{L} \leq \frac{1}{2} \cdot \frac{t^2 - G^2}{G^2}. \tag{2.15}$$

PROOF   If we choose in (2.6) $f(x) = \frac{1}{x}$, then

$$\frac{1}{b-a}\int_a^b \frac{dx}{x} \leq \frac{1}{2t} - \frac{1}{2} \cdot \frac{t\left(\frac{1}{b}-\frac{1}{a}\right)}{b-a} = \frac{1}{2t} + \frac{t}{2ab},$$

that is,

$$\frac{1}{L} \leq \frac{1}{2t} + \frac{t}{2ab} \quad \text{or} \quad \frac{1}{L} - \frac{1}{t} \leq \frac{t}{2ab} - \frac{1}{2t},$$

which is equivalent to (2.15). $\square$

The following are special cases of (2.15):

$$0 \leq \frac{L_p - L}{L} \leq \frac{1}{2} \cdot \frac{L_p^2 - G^2}{G^2} \quad (p \geq 1),$$

$$0 \leq \frac{A - L}{L} \leq \frac{1}{2} \cdot \frac{A^2 - G^2}{G^2}, \quad 0 \leq \frac{I - L}{L} \leq \frac{1}{2} \cdot \frac{I^2 - G^2}{G^2},$$

and

$$0 \leq \frac{1}{2} \cdot \frac{G^2 - H^2}{G^2} \leq \frac{L - H}{L}.$$

(**f**) We also have the following application of (2.6) [103].

Let $0 < a < b$. Then one has the inequality

$$\frac{L - t}{L} \leq \ln I - \ln t, \tag{2.16}$$

for all $t \in [a, b]$.

PROOF   If in (2.6) we choose $f(x) = -\ln x, x \in [a, b]$ we get

$$-\frac{1}{b-a}\int_a^b \ln x\, dx \leq \frac{\ln t}{2} - \frac{1}{2} \cdot \frac{b\ln b - a\ln a - t(\ln b - \ln a)}{b-a}$$

$$= -\frac{\ln t}{2} - \frac{1}{2}\ln\left(\frac{b^b}{a^a}\right)^{\frac{1}{b-a}} + \frac{1}{2}t\left(\frac{\ln b - \ln a}{b-a}\right)$$

$$= -\frac{\ln t}{2} - \frac{1}{2}\ln[eI] + \frac{t}{2L}.$$

This gives us

$$-\ln I \leq -\frac{\ln t}{2} - \frac{1}{2}[1 + \ln I] + \frac{t}{2L},$$

that is,

$$-2\ln I \leq -\ln t - 1 - \ln I + \frac{t}{L},$$

which is equivalent to

$$1 - \frac{t}{L} \leq \ln I - \ln t,$$

and proves the inequality.   ⬚

From the above inequality (2.16) we deduce the following particular inequalities involving the logarithmic, geometric, identric, harmonic, and arithmetic means:

$$0 \le \frac{L - G}{L} \le \ln I - \ln G, \quad 0 \le \frac{L - H}{L} \le \ln I - \ln H$$

and

$$\frac{A - L}{L} \ge \ln A - \ln I \ge 0.$$

In what follows we shall point out some natural further applications of (2.9) [103].

(g) *Let* $r \in (-\infty, 0) \cup [1, \infty) \setminus \{-1\}$ *and* $[a, b] \subset [0, \infty)$. *If* $x_i \in [a, b]$, $p_i \ge 0$ *with* $P_n := \sum_{i=1}^{n} p_i > 0$ $(i = \overline{1, n})$, *we have the inequality:*

$$L_r^r (a, b) - \left[M_n^{[r]} (x, p) \right]^r \tag{2.17}$$

$$\le r \left[[L_r (a, b)]^r - A_n (x, p) [L_{r-1} (a, b)]^{r-1} \right]$$

$$\le 2A (b^r, a^r) - [L_r (a, b)]^r - \left[M_n^{[r]} (x; p) \right]^r$$

where

$$A_n (x, p) := \frac{1}{P_n} \sum_{i=1}^{n} p_i x_i$$

is the weighted arithmetic mean and

$$M_n^{[r]} (x; p) := \left(\frac{1}{P_n} \sum_{i=1}^{n} p_i x_i^r \right)^{\frac{1}{r}}$$

is the r-power mean.

PROOF   In (2.9) we choose $f(x) = x^r$, $x \in [a, b]$, to get

$$\frac{1}{b - a} \int_a^b x^r \, dx \le \frac{1}{2} \left[\left(M_n^{[r]} (x; p) \right)^r + \frac{b^{r+1} - a^{r+1}}{b - a} - A_n (x, p) \cdot \frac{b^r - a^r}{b - a} \right]$$

$$\le A (b^r, a^r).$$

Since

$$\frac{b^{r+1} - a^{r+1}}{b - a} = (r + 1) [L_r (a, b)]^r$$

and

$$\frac{b^r - a^r}{b - a} = r [L_{r-1} (a, b)]^{r-1},$$

we have:

$$L_r^r (a, b) \le \frac{1}{2} \left[\left[M_n^{[r]} (x, p) \right]^r + (r + 1) \left[L_r (a, b) \right]^r - r A_n (x, p) \left[L_{r-1} (a, b) \right]^{r-1} \right]$$
$$\le A (b^r, a^r)$$

which proves (2.17).   ⧠

If in (2.17) we choose $x_i = t$, $i = \overline{1, n}$ we get

$$L_r^r - t^r \le r \left[L_r^r - t L_{r-1}^{r-1} \right] \le 2A (b^r, a^r) - L_r^r - t^r, \quad t \in [a, b], \qquad (2.18)$$

which provides a counterpart for the inequality (2.14) (for $r = p$). The second inequality in (2.18) produces the particular inequalities:

$$0 \le r \left[L_r^r - A L_{r-1}^{r-1} \right] \le 2A (b^r, a^r) - L_r^r - A^r,$$
$$0 \le r \left[L_r^r - L L_{r-1}^{r-1} \right] \le 2A (b^r, a^r) - L_r^r - L^r,$$
$$0 \le r \left[L_r^r - I L_{r-1}^{r-1} \right] \le 2A (b^r, a^r) - L_r^r - I^r,$$

and

$$0 \le r \left[L_r^r - G L_{r-1}^{r-1} \right] \le 2A (b^r, a^r) - L_r^r - G^r.$$

(**h**) The following proposition holds [103]:

Let $0 < a < b$, $x_i \in [a, b]$, $i = \overline{1, n}$, $p_i > 0$, $i = \overline{1, n}$. Then one has the inequality:

$$\frac{H_n (p, x) - L (a, b)}{L (a, b)} \le \frac{1}{2} \cdot \frac{A_n (p, x) H_n (p, x) - G^2 (a, b)}{G^2 (a, b)} \qquad (2.19)$$
$$\le \frac{A (a, b) H_n (p, x) - G^2 (a, b)}{G^2 (a, b)},$$

where $H_n (p, x)$ is the weighted harmonic mean, that is,

$$H_n (p, x) = \left[M_n^{[-1]} (x, p) \right]^{-1} = \frac{P_n}{\sum_{i=1}^{n} \frac{p_i}{x_i}}.$$

PROOF   If in (2.9) we choose $f (x) = \frac{1}{x}$, $x \in [a, b]$, we get:

$$\frac{1}{b - a} \int_a^b \frac{dx}{x} \le \frac{1}{2} \cdot \left[\frac{1}{P_n} \sum_{i=1}^{n} \frac{p_i}{x_i} + \frac{\frac{b}{b} - \frac{a}{a}}{b - a} - A_n (x, p) \frac{\frac{1}{b} - \frac{1}{a}}{b - a} \right] \le \frac{\frac{1}{a} + \frac{1}{b}}{2},$$

that is,

$$\frac{1}{L (a, b)} \le \frac{1}{2} \cdot \left[\frac{1}{H_n (p, x)} + \frac{H_n (p, x)}{G^2 (a, b)} \right] \le \frac{A (a, b)}{G^2 (a, b)}.$$

This gives

$$\frac{1}{L\left(a,b\right)} - \frac{1}{H_n\left(p,x\right)} \leq \frac{1}{2} \cdot \left[\frac{A_n\left(x,p\right)}{G^2\left(a,b\right)} - \frac{1}{H_n\left(p,x\right)}\right] \leq \frac{A\left(a,b\right)}{G^2\left(a,b\right)} - \frac{1}{H_n\left(p,x\right)},$$

which is equivalent to

$$\frac{H_n\left(p,x\right) - L\left(a,b\right)}{L\left(a,b\right) H_n\left(p,x\right)} \leq \frac{1}{2} \cdot \left[\frac{A_n\left(x,p\right) H_n\left(p,x\right) - G^2\left(a,b\right)}{G^2\left(a,b\right) H_n\left(p,x\right)}\right]$$

$$\leq \frac{A\left(a,b\right) H_n\left(p,x\right) - G^2\left(a,b\right)}{G^2\left(a,b\right) H_n\left(p,x\right)}$$

and proves the inequality (2.19). □

If in (2.19) we choose $x_i = t$, $i = \overline{1,n}$ we get

$$\frac{t-L}{L} \leq \frac{1}{2} \cdot \frac{t^2 - G^2}{G^2} \leq \frac{tA - G^2}{G^2}, \quad t \in [a,b], \tag{2.20}$$

which provides a counterpart for the inequality (2.15). The second inequality (2.20) gives us the following particular results:

$$0 \leq \frac{1}{2} \cdot \frac{I^2 - G^2}{G^2} \leq \frac{IA - G^2}{G^2}, \quad 0 \leq \frac{1}{2} \cdot \frac{L^2 - G^2}{G^2} \leq \frac{LA - G^2}{G^2}.$$

(i) Finally, we also have the following result [103].

Let $0 < a < b$ and $x_i \in [a,b]$, $p_i \geq 0$ $(i = \overline{1,n})$ with $P_n > 0$. Then we have the inequality:

$$\ln I\left(a,b\right) - \ln G_n\left(p,x\right) \geq \frac{L\left(a,b\right) - A_n\left(x,p\right)}{L\left(a,b\right)} \tag{2.21}$$

$$\geq \ln G^2\left(a,b\right) - \ln I\left(a,b\right) - \ln G_n\left(p,x\right),$$

where $G_n\left(p,x\right)$ is the weighted geometric mean. That is:

$$G_n\left(p,x\right) := \left(\prod_{i=1}^{n} x_i^{p_i}\right)^{\frac{1}{P_n}}.$$

PROOF If in (2.9) we choose $f\left(x\right) = -\ln x$, $x \in [a,b]$ we obtain:

$$\frac{1}{b-a} \int_a^b \left(-\ln x\right) dx$$

$$\leq -\frac{1}{2} \cdot \left[\frac{1}{P_n} \sum_{i=1}^{n} p_i \ln x_i + \frac{b \ln b - a \ln a}{b - a} - A_n\left(x,p\right) \frac{\ln b - \ln a}{b - a}\right]$$

$$\leq -\frac{\ln b - \ln a}{2}.$$

This is equivalent to

$$\frac{1}{b-a}\int_a^b \ln x\, dx \geq \frac{1}{2}\left[\ln G_n\left(p,x\right) + \ln\left[e \cdot I\left(a,b\right)\right] - \frac{A_n\left(x,p\right)}{L\left(a,b\right)}\right] \geq \ln G\left(a,b\right)$$

or

$$\ln I\left(a,b\right) \geq \frac{1}{2}\ln G_n\left(p,x\right) + \frac{1}{2} + \frac{1}{2}\ln I\left(a,b\right) - \frac{A_n\left(x,p\right)}{2L\left(a,b\right)} \geq \ln G\left(a,b\right).$$

Thus,

$$\frac{1}{2}\ln I\left(a,b\right) \geq \frac{1}{2}\left[\ln G_n\left(p,x\right) + 1 - \frac{A_n\left(x,p\right)}{L\left(a,b\right)}\right] \geq \ln G\left(a,b\right) - \frac{1}{2}\ln I\left(a,b\right)$$

or, additionally,

$$\frac{1}{2}\left[\ln I\left(a,b\right) - \ln G_n\left(p,x\right)\right] \geq \frac{1}{2}\left[\frac{L\left(a,b\right) - A_n\left(x,p\right)}{L\left(a,b\right)}\right]$$
$$\geq \ln G\left(a,b\right) - \frac{1}{2}\ln I\left(a,b\right) - \frac{1}{2}\ln G_n\left(p,x\right).$$

This proves the inequality (2.21).   ⬜

If in (2.21) we put $x_i = t, i = \overline{1,n}$ we get

$$\ln I - \ln t \geq \frac{L-t}{L} \geq \ln G^2 - \ln I - \ln t,\ t \in [a,b],\qquad(2.22)$$

which counterparts the inequality (2.16). This last inequality also gives us the following particular inequalities:

$$0 \leq \frac{A-L}{L} \leq -\ln G^2 + \ln I + \ln A$$

and

$$0 \leq \frac{I-L}{L} \leq -\ln G^2 + \ln I^2,$$

which are equivalent to

$$0 \leq \frac{A-L}{L} \leq \ln\left(\frac{IA}{G^2}\right) \quad \text{and} \quad 0 \leq \frac{I-L}{L} \leq \ln\left(\frac{I^2}{G^2}\right).$$

Furthermore,

$$1 \leq \exp\left(\frac{A}{L}-1\right) \leq \frac{IA}{G^2} \quad \text{and} \quad 2 \leq \exp\left(\frac{1}{L}-1\right) \leq \frac{I^2}{G^2}.$$

2.3 Hermite-Hadamard Inequality for Log-Convex Mappings

Let $f : I \to (0, \infty)$ be a *log-convex mapping on* I, *so that it satisfies*

$$f(tx+(1-t)y) \leq [f(x)]^t [f(y)]^{1-t}, \quad \text{for all } x, y \in I, \text{ for all } t \in [0, 1]. \quad (2.23)$$

If $a, b \in I$, $a < b$, *then we have the inequality* [99]

$$f\left(\frac{a+b}{2}\right) \leq \exp\left[\frac{1}{b-a} \int_a^b \ln f(x)\, dx\right] \quad (2.24)$$

$$\leq \frac{1}{b-a} \int_a^b G(f(x), f(a+b-x))\, dx$$

$$\leq \frac{1}{b-a} \int_a^b f(x)\, dx$$

$$\leq L(f(a), f(b)),$$

where $G(p, q)$ *is the geometric mean of* p, q.

PROOF Writing the Hermite-Hadamard inequality for $\ln f$, which is convex, produces

$$\ln f\left(\frac{a+b}{2}\right) \leq \frac{1}{b-a} \int_a^b \ln f(x)\, dx$$

from which the first inequality in (2.24) is obtained.

Integrating the identity

$$G(f(x), f(a+b-x)) = \exp[\ln(G(f(x), f(a+b-x)))]$$

on $[a, b]$ and using the well-known Jensen's integral inequality for the convex mapping $\exp(\cdot)$, we have

$$\frac{1}{b-a} \int_a^b G(f(x), f(a+b-x))\, dx$$

$$= \frac{1}{b-a} \int_a^b \exp[\ln(G(f(x), f(a+b-x)))]\, dx$$

$$\geq \exp\left[\frac{1}{b-a} \int_a^b \ln[G(f(x), f(a+b-x))]\, dx\right]$$

$$= \exp\left[\frac{1}{b-a} \int_a^b \left(\frac{\ln f(x) + \ln f(a+b-x)}{2}\right) dx\right]$$

$$= \exp\left[\frac{1}{b-a} \int_a^b \ln f(x)\, dx\right],$$

since obviously

$$\int_a^b \ln f(x)\, dx = \int_a^b \ln f(a+b-x)\, dx.$$

Thus, the second inequality in (2.24) is proved.

By the arithmetic mean–geometric mean inequality, we have that

$$G(f(x), f(a+b-x)) \le \frac{f(x) + f(a+b-x)}{2}, \quad x \in [a, b]$$

from which, upon integration, we get

$$\frac{1}{b-a} \int_a^b G(f(x), f(a+b-x))\, dx \le \frac{1}{b-a} \int_a^b f(x)\, dx.$$

This proves the third inequality in (2.24).

To prove the last inequality, we observe from the assumption of log-convexity of f that

$$f(ta + (1-t)b) \le [f(a)]^t [f(b)]^{1-t}, \quad \text{for all } t \in [a, b]. \tag{2.25}$$

Integrating (2.25) over t in $[0, 1]$, we have

$$\int_0^1 f(ta + (1-t)b)\, dt \le \int_0^1 [f(a)]^t [f(b)]^{1-t}\, dt.$$

Now, since

$$\int_0^1 f(ta + (1-t)b)\, dt = \frac{1}{b-a} \int_a^b f(x)\, dx$$

and

$$\int_0^1 [f(a)]^t [f(b)]^{1-t}\, dt = f(b) \int_0^1 \left(\frac{f(a)}{f(b)}\right)^t dt = f(b) \left[\frac{\left(\frac{f(a)}{f(b)}\right)^t}{\ln\left(\frac{f(a)}{f(b)}\right)} \right]_0^1$$

$$= f(b) \left[\frac{\frac{f(a)}{f(b)} - 1}{\ln\left(\frac{f(a)}{f(b)}\right)} \right] = L(f(a), f(b)),$$

the inequality is thus proved. ⬜

Comments

Other inequalities for log-convex functions can be found in Dragomir and Mond [99].

2.4 Hermite-Hadamard Inequality for the Godnova-Levin Class of Functions

In 1985, Godnova and Levin (see Dragomir, Pečarić, and Persson [108] or Mitrinović, Pečarić, and Fink [141, pp. 410–433]) introduced the following class of functions.

A mapping $f : I \to \mathbb{R}$ is said to belong to the class $Q(I)$ if it is nonnegative and for all $x, y \in I$ and $\lambda \in (0, 1)$, satisfies the inequality

$$f(\lambda x + (1 - \lambda) y) \le \frac{f(x)}{\lambda} + \frac{f(y)}{1 - \lambda}. \tag{2.26}$$

Godnova and Levin also noted that all nonnegative monotonic and nonnegative convex functions belong to this class, and proved the following motivating result:

If $f \in Q(I)$ and $x, y, z \in I$, then

$$f(x)(x - y)(x - z) + f(y)(y - x)(y - z) + f(z)(z - x)(z - y) \ge 0. \tag{2.27}$$

In fact, (2.27) is equivalent to (2.26). Therefore it can alternatively be used in the definition of the class $Q(I)$.

For the case $f(x) = x^r$, $r \in \mathbb{R}$, the inequality (2.27) obviously coincides with the well-known *Schur inequality*.

The following result of the Hermite-Hadamard type holds [108].

Let $f \in Q(I)$, $a, b \in I$ with $a < b$ and $f \in L_1[a, b]$. Then one has the inequalities:

$$f\left(\frac{a + b}{2}\right) \le \frac{4}{b - a} \int_a^b f(x)\, dx \tag{2.28}$$

and

$$\frac{1}{b - a} \int_a^b p(x) f(x)\, dx \le \frac{f(a) + f(b)}{2}, \tag{2.29}$$

where $p(x) = \frac{(b-x)(x-a)}{(b-a)^2}$, $x \in [a, b]$. The constant 4 in (2.28) is the best possible.

PROOF   Since $f \in Q(I)$, we have, for all $x, y \in I$ (with $\lambda = \frac{1}{2}$ in Equation 2.26)

$$2(f(x) + f(y)) \ge f\left(\frac{x + y}{2}\right).$$

Choosing $x = ta + (1 - t) b$, $y = (1 - t) a + tb$, we get

$$2(f(ta + (1 - t) b) + f((1 - t) a + tb)) \ge f\left(\frac{a + b}{2}\right).$$

Further, integrating for $t \in [0, 1]$, we have

$$2 \left[\int_0^1 f\left(ta + (1-t)\,b\right)dt + \int_0^1 f\left((1-t)\,a + tb\right)dt \right] \geq f\left(\frac{a+b}{2}\right). \quad (2.30)$$

Since each of the integrals is equal to $\frac{1}{b-a}\int_a^b f(x)\,dx$, we obtain the inequality (2.28) from (2.30).

For the proof of (2.29), we first note that if $f \in Q(I)$, then from (2.26) for all $a, b \in I$ and $\lambda \in [0, 1]$, we have

$$\lambda(1-\lambda)\,f\left(\lambda a + (1-\lambda)\,b\right) \leq (1-\lambda)\,f(a) + \lambda f(b)$$

and

$$\lambda(1-\lambda)\,f\left((1-\lambda)\,a + \lambda b\right) \leq \lambda f(a) + (1-\lambda)\,f(b).$$

By adding these inequalities and integrating, we find that

$$\int_0^1 \lambda(1-\lambda)\left[f\left(\lambda a + (1-\lambda)\,b\right) + f\left((1-\lambda)\,a + \lambda b\right)\right]d\lambda$$

$$\leq f(a) + f(b). \quad (2.31)$$

Moreover,

$$\int_0^1 \lambda(1-\lambda)f\left(\lambda a + (1-\lambda)\,b\right)d\lambda \quad (2.32)$$

$$= \int_0^1 \lambda(1-\lambda)\,f\left((1-\lambda)\,a + \lambda b\right)d\lambda$$

$$= \frac{1}{b-a}\int_a^b \frac{(b-x)(x-a)}{(b-a)^2}f(x)\,dx.$$

We get (2.29) by combining (2.31) with (2.32); and the proof is complete.

The constant 4 in (2.28) is the best possible because this inequality reduces to an equality for the function

$$f(x) = \begin{cases} 1, & a \leq x < \frac{a+b}{2} \\ 4, & x = \frac{a+b}{2} \\ 1, & \frac{a+b}{2} < x \leq b. \end{cases}$$

Additionally, this function is in the class $Q(I)$ because

$$\frac{f(x)}{\lambda} + \frac{f(y)}{1-\lambda} \geq \frac{1}{\lambda} + \frac{1}{1-\lambda} = g(\lambda)$$

$$\geq \min_{0<\lambda<1} g(\lambda) = g\left(\frac{1}{2}\right)$$

$$= 4 \geq f\left(\lambda x + (1-\lambda)\,y\right)$$

for all $x, y \in [a, b]$ and $\lambda \in (0, 1)$.

The proof is thus complete. ☐

Next, we shall restrict the Godnova-Levin class of functions and point out a sharp version of the Hermite-Hadamard inequality in this class. More precisely, we say that a mapping $f : I \to \mathbb{R}$ belongs to the class $P(I)$ if it is nonnegative and, for all $x, y \in I$ and $\lambda \in [0, 1]$, satisfies the following inequality:

$$f(\lambda x + (1 - \lambda) y) \leq f(x) + f(y). \qquad (2.33)$$

Obviously, $Q(I) \supset P(I)$. It is important for applications to note that $P(I)$ also consists of only nonnegative monotonic, convex, and quasi-convex functions, that is, nonnegative functions satisfying

$$f(\lambda x + (1 - \lambda) y) \leq \max \{f(x), f(y)\}.$$

The following result of the Hermite-Hadamard type holds [108]:

Let $f \in P(I)$, $a, b \in I$ with $a < b$, and $f \in L_1[a, b]$. Then one has the inequality

$$f\left(\frac{a+b}{2}\right) \leq \frac{2}{b-a} \int_a^b f(x)\,dx \leq 2(f(a) + f(b)). \qquad (2.34)$$

Both inequalities are the best possible.

PROOF   By employing (2.33) with the choice of $x = ta + (1 - t)b$, $y = (1 - t)a + tb$, and $\lambda = \frac{1}{2}$, we find that

$$f\left(\frac{a+b}{2}\right) \leq f(ta + (1 - t)b) + f((1 - t)a + tb)$$

for all $t \in [0, 1]$. Thus, by integrating on $[0, 1]$, we obtain

$$f\left(\frac{a+b}{2}\right) \leq \int_0^1 [f(ta + (1 - t)b) + f((1 - t)a + tb)]\,dt$$

$$= \frac{2}{b-a} \int_a^b f(x)\,dx$$

and prove the first inequality.

The proof of the second inequality follows by using (2.33) with $x = a$ and $y = b$ and integrating with respect to λ over $[0, 1]$.

The first inequality in (2.34) reduces to an equality for the nondecreasing function

$$f(x) = \begin{cases} 0, & a \leq x < \frac{a+b}{2}, \\ 1, & \frac{a+b}{2} \leq x \leq b. \end{cases}$$

The second inequality reduces to an equality for the nondecreasing function

$$f\left(x\right) = \begin{cases} 0, \, x = a, \\ 1, \, a < x \leq b. \end{cases}$$

The proof is thus complete.   □

Comments

For other results on Hermite-Hadamard type inequalities, see Dragomir, Pečarić, and Persson [108].

2.5   The Hermite-Hadamard Inequality for Quasi-Convex Functions

We shall start with the following definition:

The mapping $f : I \to \mathbb{R}$ is said to be Jensen (or J)-quasi-convex if

$$f\left(\frac{x+y}{2}\right) \leq \max\left\{f\left(x\right), f\left(y\right)\right\} \tag{2.35}$$

for all $x, y \in I$,

Note that the class $JQC\left(I\right)$ of J-quasi-convex functions on I contains the class $J\left(I\right)$ of J-convex functions on I. We recall that J-convex functions are those which satisfy the condition

$$f\left(\frac{x+y}{2}\right) \leq \frac{f\left(x\right) + f\left(y\right)}{2} \text{ for all } x, y \in I. \tag{2.36}$$

The following inequality of the Hermite-Hadamard type holds [102]:

Suppose $a, b \in I \subseteq \mathbb{R}$ and $a < b$. If $f \in JQC\left(I\right) \cap L_1\left[a, b\right]$, then

$$f\left(\frac{a+b}{2}\right) \leq \frac{1}{b-a} \int_a^b f\left(x\right) dx + I\left(a, b\right), \tag{2.37}$$

where

$$I\left(a, b\right) := \frac{1}{2\left(b-a\right)} \int_a^b \left|f\left(x\right) - f\left(a+b-x\right)\right| dx.$$

Furthermore, $I(a, b)$ satisfies the inequalities:

$$0 \le I(a, b) \tag{2.38}$$

$$\le \frac{1}{b-a} \min \left\{ \int_a^b |f(x)|\, dx \,, \right.$$

$$\left. \frac{1}{\sqrt{2}} \left((b-a) \int_a^b f^2(x)\, dx - J(a, b) \right)^{\frac{1}{2}} \right\},$$

where

$$J(a, b) := (b-a) \int_a^b f(x) f(a+b-x)\, dx.$$

PROOF   Since f is J-quasi-convex on I, we have the following from the well-known definition of max, for all $x, y \in I$:

$$f\left(\frac{x+y}{2}\right) \le \frac{f(x) + f(y) + |f(x) - f(y)|}{2}.$$

For $t \in [0, 1]$, put $x = ta + (1-t)b$, $y = (1-t)a + tb \in I$. Then

$$f\left(\frac{a+b}{2}\right) \le \frac{1}{2}[f(ta + (1-t)b) + f((1-t)a + tb)$$

$$+ |f(ta + (1-t)b) - f((1-t)a + tb)|].$$

Integrating this inequality over $[0, 1]$ gives

$$f\left(\frac{a+b}{2}\right) \le \frac{1}{2}\left[\int_0^1 f(ta + (1-t)b)\, dt + \int_0^1 f((1-t)a + tb)\, dt\right]$$

$$+ \frac{1}{2}\int_0^1 |f(ta + (1-t)b) - f((1-t)a + tb)|\, dt.$$

Since

$$\int_0^1 f(ta + (1-t)b)\, dt = \int_0^1 f((1-t)a + tb)\, dt = \frac{1}{b-a}\int_a^b f(x)\, dx,$$

and by using the change of variable $x = ta + (1-t)b$, we have

$$\frac{1}{2}\int_0^1 |f(ta + (1-t)b) - f((1-t)a + tb)|\, dt$$

$$= \frac{1}{2(b-a)}\int_a^b |f(x) - f(a+b-x)|\, dx$$

and the inequality (2.37) is proved.

We now observe that

$$0 \leq I(a,b) \leq \frac{1}{2(b-a)} \left[\int_a^b |f(x)| \, dx - \int_a^b |f(a+b-x)| \, dx \right]$$

$$= \frac{1}{b-a} \int_a^b |f(x)| \, dx.$$

On the other hand, by the Cauchy-Bunyakovsky-Schwarz inequality, we have

$$\frac{1}{2(b-a)} \int_a^b |f(x) - f(a+b-x)| \, dx$$

$$\leq \frac{1}{2} \left[\frac{1}{b-a} \int_a^b |f(x) - f(a+b-x)|^2 \, dx \right]^{\frac{1}{2}}$$

$$= \frac{1}{2} \left[\frac{1}{b-a} \int_a^b \left(f^2(x) - 2f(x) f(a+b-x) + f^2(a+b-x) \right) dx \right]^{\frac{1}{2}}$$

$$= \frac{1}{2} \left[\frac{2}{b-a} \int_a^b f^2(x) \, dx - \frac{2}{b-a} \int_a^b f(x) f(a+b-x) \, dx \right]^{\frac{1}{2}}$$

$$= \frac{\sqrt{2}}{2(b-a)} \left[(b-a) \int_a^b f^2(x) \, dx - (b-a) \int_a^b f(x) f(a+b-x) \, dx \right]^{\frac{1}{2}}$$

and the inequality (2.38) is proved. □

Wright introduced an interesting class of functions [167], generalising the concept of convexity.

Namely, we say that $f : I \to \mathbb{R}$ is a *Wright-convex* function on $I \subseteq \mathbb{R}$ if, for each $y > x$ and $\delta > 0$ with $y + \delta, x \in I$ we have

$$f(x+\delta) - f(x) \leq f(y+\delta) - f(y). \tag{2.39}$$

The following characterisation holds for W-convex functions [167]:

Suppose $I \subseteq \mathbb{R}$. Then the following statements are equivalent for a function $f : I \to \mathbb{R}$:

(i) *f is W-convex on I;*

(ii) *For all $a, b \in I$ and $t \in [0,1]$, we have the inequality:*

$$f((1-t)a + tb) + f(ta + (1-t)b) \leq f(a) + f(b). \tag{2.40}$$

PROOF   For "(i)$\Rightarrow$(ii)", let $a, b \in I$ and $t \in [0,1]$. First, suppose $a < b$. If f is W-convex on I, then for all $y > x$ and $\delta > 0$ with $y + \delta, x \in I$ we have

$$f(x+\delta) - f(x) \leq f(y+\delta) - f(y). \tag{2.41}$$

Choose $x = a$, $y = ta + (1 - t) b > 0$ and $\delta := b - (ta + (1 - t) b) > 0$. Then $x + \delta = (1 - t) a + tb$, $y + \delta = b$. Thus, by (2.41) we obtain

$$f((1 - t) a + tb) - f(a) \leq f(b) - f(ta + (1 - t) b);$$

and hence we have (2.40).

The proof is similar for the case $a > b$.

For "(ii)$\Rightarrow$(i)", let $y > x$ and $\delta > 0$ with $y + \delta, x \in I$. In (2.41) choose $a = x$, $b > a$, and $t \in [0, 1]$ with $ta + (1 - t) b = y$ and $b - (ta + (1 - t) b) = \delta$. We have $y + \delta = b \in I$, $x \in I$, and $x + \delta = (1 - t) a + tb$. From (2.40) we derive

$$f(x) + f(y + \delta) \geq f(y) + f(x + \delta),$$

which shows that the map is W-convex on I. $\quad\square$

The equivalence motivates the introduction of the following class of functions [167]:

For $I \subseteq \mathbb{R}$, the mapping $f : I \to \mathbb{R}$ is Wright-quasi-convex if, for all $x, y \in I$ and $t \in [0, 1]$, one has the inequality

$$\frac{1}{2}\left[f(tx + (1 - t) y) + f((1 - t) x + ty)\right] \leq \max\{f(x), f(y)\}, \qquad (2.42)$$

or, equivalently,

$$\frac{1}{2}\left[f(y) + f(\delta)\right] \leq \max\{f(x), f(y + \delta)\}$$

for every $x, y + \delta \in I$ with $x < y$ and $\delta > 0$.

We show that the following inequality of the Hermite-Hadamard type holds [167].

Let $f : I \to \mathbb{R}$ be a W-quasi-convex map on I. Suppose $a, b \in I \subseteq \mathbb{R}$ with $a < b$ and $f \in L_1[a, b]$. Then we have the inequality

$$\frac{1}{b - a} \int_a^b f(x)\, dx \leq \max\{f(a), f(b)\}. \qquad (2.43)$$

PROOF For all $t \in [0, 1]$ we have

$$\frac{1}{2}\left[f(ta + (1 - t) b) + f((1 - t) a + tb)\right] \leq \max\{f(a), f(b)\}.$$

On integrating this inequality over $[0, 1]$, we obtain the desired inequality. $\quad\square$

Comments

We now introduce the notion of a *quasi-monotone* function.

For $I \subseteq \mathbb{R}$, the mapping $f : I \to \mathbb{R}$ is quasi-monotone on I if it is either monotone on $I = [c, d]$ or monotone nonincreasing on a proper subinterval $[c, c'] \subset I$ and monotone nondecreasing on $[c', d]$.

The class $QM(I)$ of quasi-monotone functions on I provides an immediate characterisation of quasi-convex functions [167].

Suppose $I \subseteq \mathbb{R}$. Then the following statements are equivalent for a function $f : I \to \mathbb{R}$:
  (a) $f \in QM(I)$;
  (b) *On any subinterval of I, f achieves a supremum at an end point;*
  (c) $f \in QC(I)$.

PROOF   That (a) implies (b) is immediate from the definition of quasi-monotonicity. For the reverse implication, suppose it is possible that (b) holds but $f \notin QM(I)$. Then there must exist points $x, y, z \in I$ with $x < y < z$ and $f(y) > \max\{f(x), f(z)\}$, contradicting (b) for the subinterval $[x, z]$. The equivalence of (b) and (c) is simply the definition of quasi-convexity.   ∎

The following inclusion results hold [167]:

Let $WQC(I)$ denote the class of Wright-quasi-convex functions on $I \subseteq \mathbb{R}$. Then

$$QC(I) \subset WQC(I) \subset JQC(I). \tag{2.44}$$

Both inclusions are proper.

PROOF   Let $f \in QC(I)$. Then, for all $x, y \in I$ and $t \in [0, 1]$ we have

$$f(tx + (1-t)y) \leq \max\{f(x), f(y), f((1-t)x + ty)\} \leq \max\{f(x), f(y)\}$$

which, by addition, gives

$$\frac{1}{2}[f(tx + (1-t)y) + f((1-t)x + ty)] \leq \max\{f(x), f(y)\} \tag{2.45}$$

for all $x, y \in I$ and $t \in [0, 1]$, which means that $f \in WQC(I)$. The second inclusion becomes obvious on choosing $t = \frac{1}{2}$ in (2.45).

Let H be a Hamel basis over the rationals. Then each real number u has the following unique representation,

$$u = \sum_{h \in H} r_{u,h} \cdot h,$$

in which only finitely many of the coefficients $r_{u,h}$ are nonzero. Define a mapping $f : I \to \mathbb{R}$ by

$$f(u) = \sum_{h \in H} r_{u,h}.$$

Then

$$\frac{1}{2}\left[f\left(y\right)+f\left(x+\delta\right)\right]=\frac{1}{2}\left[\sum_h r_{y,h}+\sum_h\left(r_{x,h}+r_{\delta,h}\right)\right]$$

$$=\frac{1}{2}\left[\sum_h r_{x,h}+\sum_h\left(r_{y,h}+r_{\delta,h}\right)\right]$$

$$\leq\max\left[\sum_h r_{x,h},\sum_h\left(r_{y,h}+r_{\delta,h}\right)\right]$$

$$=\max\left\{f\left(x\right),f\left(y+\delta\right)\right\}$$

so that $f\in WQC\left(I\right)$.

We now demonstrate that H can be selected so that $f\notin QC\left(I\right)$. Choose $\delta>0$ and $x\neq0$ to be rational and $y+\delta$ to be irrational. We may choose H such that $y+\delta,-\left|x\right|\in H$. Then $f\left(\delta\right)<0$, $f\left(x\right)=-\operatorname{sgn}\left(x\right)$ and $f\left(y+\delta\right)=1$. The mapping f is additive. Therefore,

$$f\left(y\right)=f\left(y+\delta\right)-f\left(\delta\right)>f\left(y+\delta\right)=1=\max\left\{f\left(x\right),f\left(y+\delta\right)\right\},$$

and so $f\notin QC\left(I\right)$.

For the second inequality in (2.44), consider the Dirichlet map $f:I\to\mathbb{R}$ defined by

$$f\left(u\right)=\begin{cases}1 & \text{for } u \text{ irrational}\\ 0 & \text{for } u \text{ rational.}\end{cases}$$

If x and y are both rational, then so is $\frac{\left(x+y\right)}{2}$, so that, in this case

$$f\left(\frac{x+y}{2}\right)=\max\left\{f\left(x\right),f\left(y\right)\right\}. \qquad (2.46)$$

If either x or y is rational and the other irrational, then $\frac{\left(x+y\right)}{2}$ is irrational and so, again, (2.46) holds. If both x and y are irrational, then $\max\left\{f\left(x\right),f\left(y\right)\right\}=1$, so that

$$f\left(\frac{x+y}{2}\right)\leq\max\left\{f\left(x\right),f\left(y\right)\right\}.$$

Hence $f\in JQC\left(I\right)$. However, if x and y are distinct rationals, there are uncountably many values of $t\in\left(0,1\right)$ for which $tx+\left(1-t\right)y$ and $\left(1-t\right)x+ty$ are both irrational. For each such t

$$\frac{1}{2}\left[f\left(tx+\left(1-t\right)y\right)+f\left(\left(1-t\right)x+ty\right)\right]>\max\left\{f\left(x\right),f\left(y\right)\right\}$$

so that $f\notin WQC\left(I\right)$. Hence $WQC\left(I\right)$ is a proper subset of $JQC\left(I\right)$. $\qquad\Box$

We also have the following result [167]:

We have the inclusions

$$W\left(I\right)\subset WQC\left(I\right),\ C\left(I\right)\subset QC\left(I\right),\ J\left(I\right)\subset JQC\left(I\right).$$

Each inclusion is proper. Note that $C\left(I\right)$, $W\left(I\right)$, and $J\left(I\right)$ are sets of the convex, W-convex, and J-convex functions on I, respectively.

PROOF   In view of the above results, we have for $f\in W\left(I\right)$ that

$$\frac{1}{2}\left[f\left(\left(1-t\right)a+tb\right)+f\left(ta+\left(1-t\right)b\right)\right]\leq\frac{f\left(a\right)+f\left(b\right)}{2}$$

for all $a,b\in I$ and $t\in\left[0,1\right]$.
  Since

$$\frac{f\left(a\right)+f\left(b\right)}{2}\leq\max\left\{f\left(a\right),f\left(b\right)\right\}\quad\text{for all }a,b\in I,$$

the inequality (2.42) is satisfied, that is, $f\in WQC\left(I\right)$ and the first inclusion is thus proved.
  Similar proofs hold for the other two.
  As

$$C\left(I\right)\subset W\left(I\right)\subset J\left(I\right)\tag{2.47}$$

and each inclusion is proper [127], [128], and (2.44), in order for each inclusion to be proper, it is sufficient to show that there exists a function f with $f\in QC\left(I\right)$ but $f\in J\left(I\right)$. Clearly, any strictly concave monotonic function suffices. $\qquad\square$

2.6   The Hermite-Hadamard Inequality of s-Convex Functions in the Orlicz Sense

The following concept was introduced by Orlicz [145] and was used in the theory of Orlicz spaces [136, 144]:

Let $0<s\leq1$. A function $f:\mathbb{R}_+\to\mathbb{R}$ where $\mathbb{R}_+:=\left[0,\infty\right)$ is said to be s-convex in the first sense if:

$$f\left(\alpha u+\beta v\right)\leq\alpha^s f\left(u\right)+\beta^s f\left(v\right)\tag{2.48}$$

for all $u,v\in\mathbb{R}_+$ and $\alpha,\beta\geq0$ with $\alpha^s+\beta^s=1$. We denote this class of real functions by K_s^1.

We shall present some results from Hudzik and Malingranda [123] referring to the s-convex functions in the first sense.

Let $0 < s < 1$. If $f \in K_s^1$, then f is nondecreasing on $(0, \infty)$ and $\lim_{u \to 0+} f(u) \le f(0)$.

PROOF   We have, for $u > 0$ and $\alpha \in [0, 1]$,

$$f\left[\left(\alpha^{\frac{1}{s}} + (1-\alpha)^{\frac{1}{s}}\right) u\right] \le \alpha f(u) + (1-\alpha) f(u) = f(u).$$

The function

$$h(\alpha) = \alpha^{\frac{1}{s}} + (1-\alpha)^{\frac{1}{s}}$$

is continuous on $[0, 1]$, decreasing on $\left[0, \frac{1}{2}\right]$, increasing on $\left[\frac{1}{2}, 1\right]$, and $h([0, 1]) = \left[h\left(\frac{1}{2}\right), h(1)\right] = \left[2^{1-\frac{1}{s}}, 1\right]$. This implies that

$$f(tu) \le f(u) \text{ for all } u > 0, \ t \in \left[2^{1-\frac{1}{s}}, 1\right]. \tag{2.49}$$

If $t \in \left[2^{1-\frac{1}{s}}, 1\right]$, then $t^{\frac{1}{2}} \in \left[2^{1-\frac{1}{s}}, 1\right]$, and therefore, by the fact that (2.49) holds for all $u > 0$, we get

$$f(tu) = f\left(t^{\frac{1}{2}}\left(t^{\frac{1}{2}}u\right)\right) \le f\left(t^{\frac{1}{2}}\right) \le f(u)$$

for all $u > 0$. By induction, we therefore obtain that

$$f(tu) \le f(u) \text{ for all } u > 0, \ t \in (0, 1]. \tag{2.50}$$

Hence, by taking $0 < u \le v$ and applying (2.50), we get

$$f(u) = f(u(v)v) \le f(v);$$

which means that f is nondecreasing on $(0, \infty)$.

The second part can be proved in the following manner. For $u > 0$ we have

$$f(\alpha u) = f(\alpha u + \beta 0) \le \alpha^s f(u) + \beta^s f(0)$$

and taking $u \to 0^+$, we obtain

$$\lim_{u \to 0+} f(u) \le \lim_{u \to 0+} f(\alpha u) \le \alpha^s \lim_{u \to 0+} f(u) + \beta^s f(0)$$

and hence

$$\lim_{u \to 0+} f(u) \le f(0).$$

⬜

The above results generally do not hold in the case of convex functions, that is, when $s = 1$. The reason for this is that a convex function $f : \mathbb{R}_+ \to \mathbb{R}$ may not necessarily be nondecreasing on $(0, \infty)$.

If $0 < s < 1$, then the function $f \in K_s^1$ is nondecreasing on $(0, \infty)$ but not necessarily on $[0, \infty)$ [123].

Let $0 < s < 1$ and $a, b, c, \in \mathbb{R}$. By defining for $u \in \mathbb{R}_+$, the function

$$f(u) = \begin{cases} a & \text{if } u = 0, \\ bu^s + e & \text{if } u > 0, \end{cases}$$

we have:

(i) If $b \geq 0$ and $c \leq a$, then $f \in K_s^1$.

(ii) If $b \geq 0$ and $c < a$, then f is nondecreasing on $(0, \infty)$ but not on $[0, \infty)$.

From the known examples of the s-convex functions we can build up other s-convex functions using the following composition property [123]:

Let $0 < s \leq 1$. If $f, g \in K_s^1$ and if $F : \mathbb{R}^2 \to \mathbb{R}$ is a nondecreasing convex function, then the function $h : \mathbb{R}_+ \to \mathbb{R}$ defined by $h(u) := F(f(u), g(u))$ is s-convex. In particular, if $f, g \in K_s^1$, then $f + g$ and $\max(f, g) \in K_s^1$.

PROOF If $u, v \in \mathbb{R}_+$, then for all $\alpha, \beta \geq 0$ with $\alpha^s + \beta^s = 1$ we have

$$\begin{aligned} h(\alpha u + \beta v) &= F(f(\alpha u + \beta v), g(\alpha u + \beta v)) \\ &\leq F(\alpha^s f(u) + \beta^s f(v), \alpha^s g(u) + \beta^s g(v)) \\ &\leq \alpha^s F(f(u), g(u)) + \beta^s F(f(v), g(v)) \\ &= \alpha^s h(u) + \beta^s h(v). \end{aligned}$$

Since $F(u, v) = u + v$ and $F(u, v) = \max(u, v)$ are particular examples of nondecreasing convex functions on $\mathbb{R}^2$, we get particular cases of our results.

☐

It is important to note that the condition $\alpha^s + \beta^s = 1$ in the definition of K_s^1 can be equivalently replaced by $\alpha^s + \beta^s \leq 1$ [123].

Let $f \in K_s^1$ and $0 < s \leq 1$. Then inequality (2.48) holds for all $u, v \in \mathbb{R}_+$, and all $\alpha, \beta \geq 0$ with $\alpha^s + \beta^s \leq 1$ if and only if $f(0) \leq 0$.

PROOF Necessity is obvious by taking $u = v = 0$ and $\alpha = \beta = 0$. Therefore, we may assume that $u, v \in \mathbb{R}_+$, $\alpha, \beta \geq 0$ and $0 < \gamma = \alpha^s + \beta^s < 1$. If we let $a = \alpha \gamma^{-\frac{1}{s}}$ and $b = \beta \gamma^{-\frac{1}{s}}$, then $a^s + b^s = \frac{\alpha^s}{\gamma} + \frac{\beta^s}{\gamma} = 1$ and hence

$$\begin{aligned} f(\alpha u + \beta v) &= f\left(a\gamma^{\frac{1}{s}} u + b\gamma^{\frac{1}{s}} v\right) \\ &\leq a^s f\left(\gamma^{\frac{1}{s}} u\right) + b^s f\left(\gamma^{\frac{1}{s}} v\right) \end{aligned}$$

$$= a^s f \left[\gamma^{\frac{1}{s}} u + (1 - \gamma)^{\frac{1}{s}} 0 \right] + b^s f \left[\gamma^{\frac{1}{s}} v + (1 - \gamma)^{\frac{1}{s}} 0 \right]$$

$$\leq a^s \left[\gamma f(u) + (1 - \gamma) f(0) \right] + b^s \left[\gamma f(v) + (1 - \gamma) f(0) \right]$$

$$= a^s \gamma f(u) + b^s \gamma f(v) + (1 - \gamma) f(0)$$

$$\leq a^s f(u) + b^s f(v).$$

This completes the proof. □

Using the above result we can compare both definitions of the s-convexity [123].

Let $0 < s_1 \leq s_2 \leq 1$. If $f \in K_{s_2}^1$ and $f(0) \leq 0$, then $f \in K_{s_1}^1$.

PROOF   Assume that $f \in K_{s_2}^1$ and $u, v \geq 0, \alpha, \beta \geq 0$ with $\alpha^{s_1} + \beta^{s_1} = 1$. Then $\alpha^{s_2} + \beta^{s_2} \leq \alpha^{s_1} + \beta^{s_1} = 1$. From the above results, we have

$$f(\alpha u + \beta v) \leq \alpha^{s_2} f(u) + \beta^{s_2} f(v) \leq \alpha^{s_1} f(u) + \beta^{s_1} f(v),$$

which means that $f \in K_{s_1}^1$. □

Let us note that if f is a nonnegative function in K_s^1 and $f(0) = 0$, then f is right continuous at 0, i.e., $f(0^+) = f(0) = 0$.

We now prove the following result which contains some interesting examples of s-convex functions [123].

Let $0 < s < 1$ and $p : \mathbb{R}_+ \to \mathbb{R}_+$ be a nondecreasing function. Then the function f defined for $u \in \mathbb{R}_+$ by

$$f(u) = u^{\frac{s}{(1-s)}} p(u) \tag{2.51}$$

belongs to K_s^1.

PROOF   Let $v \geq u \geq 0$ and $\alpha, \beta \geq 0$ with $\alpha^s + \beta^s = 1$. We shall consider two cases.

(i) Let $\alpha u + \beta v \leq u$. Then

$$f(\alpha u + \beta v) \leq f(u) = (\alpha^s + \beta^s) f(u) \leq \alpha^s f(u) + \beta^s f(v).$$

(ii) Let $\alpha u + \beta v > u$. This yields $\beta v > (1 - \alpha) u$ and so $\beta > 0$.
Since $\alpha \leq \alpha^s$ for $\alpha \in [0, 1]$, we obtain $\alpha - \alpha^{s+1} \leq \alpha^s - \alpha^{s+1}$ and so

$$\frac{\alpha}{(1 - \alpha)} \leq \frac{\alpha^s}{(1 - \alpha^s)} = \frac{(1 - \beta^s)}{\beta^s},$$

giving

$$\frac{\alpha \beta}{(1 - \alpha)} + \beta v \leq \beta^{1-s} - \beta. \tag{2.52}$$

We also have

$$\alpha u + \beta v \leq (\alpha + \beta) v \leq (\alpha^s + \beta^s) v = v;$$

and, in view of (2.52),

$$\alpha u + \beta v \leq \frac{\alpha \beta v}{(1 - \alpha)} \leq (\beta^{1-s} - \beta) v + \beta v = \beta^{1-s} v,$$

whence

$$(\alpha u + \beta v)^{\frac{s}{(1-s)}} \leq \beta^s v^{\frac{s}{(1-s)}}. \tag{2.53}$$

By applying (2.53) and the monotonicity of p, we arrive at

$$\begin{aligned}
f(\alpha u + \beta v) &= (\alpha u + \beta v)^{\frac{s}{(1-s)}} p(\alpha u + \beta v) \\
&\leq \beta^s v^{\frac{s}{(1-s)}} p(\alpha u + \beta v) \leq \beta^s v^{\frac{s}{(1-s)}} p(v) \\
&= \beta^s f(v) \leq \alpha^s f(u) + \beta^s f(v).
\end{aligned}$$

The proof is thus completed.

☐

The following result contains some other examples of s-convex functions in the first sense [123]:

Let $f \in K^1_{s_1}$ and $g \in K^1_{s_2}$, where $0 < s_1, s_2 \leq 1$.

(i) *If f is a nondecreasing function and g is a nonnegative function such that $f(0) \leq 0 = g(0)$, then the composition $f \circ g$ belongs to K^1_s, where $s = s_1 \cdot s_2$.*

(ii) *Assume that $0 < s_1, s_2 < 1$. If f and g are nonnegative functions such that either $f(0) = 0$ or $g(0) = 0$, then the product $f \cdot g$ belongs to K^1_s, where $s = \min(s_1, s_2)$.*

PROOF

(i) Let $u, v \in \mathbb{R}_+$ and $\alpha, \beta \geq 0$ with $\alpha^s + \beta^s = 1$, where $s = s_1 \cdot s_2$. Since $\alpha^{s_i} + \beta^{s_i} \leq \alpha^{s_1 s_2} + \beta^{s_1 s_2} = 1$ for $i = 1, 2$, we have

$$\begin{aligned}
(f \circ g)(\alpha u + \beta v) &= f(g(\alpha u + \beta v)) \\
&\leq f(\alpha^{s_2} g(u) + \beta^{s_2} g(v)) \\
&\leq \alpha^{s_1 s_2} f(g(u)) + \beta^{s_1 s_2} f(g(u)) \\
&\leq \alpha^s (f \circ g)(u) + \beta^s (f \circ g)(v),
\end{aligned}$$

which means that $f \circ g \in K^1_s$.

(ii) Observe that both functions f and g are nondecreasing on $(0, \infty)$. Therefore

$$(f(u) - f(v))(g(u) - g(v)) \leq 0,$$

or, equivalently,

$$f(u) g(v) + f(v) g(u) \leq f(u) g(u) + f(v) g(v) \qquad (2.54)$$

for all $v \geq u > 0$. If $v > u = 0$, then inequality (2.54) is still valid as f and g are nonnegative; and $f(0) = g(0) = 0$.

Now, let $u, v \in \mathbb{R}_+$ and $\alpha, \beta \geq 0$ with $\alpha^s + \beta^s = 1$, where $s = \min(s_1, s_2)$. Then $\alpha^{s_i} + \beta^{s_i} \leq \alpha^s + \beta^s = 1$ for $i = 1, 2$, and by inequality (2.54) we have

$$
\begin{aligned}
&f(\alpha u + \beta v) g(\alpha u + \beta v) \\
&\leq (\alpha^{s_1} f(u) + \beta^{s_1} f(v))(\alpha^{s_2} g(u) + \beta^{s_2} g(v)) \\
&= \alpha^{s_1 + s_2} f(u) g(u) + \alpha^{s_1} \beta^{s_2} f(u) g(v) \\
&\quad + \alpha^{s_2} \beta^{s_1} f(v) g(u) + \beta^{s_1 + s_2} f(v) g(v) \\
&\leq \alpha^{2s} f(u) g(u) + \alpha^s \beta^s (f(u) g(v) + f(v) g(u)) + \beta^{2s} f(v) g(v) \\
&= \alpha^s f(u) g(u) + \beta^s f(v) g(v)
\end{aligned}
$$

which means that $f, g \in K_s^1$.

⬜

The following particular case may be stated:

If ϕ is a convex ψ-function, namely, $\phi(0) = 0$ and ϕ is nondecreasing and continuous on $[0, \infty)$, and g is a ψ-function in K_s^1, then the composition $\phi \circ g$ belongs to K_s^1. In particular, the ψ-function $h(u) = \phi(u^s)$ belongs to K_s^1.

Finally, we also have [123]:

Let f be a ψ-function and $f \in K_s^1$ $(0 < s < 1)$. Then there exists a convex ψ-function Φ such that the ψ-function Ψ defined for $u \geq 0$ by $\Psi(u) = \Phi(u^s)$ is equivalent to f.

PROOF By the s-convexity of the function f and by $f(0) = 0$, we obtain $f(\alpha u) \leq \alpha^s f(u)$ for all $u \geq 0$ and $\alpha \in [0, 1]$.

Assume that $v > u \geq 0$. Then $f\left(u^{\frac{1}{s}}\right) \leq f\left(\left(\frac{u}{v}\right)^{\frac{1}{s}} v^{\frac{1}{s}}\right) \leq \left(\frac{u}{v}\right) f\left(v^{\frac{1}{s}}\right)$, that is,

$$\frac{f\left(u^{\frac{1}{s}}\right)}{u} \leq \frac{f\left(v^{\frac{1}{s}}\right)}{v}. \qquad (2.55)$$

Inequality (2.55) implies that the function $\dfrac{f\left(u^{\frac{1}{s}}\right)}{u}$ is nondecreasing on $(0,\infty)$. Define

$$\Phi(u) := \begin{cases} 0 & \text{for } u = 0 \\ \int_0^u \dfrac{f\left(t^{\frac{1}{s}}\right)}{t}\,dt & \text{for } u > 0. \end{cases}$$

Then Φ is a convex ψ-function,

$$\Phi(u^s) = \int_0^{u^s} \dfrac{f\left(t^{\frac{1}{s}}\right)}{t}\,dt \le \left[\dfrac{f(u^s)^{\frac{1}{s}}}{u^s}\right] u^s = f(u)$$

and

$$\Phi(u^s) = \int_{\frac{u^s}{2}}^{u^s} \dfrac{f\left(t^{\frac{1}{s}}\right)}{t}\,dt \le \left[\dfrac{f\left(\frac{u^s}{2}\right)^{\frac{1}{s}}}{\left(\frac{u^s}{2}\right)}\right]\dfrac{u^s}{2} = f\left(2^{-\frac{1}{s}}u\right).$$

Therefore,

$$f\left(2^{-\frac{1}{s}}u\right) \le \Phi(u^s) \le f(u)$$

for all $u \ge 0$, which means that ψ is equivalent to f (this sense of equivalence is taken from the theory of Orlicz spaces [144]). The proof is completed. $\square$

With these results, we are able to point out some inequalities of the Hermite-Hadamard type for s-convex functions in the first sense [92].

Let $f : \mathbb{R}_+ \to \mathbb{R}$ be an s-*convex mapping in the first sense with* $s \in (0,1)$. *If* $a, b \in \mathbb{R}$ *with* $a < b$, *then one has the inequality:*

$$f\left(\dfrac{a+b}{2^{\frac{1}{s}}}\right) \le \dfrac{1}{b-a}\int_a^b f(x)\,dx. \tag{2.56}$$

PROOF   If in (2.48) we choose $\alpha = \frac{1}{2^{\frac{1}{s}}}$, $\beta = \frac{1}{2^{\frac{1}{s}}}$, then we have that $\alpha^s + \beta^s = 1$. Thus, for all $x, y \in [0, \infty)$,

$$f\left(\dfrac{x+y}{2^{\frac{1}{s}}}\right) \le \dfrac{f(x)+f(y)}{2}.$$

If we choose $x = ta + (1-t)b$, $y = (1-t)a + tb$, $t \in [0,1]$, then we arrive at

$$f\left(\dfrac{a+b}{2^{\frac{1}{s}}}\right) \le \dfrac{1}{2}\left[f(ta+(1-t)b)+f((1-t)a+tb)\right] \text{ for all } t \in [0,1].$$

Since f is monotonic nondecreasing on $[0,\infty)$, it is integrable on $[a,b]$. Thus, we can integrate over t in the above inequality. Taking into account that

$$\int_0^1 f(ta+(1-t)b)\,dt = \int_0^1 f((1-t)a+tb)\,dt = \dfrac{1}{b-a}\int_a^b f(x)\,dx,$$

the inequality (2.56) is proved. ⬜

The second result, which is similar, in a sense, with the second part of the Hermite-Hadamard inequality for general convex mappings, is embodied in the next statement [92].

With the above assumptions for f and s, one has the inequality:

$$\int_0^1 f\left(ta + (1 - t^s)^{\frac{1}{s}} b\right) \psi(t)\, dt \le \frac{f(a) + f(b)}{2} \tag{2.57}$$

where

$$\psi(t) := \frac{1}{2}\left[1 + (1 - t^s)^{\frac{1}{s} - 1} t^{s-1}\right],\ t \in (0, 1].$$

PROOF If we choose in (2.48) $\alpha = t, \beta = (1 - t^s)^{\frac{1}{s}}, t \in [0, 1]$, then we have $\alpha^s + \beta^s = 1$ for all $t \in [0, 1]$ and

$$f\left(ta + (1 - t^s)^{\frac{1}{s}} b\right) \le t^s f(a) + (1 - t^s) f(b)$$

for all $t \in [0, 1]$. Similarly, we have

$$f\left((1 - t^s)^{\frac{1}{s}} a + tb\right) \le (1 - t^s) f(a) + t^s f(b)$$

for all $t \in [0, 1]$.

If we add the above two inequalities, then we obtain

$$\frac{1}{2}\left[f\left(ta + (1 - t^s)^{\frac{1}{s}} b\right) + f\left((1 - t^s)^{\frac{1}{s}} a + tb\right)\right] \le \frac{f(a) + f(b)}{2}$$

for all $t \in [0, 1]$.

If we integrate this inequality over t on $[0, 1]$, then

$$\frac{1}{2}\left[\int_0^1 f\left(ta + (1 - t^s)^{\frac{1}{s}} b\right) dt + \int_0^1 f\left((1 - t^s)^{\frac{1}{s}} a + tb\right) dt\right]$$

$$\le \frac{f(a) + f(b)}{2}. \tag{2.58}$$

Let us denote $u := (1 - t^s)^{\frac{1}{s}}, t \in [0, 1]$. Then $t = (1 - u^s)^{\frac{1}{s}}$ and $dt = -(1 - u^s)^{\frac{1}{s} - 1} u^{s-1}$, where $u \in (0, 1]$. We have the following with the change of variable in (2.58):

$$\int_0^1 f\left((1 - t^s)^{\frac{1}{s}} a + tb\right) dt$$

$$= -\int_1^0 f\left(ua + (1 - u^s)^{\frac{1}{s}} b\right)(1 - u^s)^{\frac{1}{s} - 1} u^{s-1} du$$

$$= \int_0^1 f\left(ta + (1 - t^s)^{\frac{1}{s}} b\right)(1 - t^s)^{\frac{1}{s} - 1} t^{s-1} dt.$$

Using the inequality (2.58) we deduce that

$$\int_0^1 f\left(ta + (1-t^s)^{\frac{1}{s}} b\right)\left[\frac{1 + (1-t^s)^{\frac{1}{s}-1} t^{s-1}}{2}\right] dt \le \frac{f(a) + f(b)}{2}$$

and the inequality (2.57) is proved. □

Another result of the Hermite-Hadamard type holds [92].

With the above assumptions, we have the inequality:

$$f\left(\frac{a+b}{2^{\frac{1}{s}}}\right) \le \int_0^1 f\left(\frac{a+b}{2^{\frac{1}{s}}}\left[t + (1-t^s)^{\frac{1}{s}}\right]\right) dt \qquad (2.59)$$

$$\le \int_0^1 f\left(ta + (1-t^s)^{\frac{1}{s}} b\right) \psi(t)\, dt,$$

where ψ is as defined above.

PROOF As $\frac{1}{s} > 1$, we have the following by the convexity of the mapping $g : [0, \infty) \to \mathbb{R}$, $g(x) = x^{\frac{1}{s}}$:

$$\frac{(t^s)^{\frac{1}{s}} + (1-t^s)^{\frac{1}{s}}}{2} \ge \left(\frac{t^s + 1 - t^s}{2}\right)^{\frac{1}{s}} = \frac{1}{2^{\frac{1}{s}}}$$

and

$$\frac{a+b}{2^{\frac{1}{s}}} \cdot \frac{t + (1-t^s)^{\frac{1}{s}}}{2} \ge \frac{a+b}{2^{\frac{1}{s}}} \cdot \frac{1}{2^{\frac{1}{s}}}$$

from which we obtain

$$\frac{a+b}{2^{\frac{1}{s}}} \cdot \left[t + (1-t^s)^{\frac{1}{s}}\right] \ge \frac{a+b}{2^{\frac{2}{s}-1}}.$$

As the mapping f is monotonic nondecreasing on $(0, \infty)$, we get

$$f\left(\frac{a+b}{2^{\frac{1}{s}}}\left[t + (1-t^s)^{\frac{1}{s}}\right]\right) \ge f\left(\frac{a+b}{2^{\frac{2}{s}-1}}\right) \quad \text{for all } t \in [0, 1],$$

which by integration on $[0, 1]$ produces the first inequality in (2.59). As f is s-convex in the first sense, we have that

$$f\left(\frac{x+y}{2^{\frac{1}{s}}}\right) \le \frac{f(x) + f(y)}{2}$$

for all $x, y \in [0, \infty)$.

Let $x = ta + (1 - t^s)^{\frac{1}{s}} b$, $y = (1 - t^s)^{\frac{1}{s}} a + tb$, $t \in [0, 1]$. Then we have the inequality

$$\frac{1}{2} \left[f \left(ta + (1 - t^s)^{\frac{1}{s}} b \right) + f \left((1 - t^s)^{\frac{1}{s}} a + tb \right) \right] \geq f \left(\frac{a + b}{2^{\frac{1}{s}}} \left[t + (1 - t^s)^{\frac{1}{s}} \right] \right)$$

for all $t \in [0, 1]$.

If we integrate this inequality on $[0, 1]$ over t and take into account the change of variable we used in the proof of the preceding result, then we obtain the desired inequality (2.59). $\qquad\square$

Comments

Some other inequalities of H.-H. type for s-convex mappings in the first sense are embodied in the following [92]:

Let $f : [0, \infty) \to \mathbb{R}$ be an s-convex mapping in the first sense with $s \in (0, 1)$. If $a, b \in \mathbb{R}_+$ with $a < b$, then one has the inequality:

$$f \left(\frac{a + b}{2^{\frac{2}{s} - 1}} \right) \leq \int_0^1 f \left(\frac{a + b}{2^{\frac{1}{s}}} \left[t^{\frac{1}{s}} + (1 - t)^{\frac{1}{s}} \right] \right) dt \qquad (2.60)$$

$$\leq \int_0^1 f \left(a t^{\frac{1}{s}} + b (1 - t)^{\frac{1}{s}} \right) dt$$

$$\leq \frac{f(a) + f(b)}{2}.$$

PROOF   By the convexity of the mapping $g(x) = x^{\frac{1}{s}}$, $s \in (0, 1)$, we have

$$\frac{t^{\frac{1}{s}} + (1 - t)^{\frac{1}{s}}}{2} \geq \left(\frac{t + 1 - t}{2} \right)^{\frac{1}{s}} = \frac{1}{2^{\frac{1}{s}}} \quad \text{for all } t \in [0, 1].$$

By using the monotonicity of f, we get

$$f \left(\frac{a + b}{2^{\frac{1}{s}}} \left[t^{\frac{1}{s}} + (1 - t)^{\frac{1}{s}} \right] \right) \geq f \left(\frac{a + b}{2^{\frac{1}{s}}} \cdot \frac{2}{2^{\frac{1}{s}}} \right) = f \left(\frac{a + b}{2^{\frac{2}{s} - 1}} \right)$$

for all $t \in [0, 1]$; from which we obtain the first inequality in (2.60).

Since f is s-convex in the first sense, we have

$$\frac{1}{2} \left[f \left(a t^{\frac{1}{s}} + b (1 - t)^{\frac{1}{s}} \right) + f \left(a (1 - t)^{\frac{1}{s}} + b t^{\frac{1}{s}} \right) \right]$$

$$\geq f \left(\frac{a + b}{2^{\frac{1}{s}}} \left[t^s + (1 - t)^{\frac{1}{s}} \right] \right)$$

for all $t \in [0, 1]$.

By integrating over t on $[0,1]$, we obtain

$$\frac{1}{2}\left[\int_0^1 f\left(at^{\frac{1}{s}} + b(1-t)^{\frac{1}{s}}\right) dt + \int_0^1 f\left(a(1-t)^{\frac{1}{s}} + bt^{\frac{1}{s}}\right) dt\right]$$

$$\geq \int_0^1 f\left(\frac{a+b}{2^{\frac{1}{s}}}\left[t^s + (1-t)^{\frac{1}{s}}\right]\right) dt.$$

Using the change of variable $u = 1-t, t \in [0,1]$, we get

$$\int_0^1 f\left(a(1-t)^{\frac{1}{s}} + bt^{\frac{1}{s}}\right) dt = -\int_1^0 f\left(au^{\frac{1}{s}} + b(1-u)^{\frac{1}{s}}\right) du$$

$$= \int_0^1 f\left(at^{\frac{1}{s}} + b(1-t)^{\frac{1}{s}}\right) dt,$$

which implies that the second inequality in (2.60) also holds.

By the s-convexity of f on $[0,\infty)$, we have

$$f\left(t^{\frac{1}{s}}a + (1-t)^{\frac{1}{s}}b\right) \leq tf(a) + (1-t)f(b) \quad \text{for all } t \in [0,1].$$

If we integrate this inequality over t in $[0,1]$, then

$$\int_0^1 f\left(t^{\frac{1}{s}}a + (1-t)^{\frac{1}{s}}b\right) dt \leq f(a)\int_0^1 t\,dt + f(b)\int_0^1 (1-t)\,dt$$

$$= \frac{f(a) + f(b)}{2}.$$

This completes the proof. □

Finally, we have the following result which gives an upper bound for the integral mean $\frac{1}{b-a}\int_a^b f(x)\,dx$. This result is different from the one embodied in the Hermite-Hadamard inequality that holds for general convex mappings [92].

Let $f : [0,\infty) \to \mathbb{R}_+$ be an s-convex mapping in the first sense with $s \in (0,1)$. If $0 < a < b$ and the integral

$$\int_a^\infty x^{\frac{s+1}{s-1}} f(x)\,dx$$

is finite, then one has the inequality

$$\frac{1}{b-a}\int_a^b f(x)\,dx \leq \frac{s}{1-s}\left[a^{\frac{2s}{1-s}} + b^{\frac{2s}{1-s}}\right]\int_a^\infty x^{\frac{s+1}{s-1}} f(x)\,dx. \qquad (2.61)$$

PROOF By the s-convexity of f on $[0,\infty)$, we have

$$f\left(u^{\frac{1}{s}}z + (1-u)^{\frac{1}{s}}y\right) \leq uf(z) + (1-u)f(y)$$

for all $u \in [0, 1]$ and $z, y \geq 0$.

Let $z = u^{1-\frac{1}{s}} a$, $u \in (0, 1]$ and $y = (1 - u)^{1-\frac{1}{s}} b$, $u \in [0, 1)$. Then we get the inequality:

$$f\left(ua + (1 - u) b\right) \leq uf\left(u^{1-\frac{1}{s}} a\right) + (1 - u) f\left((1 - u)^{1-\frac{1}{s}} b\right) \qquad (2.62)$$

for all $u \in (0, 1)$.

Observe that by the change of variable $t = 1 - u$ ($u \in [0, 1]$), the integral

$$\int_0^1 (1 - u) f\left((1 - u)^{1-\frac{1}{s}} b\right) du$$

becomes

$$\int_0^1 tf\left(t^{1-\frac{1}{s}} b\right) dt.$$

We shall now show that the integral $\int_0^1 uf\left(u^{1-\frac{1}{s}} a\right) du$ is also finite.

If we change the variable $x = u^{1-\frac{1}{s}} a$, $u \in (0, 1]$, then

$$u = \left(\frac{x}{a}\right)^{\frac{1}{1-\frac{1}{s}}} = \left(\frac{x}{a}\right)^{\frac{s}{s-1}} = \frac{x^{\frac{s}{s-1}}}{a^{\frac{s}{s-1}}}$$

and

$$du = \frac{s}{s-1} \cdot \frac{1}{a^{\frac{s}{s-1}}} x^{\frac{s}{s-1}-1} dx = \frac{s}{s-1} \cdot \frac{1}{a^{\frac{s}{s-1}}} x^{\frac{1}{s-1}} dx.$$

Thus, we have the equality

$$\int_0^1 uf\left(u^{1-\frac{1}{s}} a\right) du = -\int_\infty^a \left[\frac{x^{\frac{s}{s-1}}}{a^{\frac{s}{s-1}}} \cdot \frac{s}{s-1} \cdot \frac{x^{\frac{1}{s-1}}}{a^{\frac{s}{s-1}}} f\left(x\right)\right] dx$$

$$= \frac{s}{1-s} \cdot a^{\frac{2s}{1-s}} \int_a^\infty x^{\frac{s+1}{s-1}} f\left(x\right) dx < \infty$$

and similarly,

$$\int_0^1 tf\left(t^{1-\frac{1}{s}} b\right) dt = \frac{s}{1-s} \cdot b^{\frac{2s}{1-s}} \int_a^\infty x^{\frac{s+1}{s-1}} f\left(x\right) dx < \infty.$$

Now, integrating the inequality (2.62) over u on $(0, 1)$, taking into account that

$$\int_0^1 f\left(ua + (1 - u) b\right) dt = \frac{1}{b - a} \int_a^b f\left(x\right) dx$$

and

$$\int_0^1 uf\left(u^{1-\frac{1}{s}} a\right) du = \frac{s}{1-s} \cdot a^{\frac{2s}{1-s}} \int_a^\infty x^{\frac{s+1}{s-1}} f\left(x\right) dx,$$

$$\int_0^1 (1 - u) f\left((1 - u)^{1-\frac{1}{s}} b\right) du = \frac{s}{1-s} \cdot b^{\frac{2s}{1-s}} \int_a^\infty x^{\frac{s+1}{s-1}} f\left(x\right) dx,$$

respectively, we deduce (2.61). $\qquad\qquad$ ⬚

2.7 The Hermite-Hadamard Inequality for s-Convex Functions in the Breckner Sense

In 1992, Hudzik and Maligranda [123] considered, among others, the following class of functions:

A function $f : \mathbb{R}_+ \to \mathbb{R}$ is said to be s-convex in the second sense or convex in the Breckner sense if

$$f(\alpha u + \beta v) \leq \alpha^s f(u) + \beta^s f(v) \tag{2.63}$$

for all $u, v \geq 0$ and $\alpha, \beta \geq 0$ with $\alpha + \beta = 1$ and s fixed in $(0,1]$. The set of all s-convex functions in the second sense is denoted by $f \in K_s^2$.

Now, we shall point out some results from Hudzik and Maligranda [123] that are connected with s-convex functions in the second sense.

If $f \in K_s^2$, then f is nonnegative on $[0, \infty)$.

PROOF   We have, for $u \in \mathbb{R}_+$,

$$f(u) = f\left(\frac{u}{2} + \frac{u}{2}\right) \leq \frac{f(u)}{2^s} + \frac{f(u)}{2^s} = 2^{1-s} f(u).$$

Therefore, $\left(2^{1-s} - 1\right) f(u) \geq 0$ and so $f(u) \geq 0$.   ◻

Example 2.1

Let $0 < s < 1$ and $a, b, c \in \mathbb{R}$. By defining for $u \in \mathbb{R}_+$ [123]:

$$f(u) := \begin{cases} a & \text{if } u = 0 \\ bu^s + c & \text{if } u > 0 \end{cases}$$

we have

(i) If $b \geq 0$ and $0 \leq c \leq a$, then $f \in K_s^2$;

(ii) If $b > 0$ and $c < 0$, then $f \notin K_s^2$.

      ◻

It is important to note that the condition $\alpha + \beta = 1$ in the definition of K_s^2 can be equivalently replaced by the condition $\alpha + \beta \leq 1$.

The following result holds [123]:

Let $f \in K_s^2$. Then inequality (2.63) holds for all $u, v \in \mathbb{R}_+$ and $\alpha, \beta \geq 0$ with $\alpha + \beta \leq 1$ if and only if $f(0) = 0$.

PROOF *Necessity.* Taking $u = v = \alpha = \beta = 0$, we obtain $f(0) \leq 0$ and as $f(0) \geq 0$, we get $f(0) = 0$.

Sufficiency. Let $u, v \in \mathbb{R}_+$ and $\alpha, \beta \geq 0$ with $0 < \gamma = \alpha + \beta < 1$. Put $a = \frac{\alpha}{\gamma}$ and $b = \frac{\beta}{\gamma}$. Then $a + b = \frac{\alpha}{\gamma} + \frac{\beta}{\gamma} = 1$ and so

$$
\begin{aligned}
f(\alpha u + \beta v) &= f(\alpha \gamma u + \beta \gamma v) \leq a^s f(\gamma u) + b^s f(\gamma v) \\
&= a^s f(\gamma u + (1 - \gamma) 0) + b^s f(\gamma v + (1 - \gamma) 0) \\
&\leq a^s \left[\gamma^s f(u) + (1 - \gamma)^s f(0) \right] + b^s \left[\gamma^s f(v) + (1 - \gamma)^s f(0) \right] \\
&= a^s \gamma^s f(u) + b^s \gamma^s f(v) + (1 - \gamma)^s f(0) \\
&= \alpha^s f(u) + \beta^s f(v).
\end{aligned}
$$

This completes the proof. □

Using the above results we can compare both definitions of the s-convexity [123].

The following statements are valid:

(i) *Let $0 < s \leq 1$. If $f \in K_s^2$ and $f(0) = 0$, then $f \in K_s^1$.*

(ii) *Let $0 < s_1 \leq s_2 \leq 1$. If $f \in K_{s_2}^2$ and $f(0) = 0$, then $f \in K_{s_1}^2$.*

PROOF

(i) Assume that $f \in K_s^2$ and $f(0) = 0$. For $u, v \in \mathbb{R}_+$ and $\alpha, \beta \geq 0$ with $\alpha^s + \beta^s = 1$, we have $\alpha + \beta \leq \alpha^s + \beta^s = 1$, and by the above result we obtain

$$
f(\alpha u + \beta v) \leq \alpha^s f(u) + \beta^s f(v),
$$

which means that $f \in K_s^1$.

(ii) Assume that $f \in K_{s_2}^2$ and that $u, v \geq 0, \alpha, \beta \geq 0$ with $\alpha + \beta = 1$. Then we have

$$
\begin{aligned}
f(\alpha u + \beta v) &\leq \alpha^{s_2} f(u) + \beta^{s_2} f(v) \\
&\leq \alpha^{s_1} f(u) + \beta^{s_1} f(v)
\end{aligned}
$$

which means that $f \in K_{s_1}^2$.

□

Using a similar argument, one can state the following results as well [123]:

Let f be a nondecreasing function in K_s^2 and g be a nonnegative convex function on $[0, \infty)$. Then the composition $f \circ g$ belongs to K_s^2.

We recall that $f : \mathbb{R}_+ \to \mathbb{R}_+$ is said to be a ψ-*function* if $f(0) = 0$, and f is nondecreasing and continuous. The following particular case holds for ψ-functions.

If ϕ is a convex ψ-function and f is a ψ-function from K_s^2, then the composition $f \circ \phi$ belongs to K_s^2. In particular, the ψ-function $h(u) = [\phi(u)]^s$ belongs to K_s^2.

Let $0 < s < 1$. Then there exists a ψ-function f in the class K_s^2 which is neither of the form $\phi(u^s)$ nor $[\phi(u)]^s$, where ϕ is a convex ψ-function.

The following inequality is the variant of the Hermite-Hadamard result for s-convex functions in the second sense [93]:

Suppose that $f : \mathbb{R}_+ \to \mathbb{R}_+$ is an *s-convex mapping in the second sense,* $s \in (0,1)$ and $a, b \in \mathbb{R}_+$ with $a < b$. If $f \in L_1[a,b]$, then one has the inequalities:

$$2^{s-1} f\left(\frac{a+b}{2}\right) \leq \frac{1}{b-a} \int_a^b f(x)\,dx \leq \frac{f(a) + f(b)}{s+1}. \qquad (2.64)$$

PROOF   As f is s-convex in the second sense, we have, for all $t \in [0,1]$,

$$f(ta + (1-t)b) \leq t^s f(a) + (1-t)^s f(b).$$

Integrating this inequality on $[0,1]$, we get

$$\int_0^1 f(ta + (1-t)b)\,dt \leq f(a) \int_0^1 t^s dt + f(b) \int_0^1 (1-t)^s\,dt$$
$$= \frac{f(a) + f(b)}{s+1}.$$

As the change of variable $x = ta + (1-t)b$ gives

$$\int_0^1 f(ta + (1-t)b)\,dt = \frac{1}{b-a} \int_a^b f(x)\,dx,$$

the second inequality in (2.64) is proved.

To prove the first inequality in (2.64), we observe that for all $x, y \in I$

$$f\left(\frac{x+y}{2}\right) \leq \frac{f(x) + f(y)}{2^s}. \qquad (2.65)$$

Now, let $x = ta + (1-t)b$ and $y = (1-t)a + tb$ with $t \in [0,1]$. Then, by (2.65) we get

$$f\left(\frac{a+b}{2}\right) \leq \frac{f(ta + (1-t)b) + f((1-t)a + tb)}{2^s} \quad \text{for all } t \in [0,1].$$

Integrating this inequality on $[0, 1]$, we deduce the first part of (2.64). □

The constant $k = \frac{1}{s+1}$ for $s \in (0, 1]$ is best possible in the second inequality in (2.64). Indeed, as the mapping $f : [0, 1] \to [0, 1]$ given by $f(x) = x^s$ is s-convex in the second sense, we obtain

$$\int_0^1 x^s dx = \frac{1}{s+1} \quad \text{and} \quad \frac{f(0) + f(1)}{s+1} = \frac{1}{s+1}.$$

Comments

(a) Suppose that f is Lebesgue integrable on $[a, b]$ and consider the mapping $H : [0, 1] \to \mathbb{R}$ given by

$$H(t) := \frac{1}{b-a} \int_a^b f\left(tx + (1-t)\frac{a+b}{2}\right) dx.$$

We are interested in pointing out some properties of this mapping as in the case of the classical convex mappings.

The following results also hold [93]:

Let $f : I \subseteq \mathbb{R}_+ \to \mathbb{R}$ be an *s-convex mapping in the second sense on I*, $s \in (0, 1]$ *and Lebesgue integrable on* $[a, b] \subset I$, $a < b$. *Then:*
(i) H *is s-convex in the second sense on* $[0, 1]$;
(ii) *We have the inequality:*

$$H(t) \geq 2^{s-1} f\left(\frac{a+b}{2}\right) \quad \text{for all } t \in [0, 1]. \tag{2.66}$$

(iii) *We have the inequality:*

$$H(t) \leq \min\{H_1(t), H_2(t)\}, t \in [0, 1] \tag{2.67}$$

where

$$H_1(t) = t^s \cdot \frac{1}{b-a} \int_a^b f(x) dx + (1-t)^s f\left(\frac{a+b}{2}\right)$$

and

$$H_2(t) = \frac{f\left(ta + (1-t)\frac{a+b}{2}\right) + f\left(tb + (1-t)\frac{a+b}{2}\right)}{s+1}$$

and $t \in (0, 1]$;
(iv) *If* $\tilde{H}(t) := \max\{H_1(t), H_2(t)\}$, $t \in [0, 1]$, *then*

$$\tilde{H}(t) \leq t^s \cdot \frac{f(a) + f(b)}{s+1} + (1-t)^s \cdot \frac{2}{s+1} f\left(\frac{a+b}{2}\right), t \in [0, 1]. \tag{2.68}$$

PROOF

(i) Let $t_1, t_2 \in [0,1]$ and $\alpha, \beta \geq 0$ with $\alpha + \beta = 1$. We have successively

$$H\left(\alpha t_1 + \beta t_2\right)$$

$$= \frac{1}{b-a} \int_a^b f\left((\alpha t_1 + \beta t_2) x + [1 - (\alpha t_1 + \beta t_2)]\frac{a+b}{2}\right) dx$$

$$= \frac{1}{b-a} \int_a^b f\left(\alpha \left[t_1 x + (1-t_1)\frac{a+b}{2}\right]\right.$$

$$\left. + \beta \left[t_2 x + (1-t_2)\frac{a+b}{2}\right]\right) dx$$

$$\leq \frac{1}{b-a} \int_a^b \left[\alpha^s f\left(t_1 x + (1-t_1)\frac{a+b}{2}\right)\right.$$

$$\left. + \beta^s f\left(\left[t_2 x + (1-t_2)\frac{a+b}{2}\right]\right)\right] dx$$

$$= \alpha^s H(t_1) + \beta^s H(t_2),$$

which shows that H is s-convex in the second sense on $[0,1]$.

(ii) Suppose that $t \in (0,1]$. Then a simple change of variable $u = tx + (1-t)\frac{a+b}{2}$ gives us

$$H(t) = \frac{1}{t(b-a)} \int_{ta+(1-t)\frac{a+b}{2}}^{tb+(1-t)\frac{a+b}{2}} f(u)\, du = \frac{1}{p-q} \int_q^p f(u)\, du$$

where $p = tb + (1-t)\frac{a+b}{2}$ and $q = ta + (1-t)\frac{a+b}{2}$. Applying the first Hermite-Hadamard inequality, we get:

$$\frac{1}{p-q} \int_q^p f(u)\, du \geq 2^{s-1}\left(\frac{p+q}{2}\right) = 2^{s-1} f\left(\frac{a+b}{2}\right),$$

giving the inequality (2.66). For the case of $t = 0$, we wish to prove

$$f\left(\frac{a+b}{2}\right) \geq 2^{s-1} f\left(\frac{a+b}{2}\right).$$

This is true since $s \in [0,1]$.

(iii) By applying the second Hermite-Hadamard inequality, we also have

$$\frac{1}{p-q} \int_q^p f(u)\, du \leq \frac{f(p) + f(q)}{r+1}$$

$$= \frac{f\left(ta + (1-t)\frac{a+b}{2}\right) + f\left(tb + (1-t)\frac{a+b}{2}\right)}{r+1}$$

for all $t \in [0, 1]$. Note that if $t = 0$, then the required inequality

$$f\left(\frac{a+b}{2}\right) = H\left(0\right) \leq H_2\left(0\right) = \frac{2}{r+1} \cdot f\left(\frac{a+b}{2}\right)$$

is true as it is equivalent to

$$\left(r - 1\right) f\left(\frac{a+b}{2}\right) \leq 0$$

and the fact that $f\left(\frac{a+b}{2}\right) \geq 0$, for $r \in (0, 1)$. On the other hand, it is obvious that

$$f\left(tx + (1 - t)\frac{a+b}{2}\right) \leq t^s f\left(x\right) + (1 - t)^s f\left(\frac{a+b}{2}\right)$$

for all $t \in [0, 1]$ and $x \in [a, b]$. Integrating this inequality on $[a, b]$ we get (2.67) for $H_1\left(t\right)$, and the statement is proved.

(iv) We have

$$H_2\left(t\right) \leq \frac{t^s f\left(a\right) + (1 - t)^s f\left(\frac{a+b}{2}\right) + t^s f\left(b\right) + (1 - t)^s f\left(\frac{a+b}{2}\right)}{s + 1}$$

$$= t^s \cdot \frac{f\left(a\right) + f\left(b\right)}{s + 1} + (1 - t)^s \cdot \frac{2}{s + 1} \cdot f\left(\frac{a+b}{2}\right)$$

for all $t \in [0, 1]$. On the other hand, we know that

$$\frac{1}{b - a} \int_a^b f\left(x\right) dx \leq \frac{f\left(a\right) + f\left(b\right)}{s + 1}$$

and

$$\left(1 - t\right)^s f\left(\frac{a+b}{2}\right) \leq (1 - t)^s \cdot \frac{2}{s + 1} \cdot f\left(\frac{a+b}{2}\right), \quad t \in [0, 1],$$

which gives us that

$$H_1\left(t\right) \leq t^s \cdot \frac{f\left(a\right) + f\left(b\right)}{2} + (1 - t)^s \cdot \frac{2}{s + 1} \cdot f\left(\frac{a+b}{2}\right).$$

This completes the proof.

$\Box$

For $s = 1$, we get the inequalities:

$$H\left(t\right) \leq \min\left\{ t \cdot \frac{1}{b - a} \int_a^b f\left(x\right) dx + (1 - t) f\left(\frac{a+b}{2}\right), \right.$$

$$\left. \frac{f\left(ta + (1 - t)\frac{a+b}{2}\right) + f\left(tb + (1 - t)\frac{a+b}{2}\right)}{2} \right\},$$

and

$$\tilde{H}(t) \le t \cdot \frac{f(a) + f(b)}{2} + (1 - t) \cdot f\left(\frac{a+b}{2}\right)$$

for all $t \in [0, 1]$.

(b) Now, assume that $f : [a, b] \to \mathbb{R}$ is Lebesgue integrable on $[a, b]$. Consider the mapping

$$F(t) := \frac{1}{(b-a)^2} \int_a^b \int_a^b f(tx + (1-t)y)\, dx dy, \ t \in [0, 1].$$

The following result contains the main properties of this mapping [93].

Let $f : I \subseteq \mathbb{R}_+ \to \mathbb{R}_+$ be an *s-convex mapping in the second sense*, $s \in (0, 1]$, $a, b \in I$ with $a < b$ and f *Lebesgue integrable on* $[a, b]$. *Then:*

(i) $F\left(s + \frac{1}{2}\right) = F\left(\frac{1}{2} - s\right)$ for all $s \in \left[0, \frac{1}{2}\right]$ and $F(t) = F(1-t)$ for all $t \in [0, 1]$;

(ii) F *is s-convex in the second sense on* $[0, 1]$;

(iii) *We have the inequality:*

$$2^{1-s} F(t) \ge F\left(\frac{1}{2}\right) = \frac{1}{(b-a)^2} \int_a^b \int_a^b f\left(\frac{x+y}{2}\right) dx dy, \ t \in [0, 1]; \qquad (2.69)$$

(iv) *We have the inequality*

$$F(t) \ge 2^{1-s} H(t) \ge 4^{s-1} f\left(\frac{a+b}{2}\right) \quad \text{for all } t \in [0, 1]; \qquad (2.70)$$

(v) *We have the inequality:*

$$F(t) \le \min\left\{ [t^s + (1-t)^s] \frac{1}{b-a} \int_a^b f(x)\, dx, \right.$$
$$\left. \frac{f(a) + f(ta + (1-t)b) + f((1-t)a + tb) + f(b)}{(s+1)^2} \right\} \qquad (2.71)$$

for all $t \in [0, 1]$.

PROOF   The proof for (i) and (ii) is obvious.

(iii) By the fact that f is s-convex in the second sense on I, we have

$$\frac{f(tx + (1-t)y) + f((1-t)x + ty)}{2^s} \ge f\left(\frac{x+y}{2}\right)$$

for all $t \in [0,1]$ and $x, y \in [a,b]$. By integrating this inequality on $[a,b]^2$ we get

$$\frac{1}{2^s} \left[\int_a^b \int_a^b f\left(tx + (1-t)\,y\right) dxdy + \int_a^b \int_a^b f\left((1-t)\,x + ty\right) dxdy \right]$$

$$\geq \int_a^b \int_a^b f\left(\frac{x+y}{2}\right) dxdy.$$

Since

$$\int_a^b \int_a^b f\left(tx + (1-t)\,y\right) dxdy = \int_a^b \int_a^b f\left((1-t)\,x + ty\right) dxdy,$$

the above inequality gives us the desired result (2.69).

(iv) First, let us observe that

$$F(t) = \frac{1}{b-a} \int_a^b \left[\frac{1}{b-a} \int_a^b f\left(tx + (1-t)\,y\right) dx \right] dy.$$

Now, for y fixed in $[a,b]$, we can consider the map $H_y : [0,1] \to \mathbb{R}$ given by

$$H_y(t) := \frac{1}{b-a} \int_a^b f\left(tx + (1-t)\,y\right) dx.$$

As shown above, for $t \in [0,1]$ we have the identity

$$H_y(t) = \frac{1}{p-q} \int_q^p f(u)\,du$$

where $p = tb + (1-t)\,y$, $q = ta + (1-t)\,y$. Applying the Hermite-Hadamard inequality we get

$$\frac{1}{p-q} \int_q^p f(u)\,du \geq 2^{s-1} f\left(\frac{p+q}{2}\right) = 2^{s-1} f\left(t \cdot \frac{a+b}{2} + (1-t)\,y\right)$$

for all $t \in (0,1)$ and $y \in [a,b]$. Integrating on $[a,b]$ over y, we may easily deduce

$$F(t) \geq 2^{s-1} H(1-t) \quad \text{for all } t \in (0,1).$$

As $F(t) = F(1-t)$, the inequality (5.83) is proved for $t \in (0,1)$. If $t = 0$ or $t = 1$, the inequality (2.70) also holds. We shall omit the details.

(v) By the definition of s-convex mappings in the second sense, we have

$$f\left(tx + (1-t)\,y\right) \leq t^s f(x) + (1-t)^s f(y)$$

for all $x, y \in [a, b]$ and $t \in [0, 1]$. Integrating this inequality on $[a, b]^2$, we deduce the first part of the inequality (2.71).

Now, let us observe, by the second part of the Hermite-Hadamard inequality, that

$$H_y(t) = \frac{1}{p-q} \int_q^p f(u) \, du \leq \frac{f(tb + (1-t)y) + f(ta + (1-t)y)}{s+1},$$

where $p = tb + (1-t)y$ and $q = ta + (1-t)y$, $t \in [0, 1]$. Integrating this inequality on $[a, b]$ over y, we deduce

$$F(t) \leq \frac{1}{s+1} \left[\frac{1}{b-a} \int_a^b f(tb + (1-t)y) \, dy \right.$$

$$\left. + \frac{1}{b-a} \int_a^b f(ta + (1-t)y) \, dy \right].$$

A simple calculation shows that

$$\frac{1}{b-a} \int_a^b f(tb + (1-t)y) \, dy$$

$$= \frac{1}{r-l} \int_l^r f(u) \, du \leq \frac{f(r) + f(l)}{2}$$

$$= \frac{f(b) + f(tb + (1-t)a)}{s+1},$$

where $r = b$, $l = tb + (1-t)a$, $t \in (0, 1)$; and similarly,

$$\frac{1}{b-a} \int_a^b f(ta + (1-t)y) \, dy \leq \frac{f(a) + f(ta + (1-t)b)}{s+1}, \quad t \in (0, 1),$$

which gives, by addition, the second inequality in (2.71). If $t = 0$ or $t = 1$, then this inequality also holds. We shall omit the details.

$\square$

2.8   Inequalities for Hadamard's Inferior and Superior Sums

Let $[a, b]$ be a compact interval of real numbers, $d := \{x_i | i = \overline{0, n}\} \subset [a, b]$, a partition of the interval $[a, b]$, given by

$$d : a = x_0 < x_1 < x_2 < \cdots < x_{n-1} < x_n = b \ (n \geq 1),$$

and let f be a bounded mapping on $[a, b]$. We consider the following sums [85]:

$$h_d(f) := \sum_{i=0}^{n-1} f\left(\frac{x_i + x_{i+1}}{2}\right)(x_{i+1} - x_i)$$

which is called *Hadamard's inferior sum*, and

$$H_d(f) := \sum_{i=0}^{n-1} \frac{f(x_i) + f(x_{i+1})}{2}(x_{i+1} - x_i)$$

which is called *Hadamard's superior sum*. We also consider *Darboux's sums*,

$$s_d(f) := \sum_{i=0}^{n-1} m_i(x_{i+1} - x_i), \qquad S_d(f) := \sum_{i=0}^{n-1} M_i(x_{i+1} - x_i),$$

where

$$m_i = \inf_{x \in [x_i, x_{i+1}]} f(x), \qquad M_i = \sup_{x \in [x_i, x_{i+1}]} f(x), \qquad i = 0, ..., n-1.$$

It is well known that f is *Riemann integrable* on $[a, b]$ iff

$$\sup_d s_d(f) = \inf_d S_d(f) = I \in \mathbb{R}.$$

In this case,

$$I = \int_a^b f(x)\, dx.$$

The following result was proved [85].

Let $f : [a, b] \to \mathbb{R}$ be a convex function on $[a, b]$. Then

(i) $h_d(f)$ increases monotonically over d, that is, for $d_1 \subseteq d_2$ one has $h_{d_1}(f) \leq h_{d_2}(f)$;

(ii) $H_d(f)$ is decreasing monotonically over d;

(iii) We have the bounds

$$\frac{1}{b-a} \inf_d h_d(f) = f\left(\frac{a+b}{2}\right), \qquad \sup_d h_d(f) = \int_a^b f(x)\, dx \qquad (2.72)$$

and

$$\inf_d H_d(f) = \int_a^b f(x)\, dx, \qquad \frac{1}{b-a} \sup_d H_d(f) = \frac{f(a) + f(b)}{2}. \qquad (2.73)$$

PROOF

(i) Without loss of generality, we can assume that $d_1 \subseteq d_2$ with $d_1 = \{x_0, ..., x_n\}$ and $d_2 = \{x_0, ..., x_k, y, x_{k+1}, ..., x_n\}$ where $y \in [x_k, x_{k+1}]$ $(0 \le k \le n-1)$. Then

$$h_{d_2}(f) - h_{d_1}(f) = f\left(\frac{x_k + y}{2}\right)(y - x_k)$$

$$+ f\left(\frac{y + x_{k+1}}{2}\right)(x_{k+1} - y) - f\left(\frac{y + x_{k+1}}{2}\right)(x_{k+1} - x_k).$$

Let us put

$$\alpha = \frac{y - x_k}{x_{k+1} - x_k}, \quad \beta = \frac{x_{k+1} - y}{x_{k+1} - x_k}$$

and

$$x = \frac{x_k + y}{2}, \quad z = \frac{y + x_{k+1}}{2}.$$

Then

$$\alpha + \beta = 1, \quad \alpha x + \beta y = \frac{x_k + x_{k+1}}{2},$$

and, by the convexity of f we deduce that $\alpha f(x) + \beta f(z) \ge f(\alpha x + \beta z)$, that is, $h_{d_2}(f) \ge h_{d_1}(f)$.

(ii) For d_1, d_2 as above, we have

$$H_{d_2}(f) - H_{d_1}(f)$$
$$= \frac{f(x_k) + f(y)}{2}(y - x_k) + \frac{f(y) + f(x_{k+1})}{2}(x_{k+1} - y)$$
$$+ \frac{f(x_k) + f(x_{k+1})}{2}(x_{k+1} - x_k)$$
$$= \frac{f(y)(x_{k+1} - x_k)}{2} - \frac{f(x_k)(x_{k+1} - y) + f(x_{k+1})(y - x_k)}{2}.$$

Now, let α, β be as above and $u = x_k, v = x_{k+1}$. Then $\alpha u + \beta v = y$; and by the convexity of f we have $\alpha f(u) + \beta f(v) \ge f(y)$, that is, $H_{d_2}(f) \le H_{d_1}(f)$, and the statement is proved.

(iii) Let $d = \{x_0, ..., x_n\}$ with $a = x_0 < x_1 < \cdots < x_n = b$. Put $p_i := x_{i+1} - x_i$, $u_i = \frac{(x_i + x_{i+1})}{2}$, $i = 0, ..., n-1$. Then, by Jensen's discrete inequality we have

$$f\left(\frac{\sum_{i=0}^{n} p_i u_i}{\sum_{i=0}^{n} p_i}\right) \le \frac{\sum_{i=0}^{n} p_i f(u_i)}{\sum_{i=0}^{n} p_i}.$$

Since

$$\sum_{i=0}^{n} p_i = b - a, \quad \sum_{i=0}^{n} p_i u_i = \frac{b^2 - a^2}{2},$$

we can deduce the inequality

$$f\left(\frac{a+b}{2}\right) \leq \frac{1}{b-a} h_d(f).$$

If $d = d_0 = \{a, b\}$, then

$$h_{d_0}(f) = (b-a) f\left(\frac{a+b}{2}\right),$$

which proves the first bound in (2.72). By the first inequality in the Hermite-Hadamard result (2.1), we have

$$f\left(\frac{x_i + x_{i+1}}{2}\right) \leq \frac{1}{x_{i+1} - x_i} \int_{x_i}^{x_{i+1}} f(x) \, dx, \quad i = 0, ..., n-1,$$

which gives, by addition,

$$h_d(f) = \sum_{i=0}^{n} f\left(\frac{x_i + x_{i+1}}{2}\right)(x_{i+1} - x_i)$$

$$\leq \sum_{i=0}^{n} \int_{x_i}^{x_{i+1}} f(x) \, dx = \int_a^b f(x) \, dx,$$

for all d a division of $[a, b]$. Since

$$s_d(f) \leq h_d(f) \leq \int_a^b f(x) \, dx,$$

where d is a division of $[a, b]$, and f is Riemann integrable on $[a, b]$, that is,

$$\sup_d s_d(f) = \int_a^b f(x) \, dx,$$

it follows that

$$\sup_d h_d(f) = \int_a^b f(x) \, dx,$$

which proves the second relation in (2.72). To prove the relation (2.73), we observe, by the second inequality in the Hermite-Hadamard result, that

$$\int_a^b f(x) \, dx = \sum_{i=0}^{n} \int_{x_i}^{x_{i+1}} f(x) \, dx$$

$$\leq \sum_{i=0}^{n-1} \frac{f(x_i) + f(x_{i+1})}{2}(x_{i+1} - x_i) = H_d(f),$$

where d is a division of $[a, b]$. Since

$$H_d(f) \le S_d(f)$$

for all d as above, and f is integrable on $[a, b]$, we conclude that

$$\inf_d H_d(f) = \int_a^b f(x)\, dx.$$

Finally, as for all d a division of $[a, b]$ we have $d \supseteq d_0 = \{a, b\}$, thus

$$\frac{1}{b-a} \sup_d H_d(f) = \frac{f(a) + f(b)}{2}.$$

This completes the proof. □

The following particular case gives an improvement of the classical Hermite-Hadamard inequality [85]:

Let $f : [a, b] \to \mathbb{R}$ be a convex mapping on $[a, b]$. Then for all $a = x_0 < x_1 < \cdots < x_n = b$ we have

$$f\left(\frac{a+b}{2}\right) \le \frac{1}{b-a} \sum_{i=0}^{n-1} f\left(\frac{x_i + x_{i+1}}{2}\right)(x_{i+1} - x_i) \tag{2.74}$$

$$\le \frac{1}{b-a} \int_a^b f(x)\, dx$$

$$\le \frac{1}{b-a} \sum_{i=0}^{n-1} \frac{f(x_i) + f(x_{i+1})}{2}(x_{i+1} - x_i)$$

$$\le \frac{f(a) + f(b)}{2}.$$

Define the sequences:

$$h_n(f) := \frac{1}{n} \sum_{i=0}^{n-1} f\left(a + \frac{2i+1}{2n}(b-a)\right)$$

and

$$H_n(f) := \frac{1}{2n} \sum_{i=0}^{n-1} \left[f\left(a + \frac{i}{n}(b-a)\right) + f\left(a + \frac{i+1}{n}(b-a)\right) \right]$$

for $n \ge 1$.

The following result holds and provides an improvement for the Hermite-Hadamard inequality [85]:

With the above assumptions, one has the inequality:

$$f\left(\frac{a+b}{2}\right) \le h_n(f) \le \frac{1}{b-a} \int_a^b f(x)\, dx \tag{2.75}$$

$$\le H_n(f) \le \frac{f(a) + f(b)}{2}.$$

Moreover, one has

$$\lim_{n\to\infty} h_n(f) = \lim_{n\to\infty} H_n(f) = \frac{1}{b-a} \int_a^b f(x)\,dx. \qquad (2.76)$$

PROOF Inequality (2.75) follows by (2.74) for the uniform partition

$$d := \left\{ x_i = a + \frac{i}{n}(b-a) \,|\, i = \overline{0,n} \right\}.$$

The relation (2.76) is obvious by the integrability of f. We shall omit the details. □

Now, let us define the sequences:

$$t_n(f) := \frac{1}{2^n} \sum_{i=0}^{n-1} f\left(a + \frac{2^i}{2^{n+1}}(b-a) \right) 2^i$$

and

$$T_n(f) := \frac{1}{2^{n+1}} \sum_{i=0}^{n-1} \left[f\left(a + \frac{2^i}{2^n}(b-a) \right) + f\left(a + \frac{2^{i+1}}{2^n}(b-a) \right) \right] 2^i$$

for $n \geq 1$.

For these sequences we may also state [85]:

Let $f : [a,b] \to \mathbb{R}$ be a convex mapping on $[a,b]$. Then we have:
(i) *$t_n(f)$ is monotonic increasing;*
(ii) *$T_n(f)$ is monotonic decreasing;*
(iii) *The following bounds hold:*

$$\lim_{n\to\infty} t_n(f) = \sup_{n\geq 1} t_n(f) = \frac{1}{b-a} \int_a^b f(x)\,dx$$

and

$$\lim_{n\to\infty} T_n(f) = \inf_{n\geq 1} T_n(f) = \frac{1}{b-a} \int_a^b f(x)\,dx.$$

Comments
Let $[a,b]$ be a compact interval of real numbers and $d \in \mathfrak{Div}[a,b]$. By this we mean $d := \{x_i \,|\, i = \overline{0,n}\} \subset [a,b]$ is a division of the interval $[a,b]$ given by $d : a = x_0 < x_1 < \cdots < x_{n-1} < x_n = b$.
Define the sums

$$h_d^{[p]} := \sum_{i=0}^{n-1} A^p(x_i, x_{i+1})(x_{i+1} - x_i)$$

and

$$H_d^{[p]} := \sum_{i=0}^{n-1} A\left(x_i^p, x_{i+1}^p\right)(x_{i+1} - x_i)$$

where $p \in (-\infty, 0) \cup [1, \infty) \setminus \{-1\}$.

For every $d \in \mathfrak{Div}\,[a, b]$ we have the inequalities:

$$A^p(a, b) \le \frac{1}{b-a} \sum_{i=0}^{n-1} A^p(x_i, x_{i+1})(x_{i+1} - x_i)$$

$$\le L_p^p(a, b)$$

$$\le \frac{1}{b-a} \sum_{i=0}^{n-1} A\left(x_i^p, x_{i+1}^p\right)(x_{i+1} - x_i)$$

$$\le A\left(a^p, b^p\right)$$

as well as the bounds

$$\frac{1}{b-a} \sup_d h_d^{[p]} = L_p^p(a, b)$$

and

$$\frac{1}{b-a} \inf_d H_d^{[p]} = L_p^p(a, b).$$

Now, let us define the sums; for the case of $0 < a < b$:

$$h_d^{[-1]} = 2 \sum_{i=0}^{n-1} \frac{x_{i+1} - x_i}{x_{i+1} + x_i}, \quad H_d^{[-1]} = \frac{1}{2} \sum_{i=0}^{n-1} \frac{x_{i+1}^2 - x_i^2}{x_i x_{i+1}}.$$

We have the inequality

$$A^{-1}(a, b) \le \frac{2}{b-a} \sum_{i=0}^{n-1} \frac{x_{i+1} - x_i}{x_{i+1} + x_i} \le L^{-1}(a, b)$$

$$\le \frac{1}{2(b-a)} \sum_{i=0}^{n-1} \frac{x_{i+1}^2 - x_i^2}{x_i x_{i+1}} \le H^{-1}(a, b)$$

for all $d \in \mathfrak{Div}\,[a, b]$ and the bounds

$$\frac{1}{b-a} \sup_d h_d^{[-1]} = L^{-1}(a, b)$$

and

$$\frac{1}{b-a} \inf_d H_d^{[-1]} = L^{-1}(a, b).$$

We can also define the sequences

$$H_d^{[0]} := \prod_{i=0}^{n-1} \left[A(x_i, x_{i+1})\right]^{(x_{i+1} - x_i)}, \quad h_d^{[0]} := \prod_{i=0}^{n-1} \left[G(x_i, x_{i+1})\right]^{(x_{i+1} - x_i)}$$

for a division d of the interval $[a, b] \subset (0, \infty)$.

Using the above results we have the inequalities:

$$A(a, b) \geq \prod_{i=0}^{n-1} [A(x_i, x_{i+1})]^{\frac{x_{i+1} - x_i}{b-a}} \geq I(a, b)$$

$$\geq \prod_{i=0}^{n-1} [G(x_i, x_{i+1})]^{\frac{x_{i+1} - x_i}{b-a}} \geq G(a, b),$$

which follows by inequality (2.74) applied to the convex mapping $f : [a, b] \to \mathbb{R}$, where $f(x) = -\ln x$.

By (2.72) we also deduce the bounds

$$\inf_d \left\{ \prod_{i=0}^{n-1} [A(x_i, x_{i+1})]^{\frac{x_{i+1} - x_i}{b-a}} \right\} = I(a, b)$$

and

$$\sup_d \left\{ \prod_{i=0}^{n-1} [G(x_i, x_{i+1})]^{\frac{x_{i+1} - x_i}{b-a}} \right\} = I(a, b).$$

2.9 A Refinement of the Hermite-Hadamard Inequality for the Modulus

We start with the following result which contains a refinement of the second part of the H.-H. inequality obtained in Dragomir [72].

Let $f : I \subseteq \mathbb{R} \to \mathbb{R}$ be a convex function on the interval of real numbers I and $a, b \in I$ with $a < b$. Then we have the following refinement of the second part of the H.-H. inequality:

$$\frac{f(a) + f(b)}{2} - \frac{1}{b-a} \int_a^b f(x) \, dx$$

$$\geq \begin{cases} \left| f(a) - \frac{1}{b-a} \int_a^b |f(x)| \, dx \right| & \text{if } f(a) = f(b) \\[2mm] \left| \frac{1}{f(b) - f(a)} \int_{f(a)}^{f(b)} |x| \, dx - \frac{1}{b-a} \int_a^b |f(x)| \, dx \right| & \text{if } f(a) \neq f(b) \end{cases} \qquad (2.77)$$

PROOF By the convexity of f on I and the continuity property of the modulus we have:

$$tf(a) + (1-t) f(b) - f(ta + (1-t) b)$$
$$= |tf(a) + (1-t) f(b) - f(ta + (1-t) b)|$$
$$\geq ||tf(a) + (1-t) f(b)| - |f(ta + (1-t) b)|| \geq 0$$

for all $a, b \in I$ and $t \in [0, 1]$. Integrating this inequality for t over $[0, 1]$ we get the inequality

$$f(a) \int_0^1 t\,dt + f(b) \int_0^1 (1-t)\,dt - \int_0^1 f(ta + (1-t)b)\,dt$$

$$\geq \left| \int_0^1 |tf(a) + (1-t)f(b)|\,dt - \int_0^1 |f(ta + (1-t)b)|\,dt \right|.$$

Since

$$\int_0^1 t\,dt = \int_0^1 (1-t)\,dt = \frac{1}{2},$$

$$\int_0^1 |tf(a) + (1-t)f(b)|\,dt = \begin{cases} f(a) & \text{if } f(b) = f(a), \\ \frac{1}{f(b)-f(a)} \int_{f(a)}^{f(b)} |x|\,dx & \text{if } f(a) \neq f(b), \end{cases}$$

and

$$\int_0^1 |f(ta + (1-t)b)|\,dt = \frac{1}{b-a} \int_a^b |f(x)|\,dx,$$

respectively, then the inequality (2.77) is proved.   ◻

The following particular result holds [85]:

With the above assumptions, and the condition that $f(a + b - x) = f(x)$ for all $x \in [a, b]$, we have the inequality:

$$f(a) - \frac{1}{b-a} \int_a^b f(x)\,dx \geq \left| |f(a)| - \frac{1}{b-a} \int_a^b |f(x)|\,dx \right| \geq 0.$$

A refinement of the first part of the Hermite-Hadamard inequality is embodied in the following result [85]:

Let $f : I \subseteq \mathbb{R} \to \mathbb{R}$ be a convex function on the interval of real numbers I and $a, b \in I$ with $a < b$. Then we have the inequality:

$$\frac{1}{b-a} \int_a^b f(x)\,dx - f\left(\frac{a+b}{2}\right)$$

$$\geq \left| \frac{1}{b-a} \int_a^b \left| \frac{f(x) + f(a + b - x)}{2} \right| dx - \left| f\left(\frac{a+b}{2}\right) \right| \right| \geq 0. \quad (2.78)$$

PROOF   By the convexity of f we have

$$\frac{f(x) + f(y)}{2} - f\left(\frac{x+y}{2}\right) \geq \left| \left| \frac{f(x) + f(y)}{2} \right| - \left| f\left(\frac{x+y}{2}\right) \right| \right|$$

for all $x, y \in I$. Let $x = ta + (1 - t) b, y = (1 - t) a + tb$ with $t \in [0, 1]$. Then we get

$$\frac{f(ta + (1 - t) b) + f((1 - t) a + tb)}{2} - f\left(\frac{a + b}{2}\right)$$

$$\geq \left\| \left| \frac{f(ta + (1 - t) b) + f((1 - t) a + tb)}{2} \right| - \left| f\left(\frac{a + b}{2}\right) \right| \right\|$$

for all $t \in [0, 1]$. Upon integration we get:

$$\frac{\int_0^1 f(ta + (1 - t) b) \, dt + \int_0^1 f((1 - t) a + tb) \, dt}{2} - f\left(\frac{a + b}{2}\right)$$

$$\geq \left| \left| \int_0^1 \left| \frac{f(ta + (1 - t) b) + f((1 - t) a + tb)}{2} \right| dt - \left| f\left(\frac{a + b}{2}\right) \right| \right| \right|.$$

However,

$$\int_0^1 f(ta + (1 - t) b) \, dt = \int_0^1 f((1 - t) a + tb) \, dt = \frac{1}{b - a} \int_a^b f(x) \, dx,$$

and denoting $x := ta + (1 - t) b, t \in [0, 1]$, we also get that:

$$\int_0^1 \left| \frac{f(ta + (1 - t) b) + f((1 - t) a + tb)}{2} \right| dt$$

$$= \frac{1}{b - a} \int_a^b \left| \frac{f(x) + f(a + b - x)}{2} \right| dx.$$

Thus, the inequality (2.78) is proved. ⬜

The following particular case also gives a refinement of the second part of the Hermite-Hadamard inequality [85]:

With the above assumptions and if the condition that the function is symmetric, namely, $f(a + b - x) = f(x)$ is satisfied for all $x \in [a, b]$, then we have the inequality:

$$\frac{1}{b - a} \int_a^b f(x) \, dx - f\left(\frac{a + b}{2}\right) \geq \left| \frac{1}{b - a} \int_a^b |f(x)| \, dx - \left| f\left(\frac{a + b}{2}\right) \right| \right| \geq 0.$$

Comments

It is well known that the following inequality holds:

$$G(a, b) \leq I(a, b) \leq A(a, b) \qquad (G - I - A)$$

where we recall that

$$G(a, b) := \sqrt{ab}$$

is the *geometric mean*,

$$I(a, b) := \frac{1}{e} \cdot \left(\frac{b^b}{a^a}\right)^{\frac{1}{b-a}}$$

is the *identric mean*, and

$$A(a, b) := \frac{a + b}{2}$$

is the *arithmetic mean* of the nonnegative real numbers $a < b$.

(a) The following refinement of the $I - G$ inequality holds [85]:

If $a \in (0, 1], b \in [1, \infty)$ with $a \neq b$, then one has the inequality:

$$\frac{I(a, b)}{G(a, b)} \geq \exp\left[\left|\frac{(\ln b)^2 + (\ln a)^2}{\ln\left(\frac{b}{a}\right)^2} - \ln\left[\left(b^b a^a e^{2-(a+b)}\right)^{\frac{1}{b-a}}\right]\right|\right] \geq 1 \quad (2.79)$$

which improves the first inequality in $(G - I - A)$.

PROOF   Let us assume that $a \in (0, 1], b \in [1, \infty)$ and $a \neq b$. Then we denote the following for the convex mapping $f(x) = -\ln x$, $x > 0$:

$$A := \frac{f(a) + f(b)}{2} - \frac{1}{b - a}\int_a^b f(x)\, dx = -\frac{\ln a + \ln b}{2} + \frac{1}{b - a}\int_a^b \ln x\, dx$$

$$= \frac{1}{b - a}[b \ln b - a \ln a - (b - a)] - \ln G(a, b) = \ln\left[\frac{I(a, b)}{G(a, b)}\right]$$

and

$$B := \left|\frac{1}{f(b) - f(a)}\int_{f(b)}^{f(a)} |x|\, dx - \frac{1}{b - a}\int_a^b |\ln x|\, dx\right|.$$

We have

$$\int_{\ln b}^{\ln a} |x|\, dx = \frac{(\ln b)^2 + (\ln a)^2}{2}$$

and

$$\int_a^b |\ln x|\, dx = \ln\left[a^a b^b e^{2-(a+b)}\right]$$

and thus

$$B = \left|\frac{(\ln b)^2 + (\ln a)^2}{\ln\left(\frac{b}{a}\right)^2} - \ln\left[\left(b^b a^a e^{2-(a+b)}\right)^{\frac{1}{b-a}}\right]\right|.$$

By using the inequality (2.77) we can state that $A \geq B \geq 0$, and thus the result is proved. $\Box$

(b) We have the following refinement of the $A - I$ inequality [85]:

Let $a, b \in (0, \infty)$ with $a \neq b$. Then one has the inequality:

$$\frac{A(a, b)}{I(a, b)} \geq \exp\left[\left|\frac{1}{b-a}\int_a^b \left|\ln\left(\sqrt{x(a+b-x)}\right)\right| dx - \left|\ln\left(\frac{a+b}{2}\right)\right|\right|\right] \quad (2.80)$$

$$\geq 1,$$

which improves the second inequality in $(G - I - A)$.

PROOF Denote for $f(x) = -\ln x, x > 0$ that

$$C := \frac{1}{b-a}\int_a^b f(x)\, dx - f\left(\frac{a+b}{2}\right) = \ln\left(\frac{a+b}{2}\right) - \ln I(a, b)$$

$$= \ln\left[\frac{A(a, b)}{I(a, b)}\right]$$

and

$$D := \left|\frac{1}{b-a}\int_a^b \left|\frac{f(x) + f(a+b-x)}{2}\right| dx - \left|f\left(\frac{a+b}{2}\right)\right|\right|$$

$$= \left|\frac{1}{b-a}\int_a^b \left|\ln\sqrt{x(a+b-x)}\right| dx - \left|\ln\left(\frac{a+b}{2}\right)\right|\right|.$$

By inequality (2.78) we have $C \geq D \geq 0$, and the result is thus proved. ☐

Chapter 3

Ostrowski and Trapezoid Type Inequalities

In recent years, an exponential growth has been noted in the development of integral inequalities of Ostrowski and trapezoid type, mostly due to their use in producing adaptive quadrature and cubature rules for approximating single or multiple integrals.

In this chapter various Ostrowski and trapezoid type inequalities are presented, together with complete proofs, for various classes of functions. These functions include those that are absolutely continuous, of bounded variation or monotonic, as well as differentiable functions of different orders whose derivatives satisfy similar conditions. Sharp bounds for the Čebyšev functional, which is of importance in approximating the integral of a product in the one-dimensional and multidimensional cases, are given as well.

The material is complemented by numerous remarks that allow opportunities for research in related fields such as probability theory and statistics.

3.1 Ostrowski's Integral Inequality for Absolutely Continuous Mappings

Let $f : [a, b] \to \mathbb{R}$ be an absolutely continuous mapping on (a, b). Then ([111], [113], [114])

$$\left| f(x) - \frac{1}{b-a} \int_a^b f(t)\, dt \right| \leq \begin{cases} \left[\frac{1}{4} + \left(\frac{x - \frac{a+b}{2}}{b-a} \right)^2 \right] (b-a) \, \|f'\|_\infty \\[2ex] \left\{ \frac{1}{q+1} \left[\left(\frac{x-a}{b-a} \right)^{q+1} + \left(\frac{b-x}{b-a} \right)^{q+1} \right] \right\}^{\frac{1}{q}} \\[1ex] \quad \times (b-a)^{\frac{1}{q}} \, \|f'\|_p, \quad p > 1, \ \frac{1}{p} + \frac{1}{q} = 1 \\[2ex] \left[\frac{1}{2} + \frac{\left| x - \frac{a+b}{2} \right|}{b-a} \right] \|f'\|_1 \end{cases} \tag{3.1}$$

for all $x \in [a, b]$, where

$$\|f'\|_s := \begin{cases} \left(\int_a^b |f'(t)|^s \, dt\right)^{\frac{1}{s}}, & \text{if } s \in [1, \infty) \\ \sup_{t \in (a,b)} |f'(t)| & \text{if } s = \infty. \end{cases}$$

PROOF   Integrating by parts gives

$$\int_a^x (t - a) f'(t) \, dt = (x - a) f(x) - \int_a^x f(t) \, dt$$

and

$$\int_x^b (t - b) f'(t) \, dt = (b - x) f(x) - \int_x^b f(t) \, dt.$$

Summing the above two results, we get Montgomery's identity,

$$(b - a) f(x) = \int_a^b f(t) \, dt + \int_a^b p(x, t) f'(t) \, dt, \tag{3.2}$$

where $p : [a, b]^2 \to \mathbb{R}$,

$$p(x, t) := \begin{cases} t - a & \text{if } t \in [a, x] \\ t - b & \text{if } t \in (x, b]. \end{cases}$$

Now, on using properties of the modulus, we have

$$\left| \int_a^b p(x, t) f'(t) \, dt \right| \leq \sup_{t \in (a,b)} |f'(t)| \int_a^b |p(x, t)| \, dt$$

$$= \|f'\|_\infty \left[\int_a^x (t - a) \, dt + \int_x^b (t - b) \, dt \right]$$

$$= \|f'\|_\infty \left[\frac{(x - a)^2 + (b - x)^2}{2} \right]$$

$$= \|f'\|_\infty \left[\frac{1}{4} + \left(\frac{x - \frac{a+b}{2}}{b - a} \right)^2 \right] (b - a)^2,$$

which proves the first part of (3.1).

Using Hölder's integral inequality for $p > 1$, $\frac{1}{p} + \frac{1}{q} = 1$, we get

$$\left| \int_a^b p(x,t) f'(t) \, dt \right| \leq \left(\int_a^b |p(x,t)|^q \, dt \right)^{\frac{1}{q}} \|f'\|_p$$

$$= \left[\int_a^x (t-a)^q \, dt + \int_x^b (t-b)^q \, dt \right]^{\frac{1}{q}} \|f'\|_p$$

$$= \left[\frac{(x-a)^{q+1} + (b-x)^{q+1}}{q+1} \right]^{\frac{1}{q}} \|f'\|_p .$$

Using (3.2), we then get the second part of (3.1).

Finally, we have

$$\left| \int_a^b p(x,t) f'(t) \, dt \right| \leq \sup_{t \in (a,b)} |p(x,t)| \int_a^b |f'(t)| \, dt = \max\{x-a, b-x\} \|f'\|_1$$

$$= \left[\frac{b-a}{2} + \left| x - \frac{a+b}{2} \right| \right] \|f'\|_1$$

from which, via (3.2), we get the last part of (3.1). Here we have used the well-known fact that $\max\{X, Y\} = \frac{X+Y}{2} + \frac{1}{2}|X - Y|$. □

Comments

(a) The constant $\frac{1}{4}$ in the first inequality is sharp in the sense that it cannot be replaced by a smaller one. This inequality in a different form is named after Ostrowski [147].

To prove the sharpness of (3.1), we choose $f(x) = x$ to get

$$\left| x - \frac{a+b}{2} \right| \leq \left[\frac{1}{4} + \left(\frac{x - \frac{a+b}{2}}{b-a} \right)^2 \right] (b-a) \tag{3.3}$$

for all $x \in [a, b]$.

If in (3.3) we choose $x = a$ or $x = b$, then equality is achieved.

(b) In Peachey, McAndrew, and Dragomir [148] it was shown that the constant $\frac{1}{2}$ in the last inequality in (3.1) is also sharp.

(c) The constant $\frac{1}{q+1}$ in the second part of (3.1) cannot be improved by a constant of the form $\frac{c}{q+1}$ with $0 < c < 1$.

Indeed, if we assume that there is such a constant, then for $f(x) = x$ we would have

$$\left| x - \frac{a+b}{2} \right| \leq \left\{ \frac{c}{q+1} \left[\left(\frac{x-a}{b-a} \right)^{q+1} + \left(\frac{b-x}{b-a} \right)^{q+1} \right] \right\}^{\frac{1}{q}} (b-a) \tag{3.4}$$

for all $x \in [a, b]$.

If in (3.4) we choose $x = a$, we get $\frac{1}{2} \leq \left(\frac{c}{q+1}\right)^{\frac{1}{q}}$, from which we conclude that $c \geq \frac{q+1}{2^q}$ for all $q > 1$. Letting $q \to +1$, we get $c \geq 1$ and the sharpness of the constant is proved.

3.2 Ostrowski's Integral Inequality for Mappings of Bounded Variation

Let $f : [a, b] \to \mathbb{R}$ be a mapping of bounded variation on $[a, b]$. Then we have the inequality [70]

$$\left| f(x) - \frac{1}{b-a} \int_a^b f(t) \, dt \right| \leq \left[\frac{1}{2} + \frac{\left| x - \frac{a+b}{2} \right|}{b-a} \right] \bigvee_a^b (f) \tag{3.5}$$

for all $x \in [a, b]$, where $\bigvee_a^b (f)$ is the total variation of f on $[a, b]$. The constant $\frac{1}{2}$ is the best possible in (3.5).

PROOF The integration by parts formula for the Riemann-Stieltjes integral gives

$$\int_a^x (t-a) \, df(t) = (x-a) f(x) - \int_a^x f(t) \, dt \tag{3.6}$$

and

$$\int_x^b (t-b) \, df(t) = (b-x) f(x) - \int_x^b f(t) \, dt, \tag{3.7}$$

for all $x \in [a, b]$.

By adding (3.6) and (3.7), we get the identity

$$\int_a^b f(t) \, dt = (b-a) f(x) + \int_a^b p(t, x) \, df(t), \tag{3.8}$$

where

$$p(t, x) := \begin{cases} t - a & \text{if } t \in [a, x] \\ t - b & \text{if } t \in (x, b] \end{cases} , \quad x \in [a, b] .$$

It is known that if the mapping $g : [a, b] \to \mathbb{R}$ is continuous on the partitions and $v : [a, b] \to \mathbb{R}$ is of bounded variation on $[a, b]$, then g is Riemann-Stieltjes integrable with respect to v and thus

$$\left| \int_a^b g(x) \, dv(x) \right| \leq \sup_{x \in [a,b]} |g(x)| \bigvee_a^b (v). \tag{3.9}$$

By applying (3.9) for p and f, we can state

$$\left| \int_a^b p\,(t,x)\,df\,(t) \right| \le \sup_{t\in[a,b]} |p\,(t,x)| \bigvee_a^b (f) \tag{3.10}$$

with

$$\sup_{t\in[a,b]} |p\,(t,x)| = \max_{x\in[a,b]} \{x-a, b-x\}$$

$$= \frac{1}{2}\,(b-a) + \left| x - \frac{a+b}{2} \right|$$

where we have used the fact that $2 \cdot \max\{A, B\} = A + B + |B - A|$.

Then by (3.10) and (3.8) we deduce the desired inequality (3.5).

To prove the sharpness of the constant $\frac{1}{2}$, assume that (3.5) holds with constant $c > 0$, that is,

$$\left| f\,(x) - \frac{1}{b-a} \int_a^b f\,(t)\,dt \right| \le \left[c + \frac{\left| x - \frac{a+b}{2} \right|}{b-a} \right] \bigvee_a^b (f) \tag{3.11}$$

for all $x \in [a, b]$.

Choose $f : [a, b] \to \mathbb{R}$ given by

$$f\,(x) = \begin{cases} 0 \text{ if } x \in \left[a, \frac{a+b}{2}\right) \cup \left(\frac{a+b}{2}, b\right], \\ 1 \text{ if } x = \frac{a+b}{2}. \end{cases}$$

Then f is of bounded variation on $[a, b]$, $\int_a^b f\,(t)\,dt = 0$ and $\bigvee_a^b (f) = 2$. With this choice of f in (3.11) and $x = \frac{a+b}{2}$, we get $1 \le 2c$. Thus, $c \ge \frac{1}{2}$ and the proof is completed. $\qquad\Box$

Comments

Let f be as above and $I_n := a = x_0 < x_1 < \cdots < x_{n-1} < x_n = b$ be a partition of $[a, b]$, $h_i := x_{i+1} - x_i$, $\xi_i \in [x_i, x_{i+1}]$, $i = \overline{0, n-1}$, and let the Riemann sum be represented by

$$\sigma\,(f, \xi, I_n) := \sum_{i=0}^{n-1} f\,(\xi_i)\,h_i.$$

We then have

$$\int_a^b f\,(x)\,dx = \sigma\,(f, \xi, I_n) + R\,(f, \xi, I_n), \tag{3.12}$$

where the remainder $R\,(f, \xi, I_n)$ satisfies the estimation

$$|R\,(f, \xi, I_n)| \le \left[\frac{1}{2}\nu\,(h) + \sup_{i=\overline{0,n-1}} \left| \xi_i - \frac{x_i + x_{i+1}}{2} \right| \right] \bigvee_a^b (f). \tag{3.13}$$

Indeed, applying (3.5) on the intervals $[x_i, x_{i+1}]$ and for $\xi_i \in [x_i, x_{i+1}]$, we get

$$\left| \int_{x_i}^{x_{i+1}} f(t)\, dt - h_i f(\xi_i) \right| \leq \left[\frac{1}{2} h_i + \left| \xi_i - \frac{x_i + x_{i+1}}{2} \right| \right] \bigvee_{x_i}^{x_{i+1}} (f).$$

Summing over i from 0 to $n-1$ we get

$$
\begin{aligned}
|R(f, \xi, I_n)| &\leq \sum_{i=0}^{n-1} \left[\frac{1}{2} h_i + \left| \xi_i - \frac{x_i + x_{i+1}}{2} \right| \right] \bigvee_{x_i}^{x_{i+1}} (f) \\
&\leq \max_{i=\overline{0,n-1}} \left[\frac{1}{2} h_i + \left| \xi_i - \frac{x_i + x_{i+1}}{2} \right| \right] \sum_{i=1}^{n} \bigvee_{x_i}^{x_{i+1}} (f) \\
&= \left[\frac{1}{2} \nu(h) + \max_{i=\overline{0,n-1}} \left| \xi_i - \frac{x_i + x_{i+1}}{2} \right| \right] \bigvee_{a}^{b} (f),
\end{aligned}
$$

where $\nu(h) = \max\limits_{i=\overline{0,n-1}} h_i$.

3.3   Trapezoid Inequality for Functions of Bounded Variation

Let $f : [a, b] \to \mathbb{R}$ be a function of bounded variation. Then we have the inequality [38]:

$$\left| \int_a^b f(x)\, dx - \frac{f(a) + f(b)}{2} (b - a) \right| \leq \frac{b-a}{2} \bigvee_a^b (f), \tag{3.14}$$

where $\bigvee_a^b (f)$ is the total variation of f on $[a, b]$.

The constant $\frac{1}{2}$ is the best possible in (3.14).

PROOF   Using the integration by parts formula for the Riemann-Stieltjes integral, we have

$$\int_a^b \left(x - \frac{a+b}{2} \right) df(x) = \frac{f(a) + f(b)}{2} (b - a) - \int_a^b f(x)\, dx. \tag{3.15}$$

Now, assume that $\Delta_n : a = x_0^{(n)} < x_1^{(n)} < \cdots < x_{n-1}^{(n)} < x_n^{(n)} = b$ is a sequence of divisions with $\nu(\Delta_n) := \max\limits_{i=\overline{0,n-1}} \left(x_{i+1}^{(n)} - x_i^{(n)} \right)$ and $\xi_i^{(n)} \in \left[x_i^{(n)}, x_{i+1}^{(n)} \right]$. If $p : [a, b] \to \mathbb{R}$ is a continuous mapping on $[a, b]$ and $v : [a, b] \to \mathbb{R}$ is of bounded variation on $[a, b]$, then

$$\left| \int_a^b p(x) \, dv(x) \right| = \left| \lim_{\nu(\Delta_n) \to 0} \sum_{i=0}^{n-1} p\left(\xi_i^{(n)}\right) \left[v\left(x_{i+1}^{(n)}\right) - v\left(x_i^{(n)}\right) \right] \right| \quad (3.16)$$

$$\leq \lim_{\nu(\Delta_n) \to 0} \sum_{i=0}^{n-1} \left| p\left(\xi_i^{(n)}\right) \right| \left| v\left(x_{i+1}^{(n)}\right) - v\left(x_i^{(n)}\right) \right|$$

$$\leq \max_{x \in [a,b]} |p(x)| \sup_{\Delta_n} \sum_{i=0}^{n-1} \left| v\left(x_{i+1}^{(n)}\right) - v\left(x_i^{(n)}\right) \right|$$

$$= \max_{x \in [a,b]} |p(x)| \bigvee_a^b (v).$$

Applying inequality (3.16), we get

$$\left| \int_a^b \left(x - \frac{a+b}{2}\right) df(x) \right| \leq \max_{x \in [a,b]} \left| x - \frac{a+b}{2} \right| \bigvee_a^b (f) \quad (3.17)$$

$$= \frac{b-a}{2} \bigvee_a^b (f)$$

and, via identity (3.15), the inequality (3.14) is proved.

For the sharpness of the constant, assume that the inequality (3.14) holds with a constant $c > 0$, that is,

$$\left| \int_a^b f(x) \, dx - \frac{f(a) + f(b)}{2} (b-a) \right| \leq c(b-a) \bigvee_a^b (f). \quad (3.18)$$

Let us consider the mapping $f : [a, b] \to \mathbb{R}$ given by

$$f(x) = \begin{cases} 1 \text{ if } x \in \{a, b\} \\ \\ 0 \text{ if } x \in (a, b). \end{cases}$$

Then f is of bounded variation, $\bigvee_a^b (f) = 2$ and $\int_a^b f(x) \, dx = 0$, giving

$$\int_a^b f(x) \, dx - \frac{f(a) + f(b)}{2} (b-a) = -(b-a)$$

and

$$(b-a) \bigvee_a^b (f) = 2(b-a).$$

Thus by inequality (3.18) we get $(b-a) \leq 2c(b-a)$, which implies that $c \geq \frac{1}{2}$ and demonstrates that the constant $\frac{1}{2}$ is the best possible in (3.14). □

Comment

If we assume that $I_n : a = x_0 < x_1 < \cdots < x_{n-1} < x_n = b$ is a partition of the interval $[a, b]$ and f is as above, then we have

$$\int_a^b f(x)\, dx = A_T(f, I_n) + R_T(f, I_n) \tag{3.19}$$

where $A_T(f, I_n)$ is the *trapezoid rule*. That is, we recall

$$A_T(f, I_n) := \frac{1}{2} \sum_{i=1}^{n} [f(x_i) + f(x_{i+1})]\, h_i, \quad h_i := x_{i+1} - x_i,$$

and the remainder $R_T(f, I_n)$ satisfies the estimation

$$|R_T(f, I_n)| \leq \frac{\nu(h)}{2} \bigvee_a^b (f), \tag{3.20}$$

where $\nu(h) := \max_{i=\overline{0,n-1}} \{h_i\}$. The constant $\frac{1}{2}$ is the best possible.

3.4   Trapezoid Inequality for Monotonic Mappings

Let $f : [a, b] \to \mathbb{R}$ be a monotonic nondecreasing mapping on $[a, b]$. Then we have the inequality [38]

$$\left| \int_a^b f(x)\, dx - \frac{f(a) + f(b)}{2} (b - a) \right| \tag{3.21}$$

$$\leq \frac{1}{2}(b - a)(f(b) - f(a)) - \int_a^b \operatorname{sgn}\left(x - \frac{a + b}{2}\right) f(x)\, dx$$

$$\leq \frac{(b - a)(f(b) - f(a))}{2}.$$

The constant $\frac{1}{2}$ is sharp in both inequalities.

PROOF   The integration by parts formula for the Riemann-Stieltjes integral gives

$$\int_a^b \left(x - \frac{a + b}{2}\right) df(x) = \frac{f(a) + f(b)}{2}(b - a) - \int_a^b f(x)\, dx. \tag{3.22}$$

Now, assume that $\Delta_n : a = x_0^{(n)} < x_1^{(n)} < \cdots < x_{n-1}^{(n)} < x_n^{(n)} = b$ is a sequence of divisions of the interval $[a, b]$ with $\nu(\Delta_n) \to 0$ as $n \to \infty$, where

$\nu\left(\Delta_n\right) := \max\limits_{i=0,n-1} \left(x_{i+1}^{(n)} - x_i^{(n)}\right)$ and $\xi_i^{(n)} \in \left[x_i^{(n)}, x_{i+1}^{(n)}\right]$. If $p : [a,b] \to \mathbb{R}$ is a continuous mapping on $[a,b]$ and v is monotonic nondecreasing on $[a,b]$, then

$$\left|\int_a^b p(x)\, dv(x)\right| = \left|\lim_{\nu(\Delta_n)\to 0} \sum_{i=0}^{n-1} p\left(\xi_i^{(n)}\right)\left[v\left(x_{i+1}^{(n)}\right) - v\left(x_i^{(n)}\right)\right]\right| \qquad (3.23)$$

$$\leq \lim_{\nu(\Delta_n)\to 0} \sum_{i=0}^{n-1} \left|p\left(\xi_i^{(n)}\right)\right|\left|v\left(x_{i+1}^{(n)}\right) - v\left(x_i^{(n)}\right)\right|$$

$$= \lim_{\nu(\Delta_n)\to 0} \sum_{i=0}^{n-1} \left|p\left(\xi_i^{(n)}\right)\right|\left(v\left(x_{i+1}^{(n)}\right) - v\left(x_i^{(n)}\right)\right)$$

$$= \int_a^b |p(x)|\, dv(x).$$

Applying (3.23), we can state:

$$\left|\int_a^b \left(x - \frac{a+b}{2}\right) df(x)\right| \qquad (3.24)$$

$$\leq \int_a^b \left|x - \frac{a+b}{2}\right| df(x)$$

$$= \int_a^{\frac{a+b}{2}} \left(\frac{a+b}{2} - x\right) df(x) + \int_{\frac{a+b}{2}}^b \left(x - \frac{a+b}{2}\right) df(x)$$

$$= \left(\frac{a+b}{2} - x\right) f(x)\Big]_a^{\frac{a+b}{2}} + \int_a^{\frac{a+b}{2}} f(x)\, dx$$

$$+ \left(x - \frac{a+b}{2}\right) f(x)\Big]_{\frac{a+b}{2}}^b + \int_{\frac{a+b}{2}}^b f(x)\, dx$$

$$= \frac{1}{2}(b-a)(f(b) - f(a)) - \int_a^b \text{sgn}\left(x - \frac{a+b}{2}\right) f(x)\, dx$$

and the first inequality in (3.21) is proved.

As f is monotonic nondecreasing on $[a,b]$, we can also state that

$$\int_a^{\frac{a+b}{2}} f(x)\, dx \leq \left(\frac{a+b}{2} - a\right) f\left(\frac{a+b}{2}\right) = \frac{b-a}{2} f\left(\frac{a+b}{2}\right)$$

and

$$\int_{\frac{a+b}{2}}^b f(x)\, dx \geq \left(b - \frac{a+b}{2}\right) f\left(\frac{a+b}{2}\right) = \frac{b-a}{2} f\left(\frac{a+b}{2}\right)$$

and so

$$\int_a^b \mathrm{sgn}\left(x - \frac{a+b}{2}\right) f\left(x\right) dx = -\int_a^{\frac{a+b}{2}} f\left(x\right) dx + \int_{\frac{a+b}{2}}^b f\left(x\right) dx$$

$$\geq \frac{b-a}{2} f\left(\frac{a+b}{2}\right) - \frac{b-a}{2} f\left(\frac{a+b}{2}\right)$$

$$= 0,$$

and the second inequality in (3.21) is also proved.

To prove the sharpness of the constant $\frac{1}{2}$, we choose the following function f:

$$f\left(x\right) = \begin{cases} 0 \text{ if } x \in [a, b) \\ \\ 1 \text{ if } x = b, \end{cases}$$

which is monotone nondecreasing on $[a, b]$ and produces equality in both inequalities in (3.21). ☐

Comments

If f is as above and I_n is a partition of $[a, b]$, then we have

$$\int_a^b f\left(x\right) dx = A_T\left(f, I_n\right) + R_T\left(f, I_n\right) \tag{3.25}$$

where $A_T\left(f, I_n\right)$ is the *trapezoid rule* and the remainder $R_T\left(f; I_n\right)$ satisfies the estimation;

$$\left|R_T\left(f, I_k\right)\right| \tag{3.26}$$

$$\leq \frac{1}{2} \sum_{i=0}^{n-1} \left[f\left(x_{i+1}\right) - f\left(x_i\right)\right] h_i - \sum_{i=0}^{n-1} \int_{x_i}^{x_{i+1}} \mathrm{sgn}\left(x - \frac{x_i + x_{i+1}}{2}\right) f\left(x\right) dx$$

$$\leq \frac{1}{2} \sum_{i=0}^{n-1} \left[f\left(x_{i+1}\right) - f\left(x_i\right)\right] h_i \leq \frac{\nu\left(h\right)}{2} \left(f\left(b\right) - f\left(a\right)\right),$$

where $h_i := x_{i+1} - x_i$ and $\nu\left(h\right) = \max_{i=0,n-1} h_i$.

3.5 Trapezoid Inequality for Absolutely Continuous Mappings

Let $f : [a, b] \to \mathbb{R}$ be an absolutely continuous mapping on $[a, b]$. Then [38]

$$\left| \int_a^b f(x) \, dx - \frac{f(a) + f(b)}{2} (b - a) \right|$$

$$\leq \begin{cases} \frac{(b-a)^2}{4} \|f'\|_\infty & \text{if } f' \in L_\infty [a, b]; \\ \frac{(b-a)^{1+\frac{1}{q}}}{2(q+1)^{\frac{1}{q}}} \|f'\|_p & \text{if } f' \in L_p [a, b], \\ & \quad p > 1, \ \frac{1}{p} + \frac{1}{q} = 1; \\ \frac{b-a}{2} \|f'\|_1 & \text{if } f' \in L_1 [a, b], \end{cases} \tag{3.27}$$

where $\|\cdot\|_p$ are the usual p-norms in the Lebesgue spaces $L_p [a, b]$. That is,

$$\|f'\|_\infty := \operatorname*{ess\ sup}_{t \in [a,b]} |f'(t)|, \quad \text{and} \quad \|f'\|_p := \left(\int_a^b |f'(t)|^p \, dt \right)^{\frac{1}{p}} \quad \text{for } p \geq 1.$$

PROOF The integration by parts formula gives

$$\int_a^b \left(t - \frac{a+b}{2} \right) f'(t) \, dt = \frac{(b-a)(f(a) + f(b))}{2} - \int_a^b f(t) \, dt. \tag{3.28}$$

Using the properties of modulus, we get from (3.28),

$$\left| \int_a^b f(t) \, dt - (b - a) \frac{f(a) + f(b)}{2} \right| \leq \int_a^b \left| t - \frac{a+b}{2} \right| |f'(t)| \, dt. \tag{3.29}$$

If $f' \in L_\infty [a, b]$, then

$$\int_a^b |f'(t)| \left| t - \frac{a+b}{2} \right| dt \leq \|f'\|_\infty \int_a^b \left| t - \frac{a+b}{2} \right| dt = \|f'\|_\infty \frac{(b-a)^2}{4}.$$

If $f' \in L_p [a, b]$, then by Hölder's integral inequality for $p > 1$, $\frac{1}{p} + \frac{1}{q} = 1$, we get

$$\int_a^b |f'(t)| \left| t - \frac{a+b}{2} \right| dt \leq \left(\int_a^b \left| t - \frac{a+b}{2} \right|^q dt \right)^{\frac{1}{q}} \left(\int_a^b |f'(t)|^p \, dt \right)^{\frac{1}{p}}$$

$$= \frac{(b-a)^{1+\frac{1}{q}}}{2(q+1)^{\frac{1}{q}}} \|f'\|_p.$$

Finally, if $f' \in L_1[a, b]$, then

$$\int_a^b |f'(t)| \left| t - \frac{a+b}{2} \right| dt \leq \sup_{t \in [a,b]} \left| t - \frac{a+b}{2} \right| \int_a^b |f'(t)| \, dt = \frac{b-a}{2} \|f'\|_1$$

and the inequality (3.27) is proved. $\qquad\qquad\qquad\qquad\qquad$ □

Comments

Let f be as above and I_n a partition of the interval $[a, b]$. Then we have

$$\int_a^b f(x) \, dx = A_T(f, I_n) + R_T(f, I_n) \tag{3.30}$$

where $A_T(f, I_n)$ is the *trapezoid rule* and the remainder $R_T(f, I_n)$ satisfies the estimation

$$|R_T(f, I_n)|$$

$$\leq \begin{cases} \frac{1}{4} \|f'\|_\infty \sum_{i=0}^{n-1} h_i^2; \\ \frac{1}{2(q+1)^{\frac{1}{q}}} \|f'\|_p \left(\sum_{i=0}^{n-1} h_i^{q+1} \right)^{\frac{1}{q}}, & \text{where } p > 1 \text{ and } \frac{1}{p} + \frac{1}{q} = 1; \\ \frac{1}{2} \|f'\|_1 \nu(h), \end{cases} \tag{3.31}$$

where $h_i := x_{i+1} - x_i$ $(i = 0, \dots, n - 1)$ and $\nu(h) := \max_{i=0,n-1} h_i$.

We provide the proof for the second inequality in (3.31). The proof for the first and third inequalities is left for readers.

Applying (3.27) on the intervals $[x_i, x_{i+1}]$, we get

$$\left| \int_{x_i}^{x_{i+1}} f(x) \, dx - \frac{f(x_i) + f(x_{i+1})}{2} h_i \right|$$

$$\leq \frac{1}{2(q+1)^{\frac{1}{q}}} h_i^{1+\frac{1}{q}} \left(\int_{x_i}^{x_{i+1}} |f'(t)|^p \, dt \right)^{\frac{1}{p}}, \quad \text{for all } i \in \{0, \dots, n-1\}.$$

Summing and using Hölder's discrete inequality, we get

$$|R_T(f, I_n)| \leq \frac{1}{2(q+1)^{\frac{1}{q}}} \sum_{i=0}^{n-1} h_i^{1+\frac{1}{q}} \left(\int_{x_i}^{x_{i+1}} |f'(t)|^p \, dt \right)^{\frac{1}{p}}$$

$$\leq \frac{1}{2(q+1)^{\frac{1}{q}}} \left[\sum_{i=0}^{n-1} \left[\left(\int_{x_i}^{x_{i+1}} |f'(t)|^p \, dt \right)^{\frac{1}{p}} \right]^p \right]^{\frac{1}{p}} \left[\sum_{i=0}^{n-1} \left(h_i^{1+\frac{1}{q}} \right)^q \right]^{\frac{1}{q}}$$

$$= \frac{1}{2(q+1)^{\frac{1}{q}}} \|f'\|_p \left(\sum_{i=0}^{n-1} h_i^{q+1} \right)^{\frac{1}{q}}$$

and the inequality (3.31) is completely proved.

3.6 Trapezoid Inequality in Terms of Second Derivatives

Let $f : [a, b] \to \mathbb{R}$ be a mapping whose derivative f' is absolutely continuous on $[a, b]$. Then we have the inequality [89]

$$\left| \int_a^b f(x)\, dx - \frac{f(a) + f(b)}{2} (b - a) \right|$$

$$\leq \begin{cases} \dfrac{\|f''\|_\infty}{12} (b - a)^3 & \text{if } f'' \in L_\infty [a, b]; \\[2ex] \dfrac{1}{2} \|f''\|_p [B(q + 1, q + 1)]^{\frac{1}{q}} (b - a)^{2 + \frac{1}{q}} & \text{if } f'' \in L_p [a, b], \\ & \quad p > 1, \ \frac{1}{p} + \frac{1}{q} = 1; \\[2ex] \dfrac{1}{8} \|f''\|_1 (b - a)^2 & \text{if } f'' \in L_1 [a, b], \end{cases} \qquad (3.32)$$

where $\|\cdot\|_p$ are the usual norms $(p \in [1, \infty])$ on $L_p [a, b]$ and $B(\cdot, \cdot)$ is the Beta function of Euler, that is,

$$B(\alpha, \beta) = \int_0^1 t^{\alpha - 1} (1 - t)^{\beta - 1}\, dt, \quad \alpha, \beta > 0. \qquad (3.33)$$

PROOF   Integrating by parts twice on $[a, b]$, we get

$$\int_a^b (x - a)(b - x) f''(x)\, dx = (b - a)(f(a) + f(b)) - 2 \int_a^b f(x)\, dx \qquad (3.34)$$

and so

$$\left| \int_a^b f(x)\, dx - \frac{f(a) + f(b)}{2} (b - a) \right|$$

$$\leq \frac{1}{2} \int_a^b (x - a)(b - x) |f''(x)|\, dx. \qquad (3.35)$$

If $f'' \in L_\infty [a, b]$, then

$$\frac{1}{2} \int_a^b (x - a)(b - x) |f''(x)|\, dx \leq \frac{1}{2} \|f''\|_\infty \int_a^b (x - a)(b - x)\, dx$$

$$= \frac{\|f''\|_\infty}{12} (b - a)^3.$$

This proves the first part of (3.32).

If $f'' \in L_p[a,b]$, then by Hölder's integral inequality, we have

$$\int_a^b (x-a)(b-x)|f''(x)|\,dx$$

$$\leq \|f''\|_p \left(\int_a^b (x-a)^q (b-x)^q \, dx \right)^{\frac{1}{q}}, \quad p > 1, \quad \frac{1}{p} + \frac{1}{q} = 1. \quad (3.36)$$

Using the transformation $x = (1-t)a + tb$, $t \in [0,1]$, we get

$$(x-a)^q (b-x)^q = (b-a)^{2q} t^q (1-t^q), \quad dx = (b-a)\,dt$$

and thus

$$\int_a^b (x-a)^q (b-x)^q \, dx = (b-a)^{2q+1} B(q+1,q+1).$$

Now, by (3.35) and (3.36) we get the second part of (3.32).
Finally, if $f'' \in L_1[a,b]$, then we have

$$\int_a^b (x-a)(b-x)|f''(x)|\,dx \leq \max_{x \in [a,b]} [(x-a)(b-x)]\,\|f''\|_1$$

$$= \frac{(b-a)^2}{4}\,\|f''\|_1$$

and the inequality (3.32) is completely proved.   ⬜

Comments
Let f be as above and I_n a partition of the interval $[a,b]$. Then we have

$$\int_a^b f(x)\,dx = A_T(f,I_n) + R_T(f,I_n) \quad (3.37)$$

where $A_T(f,I_n)$ is the trapezoid formula and the remainder $R_T(f,I_n)$ satisfies the estimate

$$|R_T(f,I_n)|$$

$$\leq \begin{cases} \dfrac{1}{12}\|f''\|_\infty \displaystyle\sum_{i=0}^{n-1} h_i^3; \\[2ex] \dfrac{1}{2}\|f''\|_p [B(q+1,q+1)]^{\frac{1}{q}} \left(\displaystyle\sum_{i=0}^{n-1} h_i^{2q+1} \right)^{\frac{1}{q}}, \quad p>1,\ \frac{1}{p}+\frac{1}{q}=1; \\[2ex] \dfrac{1}{8}\|f''\|_1 \nu^2(h), \end{cases} \quad (3.38)$$

where $h_i := x_{i+1} - x_i$ $(i = 0, \ldots, n-1)$ and $\nu(I_n) := \max_{i=0,n-1} h_i$.

3.7 Generalised Trapezoid Rule Involving nth Derivative Error Bounds

Throughout this section, we denote $\mathring{I}$ to be the interior of an interval $I \subseteq \mathbb{R}$. Using Hayashi's inequality (see Mitrinović, Pečarić, and Fink [141, pp. 311–312]), Cerone and Dragomir [30] obtained the following trapezoidal inequality for differentiable mappings where the bound is in terms of the upper and lower bounds of the first derivative.

Let $f : I \subseteq \mathbb{R} \to \mathbb{R}$ be a differentiable mapping on $\mathring{I}$ and $[a, b] \subset \mathring{I}$ with $M = \sup\limits_{x \in [a,b]} f'(x) < \infty$, $m = \inf\limits_{x \in [a,b]} f'(x) > -\infty$, and $M > m$. If f' is integrable on $[a, b]$, then the following inequalities hold:

$$\left| \int_a^b f(x)\, dx - \frac{b-a}{2} [f(a) + f(b)] \right| \leq \frac{(b-a)^2}{2(M-m)} (S-m)(M-S) \quad (3.39)$$

$$\leq \frac{M-m}{2} \left(\frac{b-a}{2} \right)^2, \quad (3.40)$$

where $S = \frac{f(b)-f(a)}{b-a}$.

The result (3.39) was also obtained previously in a similar fashion by Cerone and Dragomir [30]; however, their formulation did not reveal (3.40). A prior result obtained by Iyengar [124] (see also Mitrinović, Pečarić, and Fink [141, p. 471]) is recovered if we take in (3.39): $m = -M$.

Cerone and Dragomir [30] also obtained nonsymmetric bounds for a generalised trapezoidal rule, namely,

Let f satisfy the conditions of the results in (3.39) and (3.40), then the following result holds:

$$\beta_L \leq \int_a^b f(x)\, dx - (b-a) \left[\left(\frac{\theta - a}{b-a} \right) f(a) + \left(\frac{b - \theta}{b-a} \right) f(b) \right] \leq \beta_U, \quad (3.41)$$

where

$$\beta_U = \frac{(b-a)^2}{2(M-m)} [S(2\gamma_U - S) - mM],$$

$$\beta_L = \frac{(b-a)^2}{2(M-m)} [S(S - 2\gamma_L) + mM],$$

$$\gamma_U = \left(\frac{\theta - a}{b-a} \right) M + \left(\frac{b - \theta}{b-a} \right) m, \quad \gamma_L = M + m - \gamma_U,$$

with $S = \frac{f(b)-f(a)}{b-a}$ and $\theta \in [a, b]$.

Milovanović and Pečarić [137] proved the following specialization of a more general result in which $f^{(n-1)}$ satisfy the Lipschitz condition, namely,

Let the function $f : [a, b] \to \mathbb{R}$ have a continuous derivative of order $n - 1$ and $\left| f^{(n)}(x) \right| \leq M$ for $x \in (a, b)$.
If $f^{(k)}(a) = f^{(k)}(b) = 0$ $(k = 1, 2, \ldots, n - 1)$, then the inequality

$$\left| \int_a^b f(x)\, dx - \frac{b - a}{2} \left(f(a) + f(b) \right) \right|$$
$$\leq \frac{M(b-a)^{n+1}}{(n+1)!} \left[\zeta^{n+1} - \frac{q}{2} \left(1 + \frac{n}{2\zeta} - 1 \right) \right]$$

holds, where

$$\zeta \text{ satisfies } \zeta^n - (1 - \zeta)^n = q := \frac{n!}{M(b-a)^n} \left(f(b) - f(a) \right).$$

Generalised trapezoidal type rules involving a parameter θ are obtained in assuming that the nth derivative is bounded both above and below. Further, the restrictive assumption of vanishing lower order derivatives at the end points is not made in the current work. Some of the results are compared with those obtained in Qi [155] where a Taylor approach is utilised. It is shown that the current developments give better results than the Taylor approach used by Qi.

The following result due to Hayashi in Mitrinović, Pečarić, and Fink [141, pp. 311–312] will be required and is thus stated for convenience.

Let $h : [a, b] \to \mathbb{R}$ be a nonincreasing mapping on $[a, b]$ and $g : [a, b] \to \mathbb{R}$ an integrable mapping on $[a, b]$ with

$$0 \leq g(x) \leq A, \quad \text{for all } x \in [a, b],$$

then

$$A \int_{b-\lambda}^b h(x)\, dx \leq \int_a^b h(x) g(x)\, dx \leq A \int_a^{a+\lambda} h(x)\, dx, \qquad (3.42)$$

where

$$\lambda = \frac{1}{A} \int_a^b g(x)\, dx.$$

The result (3.42) is attributed to a generalisation of Steffensen's inequality (see Mitrinović, Pečarić, and Fink [141, pp. 311–312]) which is obtained by taking $A = 1$ in the above result. Inspection of Steffensen's original paper [162] reveals that the more general situation depicted by the following result was also treated. Thus Hayashi's result is a special case of the following result of Steffensen (see also Section 4.4).

Let $h : [a, b] \to \mathbb{R}$ be a nonincreasing mapping on $[a, b]$ and $g : [a, b] \to \mathbb{R}$ be an integrable mapping on $[a, b]$ with

$$\phi \le g(x) \le \Phi, \quad \text{for all } x \in [a, b],$$

then

$$\phi \cdot \int_a^{b-\lambda} h(x)\,dx + \Phi \int_{b-\lambda}^b h(x)\,dx$$

$$\le \int_a^b h(x)\,g(x)\,dx \le \Phi \cdot \int_a^{a+\lambda} h(x)\,dx + \phi \int_{a+\lambda}^b h(x)\,dx, \quad (3.43)$$

where

$$\lambda = \int_a^b G(x)\,dx, \quad G(x) = \frac{g(x) - \phi}{\Phi - \phi}, \quad \Phi \ne \phi. \quad (3.44)$$

We note that result (3.43) may be obtained upon simplification and by using Steffensen's more well-known result that

$$\int_{b-\lambda}^b h(x)\,dx \le \int_a^b h(x)\,G(x)\,dx \le \int_a^{a+\lambda} h(x)\,dx, \quad (3.45)$$

where λ is as given by (3.44) and $0 \le G(x) \le 1$. Contrarily, if we take $\phi = 0$ and $\Phi = 1$ we obtain (3.45) from (3.43). Also, if we take $\phi = 0$ in (3.43) then the Hayashi result (3.42) is seen to be included.

Equation (3.45) has the pleasant interpretation, as noted by Steffensen, that if we divide by λ then

$$\frac{1}{\lambda} \int_{b-\lambda}^b h(x)\,dx \le \frac{\int_a^b G(x)\,h(x)\,dx}{\int_a^b G(x)\,dx} \le \frac{1}{\lambda} \int_a^{a+\lambda} h(x)\,dx.$$

Thus, the weighted integral mean of $h(x)$ is bounded by the integral means over the end intervals of length λ, the total weight.

Further, it should be stated here that the discrete versions of (3.43) and (3.45) were also treated in Steffensen [162].

The following result gives trapezoid type rules using the above results [18]:

Let $f : I \subseteq \mathbb{R} \to \mathbb{R}$ and $f^{(n-1)}$ be absolutely continuous on $\overset{\circ}{I}$ ($\overset{\circ}{I}$ is the interior of I) and $[a, b] \subset \overset{\circ}{I}$ with $m = \inf\limits_{x \in [a,b]} f^{(n)}(x) > -\infty$, $M = \sup\limits_{x \in [a,b]} f^{(n)}(x) < \infty$, and $M > m$. If $f^{(n)}$ is integrable on $[a, b]$, then the following inequalities hold:

$$\left| \int_a^b f(x)\,dx - T_n(\theta; a, b) - P_n(\theta; a, b) \right| \le Q_n^-(\theta; a, b), \quad (3.46)$$

where

$$T_n(\theta; a, b) = \sum_{k=1}^{n} \frac{(\theta - a)^k f^{(k-1)}(a) + (-1)^{k-1}(b - \theta)^k f^{(k-1)}(b)}{k!}, \quad (3.47)$$

$$P_n(\theta; a, b) = \frac{m}{(n+1)!}\left[(\theta - a)^{n+1} + (-1)^n (b - \theta)^{n+1}\right] + Q_n^+(\theta; a, b), \quad (3.48)$$

with

$$\frac{2(n+1)!}{M-m} Q_n^{\pm}(\theta; a, b) \qquad (3.49)$$

$$= \begin{cases} \sum_{j=1}^{n+1} (-1)^{j-1} \binom{n+1}{j} (\lambda_n^0)^j \left[(\theta - a)^{n+1-j} \pm (-1)^n (b - \theta)^{n+1-j}\right], & n \text{ odd}, \\[2mm] \sum_{j=1}^{n+1} (-1)^{j-1} \binom{n+1}{j} \left[(\lambda_n^a)^j (\theta - a)^{n+1-j} + (-1)^n (\lambda_n^b)^j (b - \theta)^{n+1-j}\right] \\[1mm] \pm \left[(\lambda_n^a)^{n+1} + (-1)^n (\lambda_n^b)^{n+1}\right], & n \text{ even} \end{cases}$$

and

$$\lambda_n^0 = \lambda_n(a, b), \; \lambda_n^a = \lambda_n(a, \theta), \; \lambda_n^b = \lambda_n(\theta, b) \quad and \quad \theta \in [a, b], \quad (3.50)$$

where

$$\lambda_n(a, b) = \frac{b-a}{M-m}(S_{n-1}(a, b) - m) \qquad (3.51)$$

and

$$S_{n-1}(a, b) = \frac{f^{(n-1)}(b) - f^{(n-1)}(a)}{b-a}. \qquad (3.52)$$

PROOF Following Cerone [18], let $h(x) = \frac{(\theta-x)^n}{n!}$, $\theta \in [a, b]$ and $g(x) = f^{(n)}(x) - m$. Assume for the time being that n is odd, then $h(x)$ is a nonincreasing function, and so from Hayashi's inequality

$$L_o \leq I_n \leq U_o, \qquad (3.53)$$

where

$$I_n = I_n(\theta; a, b) = \int_a^b \frac{(\theta - x)^n}{n!}\left(f^{(n)}(x) - m\right) dx,$$

$$\lambda_n(a, b) = \frac{1}{M-m}\int_a^b \left(f^{(n)}(x) - m\right) dx = \lambda_n^0 \text{ (for } n \text{ odd)},$$

and

$$L_o = W(b - \lambda_n^0, b), \; U_o = W(a, a + \lambda_n^0)$$

with

$$W\left(l,u\right)=\left(M-m\right)\int_{l}^{u}h\left(x\right)dx=\left(M-m\right)\int_{l}^{u}\frac{\left(\theta-x\right)^{n}}{n!}dx.$$

The above expressions may be simplified to give

$$I_{n}\left(\theta;a,b\right)=\int_{a}^{b}f\left(x\right)dx-T_{n}\left(\theta;a,b\right)$$

$$-\frac{m}{\left(n+1\right)!}\left[\left(\theta-a\right)^{n+1}+\left(-1\right)^{n}\left(b-\theta\right)^{n+1}\right],\quad(3.54)$$

where λ_{n}^{0} is as given by (3.52) and (3.51), and

$$W\left(l,u\right)=\frac{M-m}{\left(n+1\right)!}\left[\left(\theta-l\right)^{n+1}-\left(\theta-u\right)^{n+1}\right] \qquad (3.55)$$

with

$$L_{o}=\frac{M-m}{\left(n+1\right)!}\left[\left(\theta-\left(b-\lambda_{n}^{0}\right)\right)^{n+1}-\left(\theta-b\right)^{n+1}\right] \qquad (3.56)$$

and

$$U_{o}=\frac{M-m}{\left(n+1\right)!}\left[\left(\theta-a\right)^{n+1}-\left(\theta-\left(a+\lambda_{n}^{0}\right)\right)^{n+1}\right]. \qquad (3.57)$$

Further, it may be noticed from (3.53) that

$$\left|I_{n}-\frac{U_{o}+L_{o}}{2}\right|\leq\frac{U_{o}-L_{o}}{2}, \qquad (3.58)$$

where

$$\frac{U_{o}\pm L_{o}}{2}=\frac{M-m}{2\left(n+1\right)!}\sum_{j=1}^{n+1}\left(-1\right)^{j-1}\binom{n+1}{j}$$

$$\times\left(\lambda_{n}^{0}\right)^{j}\left[\left(\theta-a\right)^{n+1-j}\pm\left(-1\right)^{n}\left(b-\theta\right)^{n+1-j}\right]. \qquad (3.59)$$

Combining (3.54), (3.58), and (3.59) produces the results (3.46)–(3.52) for n, odd.

Now, for the case when n is even. It should be noted that the inequality (3.42) is reversed for $h\left(x\right)$ nondecreasing and so for n even $\frac{\left(\theta-x\right)^{n}}{n!}$ is nonincreasing for $x\in\left[a,\theta\right]$ and nondecreasing for $x\in\left(\theta,b\right]$. Let a superscript of a or b represent these intervals.

Then on the interval $\left[a,\theta\right]$ we have

$$L^{a}\leq I_{n}^{a}\leq U^{a}, \qquad (3.60)$$

where

$$I_{n}^{a}=I_{n}\left(\theta;a,\theta\right),$$
$$L^{a}=W\left(\theta-\lambda_{n}^{a},\theta\right),\quad U^{a}=W\left(a,a+\lambda_{n}^{a}\right)$$

with

$$\lambda_n^a = \lambda_n(a, \theta) = \frac{\theta - a}{M - m}(S_{n-1}(a, \theta) - m).$$

Similarly, on $(\theta, b]$ we have

$$L^b \le I_n^b \le U^b, \tag{3.61}$$

where

$$I_n^b = I_n(\theta; \theta, b),$$
$$L^b = W(\theta, \theta + \lambda_n^b), \quad U^b = W(b - \lambda_n^b, b)$$

with

$$\lambda_n^b = \lambda_n(\theta, b) = \frac{b - \theta}{M - m}(S_{n-1}(\theta, b) - m).$$

Thus, combining (3.60) and (3.61) gives

$$L_e \le I_n \le U_e, \tag{3.62}$$

where

$$I_n = I_n^a + I_n^b, \tag{3.63}$$
$$L_e = L^a + L^b,$$
$$U_e = U^a + U^b.$$

That is, I_n is as given by (3.51) and, on using (3.55),

$$L_e = W(\theta - \lambda_n^a, \theta) + W(\theta, \theta + \lambda_n^b) \tag{3.64}$$
$$= \frac{M - m}{(n+1)!}\left[(\lambda_n^a)^{n+1} + (-1)^n (\lambda_n^b)^{n+1}\right]$$

and

$$U_e = W(a, a + \lambda_n^a) + W(b - \lambda_n^b, b) \tag{3.65}$$
$$= \frac{M - m}{(n+1)!} \sum_{j=1}^{n+1} (-1)^{j-1} \binom{n+1}{j}$$
$$\times \left[(\lambda_n^a)^j (\theta - a)^{n+1-j} + (-1)^n (\lambda_n^b)^j (b - \theta)^{n+1-j}\right].$$

Further, from (3.58) we have

$$\left| I_n - \frac{U_e + L_e}{2} \right| \le \frac{U_e - L_e}{2}, \tag{3.66}$$

where

$$\frac{U_e \pm L_e}{2} = \frac{M - m}{2(n+1)!}\left\{ \sum_{j=1}^{n+1} (-1)^{j-1} \binom{n+1}{j}\left[(\lambda_n^a)^j (\theta - a)^{n+1-j}\right.\right.$$
$$\left.\left. + (-1)^n (\lambda_n^b)^j (b - \theta)^{n+1-j}\right] \pm \left[(\lambda_n^a)^{n+1} + (-1)^n (\lambda_n^b)^{n+1}\right]\right\}. \tag{3.67}$$

Combining (3.54), (3.66), and (3.67) produces the results (3.47)–(3.52) for n even and thus the proof is complete. □

We may use Steffensen's inequality (3.43) with $g(x) = f^{(n)}(x)$ and $h(x) = \frac{(\theta-x)^n}{n!}$ to prove the above result. This will not be pursued here.

The following results were also obtained by Cerone [18] (see also Liu [131]):

Let the conditions leading to the results (3.46)–(3.52) be valid. Then

$$L + R \leq \int_a^b f(x)\,dx - T_n(\theta; a, b) \leq U + R \tag{3.68}$$

holds, where $T_n(\theta; a, b)$ is as given by (3.47),

$$\frac{(n+1)!}{m}R = (\theta - a)^{n+1} + (-1)^n(b - \theta)^{n+1}, \tag{3.69}$$

$$\frac{(n+1)!}{M-m}L = \begin{cases} (-1)^{n+1}\left[(b - \theta - \lambda_n^0)^{n+1} - (b - \theta)^{n+1}\right], & n\ \text{odd}, \\ (\lambda_n^a)^{n+1} + (-1)^n(\lambda_n^b)^{n+1}, & n\ \text{even} \end{cases} \tag{3.70}$$

and

$$\frac{(n+1)!}{M-m}U = \begin{cases} (\theta - a)^{n+1} - (\theta - a - \lambda_n^0)^{n+1}, & n\ \text{odd}, \\ (\theta - a)^{n+1} - (\theta - a - \lambda_n^a)^{n+1} \\ +(-1)^{n+1}\left[(b - \theta - \lambda_n^b)^{n+1} - (b - \theta)^{n+1}\right], & n\ \text{even} \end{cases} \tag{3.71}$$

with

$$\lambda_n^0 = \frac{b - a}{M - m}(S_{n-1}(a, b) - m),$$
$$\lambda_n^a = \frac{\theta - a}{M - m}(S_{n-1}(a, \theta) - m),$$
$$\lambda_n^b = \frac{b - \theta}{M - m}(S_{n-1}(\theta, b) - m),$$

and

$$S_{n-1}(a, b) = \frac{f^{(n-1)}(b) - f^{(n-1)}(a)}{b - a}.$$

Let $C_U = U + R$ and $C_L = L + R$ where R, L, and U are as defined in (3.69)–(3.71). Further, let Q_L and Q_U be as defined by (3.74) and (3.75). Consider $D_U := C_U - Q_U$ and $D_L := C_L - Q_L$, then $C_U < Q_U$ and $C_L > Q_L$.

PROOF   We have from (3.69)–(3.71) that

$$
(n+1)!D_U = \begin{cases} -(M-m)\left[(\theta - a - \lambda_n^a)^{n+1} + (b - \theta - \lambda_n^b)^{n+1}\right] & \text{for } n \text{ even} \\[2mm] -(M-m)(\theta - a - \lambda_n^0)^{n+1} & \text{for } n \text{ odd} \end{cases}
$$

and

$$
(n+1)!D_L = \begin{cases} (M-m)\left[(\lambda_n^a)^{n+1} + (\lambda_n^b)^{n+1}\right] & \text{for } n \text{ even} \\[2mm] (M-m)(b - \theta - \lambda_n^b)^{n+1} & \text{for } n \text{ odd.} \end{cases}
$$

Thus we can conclude that $D_U < 0$ and $D_L > 0$ for both n odd and even since $0 < \lambda_n^a < \theta - a$ and $0 < \lambda_n^b < b - \theta$.

$\square$

Comments

(a) It should be noticed that $U > 0$ and $L > 0$ since $0 < \lambda_n^0 < b - a$, $0 < \lambda_n^a < \theta - a$, and $0 < \lambda_n^b < b - \theta$ as $0 < \frac{S_{n-1}(a,b) - m}{M - m} < 1$. Further, $R > 0$ for n even or for $\theta > \frac{a+b}{2}$ and n odd. Now, $R < 0$ for $\theta < \frac{a+b}{2}$ and n odd.

(b) The result (3.68) gives nonsymmetric bounds for the generalised trapezoidal rule $T_n(\theta; a, b)$ as defined by (3.47) while the result (3.42) gives symmetric bounds for a perturbed trapezoidal rule. The bounds involve the upper and lower bounds M and m of $f^{(n)}(x)$, $x \in [a, b]$ and some arbitrary point $\theta \in [a, b]$. If θ is taken to be at the midpoint, that is, $\theta = \frac{a+b}{2}$, then some simplification occurs. In particular, for n odd, $P_n\left(\frac{a+b}{2}; a, b\right) = 0$; and so there is no perturbation. For n odd and $\theta = \frac{a+b}{2}$ then $R = 0$ in (3.69).

(c) If $n = 1$ in (3.46), then we recapture (3.39) on taking $\theta = \frac{a+b}{2}$. Further, (3.41) is reproduced from (3.68) on taking $n = 1$. Thus, the results of this section are an extension of the work of Cerone and Dragomir [30] to involve bounds for the generalised trapezoidal rule in terms of bounds on $f^{(n)}$. If $n = 1$ in (3.46) then

$$
\left| \int_a^b f(x)\, dx - [(\theta - a) f(a) + (b - \theta) f(b)] \right.
$$

$$
\left. - (b - a)[S_0(a, b) - m]\left(\theta - \frac{a+b}{2}\right) \right|
$$

$$
\leq \frac{(b-a)^2}{2(M-m)}(S_0(a, b) - m)(M - S_0(a, b)).
$$

Now, using the definition of $S_0(a, b) = \frac{f(b) - f(a)}{b - a}$, then the above result may be simplified to

$$\left| \int_a^b f(x)\, dx - \frac{b-a}{2} \left[f(a) + f(b) \right] + m(b-a) \left(\theta - \frac{a+b}{2} \right) \right|$$

$$\leq \frac{(b-a)^2}{2(M-m)} \left(S_0(a,b) - m \right) \left(M - S_0(a,b) \right).$$

It may be noticed that the above result is a perturbed formula which has the same bounds (3.39) independent of θ. Further, the perturbation vanishes if $\theta = \frac{a+b}{2}$.

(d) Using a Taylor series approach, Qi [155] obtained the following in our notation:

$$Q_L \leq \int_a^b f(x)\, dx - T_n(\theta; a, b) \leq Q_U; \tag{3.72}$$

where, if we define

$$Q(u, v) := u \frac{(\theta - a)^{n+1}}{(n+1)!} + (-1)^n v \frac{(b - \theta)^{n+1}}{(n+1)!}, \tag{3.73}$$

then

$$Q_L = \begin{cases} Q(m, m), & n \text{ even} \\ Q(m, M), & n \text{ odd} \end{cases} \tag{3.74}$$

and

$$Q_U = \begin{cases} Q(M, M), & n \text{ even} \\ Q(M, m), & n \text{ odd} \end{cases}, \tag{3.75}$$

where $m \leq f^{(n)}(x) \leq M$, $x \in [a, b]$.

A comparison of (3.72) with (3.68) shows that (3.72) provides better bounds (see Liu [131]). We note that R as defined in (3.69) is equivalent to $Q(m, m)$, which is the lower bound in (3.72) for n even.

3.8 A Refinement of Ostrowski's Inequality for the Čebyšev Functional

The following result holds [50]:

Let $h : [a, b] \to \mathbb{R}$ be an integrable function on $[a, b]$ such that

$$-\infty < \gamma \leq h(x) \leq \Gamma < \infty \text{ for a.e. } x \text{ on } [a, b]. \tag{3.76}$$

Then we have the inequality

$$\frac{1}{b-a}\int_a^b \left| \int_a^x h(t)\,dt - \frac{x-a}{b-a}\int_a^b h(u)\,du \right| dx \tag{3.77}$$

$$\leq \frac{1}{2}\left(\frac{1}{b-a}\int_a^b h(u)\,du - \gamma\right)\left(\Gamma - \frac{1}{b-a}\int_a^b h(u)\,du\right)\frac{b-a}{\Gamma-\gamma}$$

$$\leq \frac{1}{8}(\Gamma-\gamma)(b-a).$$

The constants $\frac{1}{2}$ and $\frac{1}{8}$ are sharp in the sense that they cannot be replaced by smaller constants.

PROOF Ostrowski [146, p. 368] proved the following result:
If $f:[a,b]\to\mathbb{R}$ is integrable and

$$-\alpha \leq f(x) \leq 1-\alpha, \quad \alpha \in [0,1] \quad \text{and} \quad \int_a^b f(x)\,dx = 0, \tag{3.78}$$

then

$$\int_a^b |F(x)|\,dx \leq \frac{\alpha(1-\alpha)}{2}(b-a)^2, \tag{3.79}$$

where $F(x) = \int_a^x f(t)\,dt$.
Define

$$f(t) = \frac{1}{\Gamma-\gamma}\left[h(t) - \frac{1}{b-a}\int_a^b h(u)\,du\right], \quad t\in[a,b], \quad (\Gamma\neq\gamma).$$

Obviously,

$$\int_a^b f(t)\,dt = 0$$

and by (3.76),

$$f(t) \leq \frac{\Gamma - \frac{1}{b-a}\int_a^b h(u)\,du}{\Gamma-\gamma} = 1 - \frac{\frac{1}{b-a}\int_a^b h(u)\,du - \gamma}{\Gamma-\gamma}, \quad t\in[a,b] \tag{3.80}$$

and

$$f(t) \geq \frac{\gamma - \frac{1}{b-a}\int_a^b h(u)\,du}{\Gamma-\gamma}, \quad t\in[a,b]. \tag{3.81}$$

Denoting

$$\alpha := \frac{\frac{1}{b-a}\int_a^b h(u)\,du - \gamma}{\Gamma-\gamma}$$

we observe, by (3.80) and (3.81), that

$$-\alpha \le f(x) \le 1 - \alpha \quad \text{for} \quad x \in [a, b].$$

We also have

$$F(x) = \int_a^x f(t)\, dt = \frac{1}{\Gamma - \gamma} \left[\int_a^x h(t)\, dt - \frac{x-a}{b-a} \int_a^b h(u)\, du \right]$$

and thus, by (3.79), we may state the following inequality:

$$\frac{1}{\Gamma - \gamma} \int_a^b \left| \int_a^x h(t)\, dt - \frac{x-a}{b-a} \int_a^b h(u)\, du \right| dx$$

$$\le \frac{1}{2} \left(\frac{\frac{1}{b-a} \int_a^b h(u)\, du - \gamma}{\Gamma - \gamma} \right) \left(\frac{\Gamma - \frac{1}{b-a} \int_a^b h(u)\, du}{\Gamma - \gamma} \right) (b-a)^2 \quad (3.82)$$

which is clearly equivalent to the first inequality in (3.77).

To prove the second inequality in (3.77), we make use of the following elementary fact:

$$\alpha\beta \le \frac{1}{4}(\alpha + \beta)^2, \alpha, \beta \in \mathbb{R}; \quad (3.83)$$

with equality if and only if $\alpha = \beta$, for the choices

$$\alpha = \frac{1}{b-a} \int_a^b h(u)\, du - \gamma, \quad \beta = \Gamma - \frac{1}{b-a} \int_a^b h(u)\, du.$$

To prove the sharpness of the constant $\frac{1}{2}$, assume that (3.77) holds with a constant $C > 0$. Namely,

$$\frac{1}{\Gamma - \gamma} \int_a^b \left| \int_a^x h(t)\, dt - \frac{x-a}{b-a} \int_a^b h(u)\, du \right| dx$$

$$\le C \left(\frac{1}{b-a} \int_a^b h(u)\, du - \gamma \right) \left(\Gamma - \frac{1}{b-a} \int_a^b h(u)\, du \right) \frac{b-a}{\Gamma - \gamma}. \quad (3.84)$$

Consider the function $h : [a, b] \to \mathbb{R}$,

$$h(t) := \begin{cases} -1 & t \in \left[a, \frac{a+b}{2}\right], \\ 1 & t \in \left(\frac{a+b}{2}, b\right]. \end{cases}$$

Obviously $\gamma = -1$, $\Gamma = 1$, and $\int_a^b h(u)\, du = 0$.

We have

$$\int_a^b \left| \int_a^x h(t)\, dt \right| = \frac{1}{b-a} \left[\int_a^{\frac{a+b}{2}} \left| \int_a^x h(t)\, dt \right| dx + \int_{\frac{a+b}{2}}^b \left| \int_a^x h(t)\, dt \right| dx \right]$$

$$= \frac{1}{b-a} \left[\int_a^{\frac{a+b}{2}} (x-a)\, dx + \int_{\frac{a+b}{2}}^b (b-x)\, dx \right] = \frac{b-a}{4},$$

and thus, by (3.84), we deduce

$$\frac{b-a}{4} \le C \cdot \frac{b-a}{2},$$

giving $C \ge \frac{1}{2}$.

The fact that the constant $\frac{1}{8}$ is best possible is obvious, and we omit the details. □

The following bounds for the Čebyšev functional $T(f,g)$ hold [50]:

Let $f,g : [a,b] \to \mathbb{R}$ be such that g is absolutely continuous on $[a,b]$ with $g' \in L_\infty[a,b]$ and f is Lebesgue integrable and such that there exist $m, M \in \mathbb{R}$ with

$$-\infty < m \le f(x) \le M < \infty \text{ for a.e. } x \text{ on } [a,b]. \tag{3.85}$$

Then we have the inequality

$$|T(f,g)| \tag{3.86}$$

$$\le \frac{1}{2} \|g'\|_\infty \frac{\left(\frac{1}{b-a} \int_a^b f(x)\, dx - m \right) \left(M - \frac{1}{b-a} \int_a^b f(x)\, dx \right)}{M-m} (b-a)$$

$$\le \frac{1}{8} (b-a)(M-m) \|g'\|_\infty .$$

The constants $\frac{1}{2}$ and $\frac{1}{8}$ are sharp in the above sense.

PROOF Integrating by parts gives

$$\frac{1}{b-a} \int_a^b \left(\int_a^x f(t)\, dt - \frac{x-a}{b-a} \int_a^b f(u)\, du \right) g'(x)\, dx$$

$$= \frac{1}{b-a} \left[\left(\int_a^x f(t)\, dt - \frac{x-a}{b-a} \int_a^b f(u)\, du \right) g(x) \Big|_a^b \right.$$

$$\left. - \int_a^b g(x) \left[f(x) - \frac{1}{b-a} \int_a^b f(u)\, du \right] dx \right]$$

$$= -\frac{1}{b-a} \int_a^b g(x) f(x)\, dx + \frac{1}{b-a} \int_a^b g(x)\, dx \cdot \frac{1}{b-a} \int_a^b f(x)\, dx$$

$$= -T(f,g).$$

Taking the modulus, we have

$$|T(f,g)| \leq \frac{1}{b-a} \int_a^b \left| \int_a^x f(t)\,dt - \frac{x-a}{b-a} \int_a^b f(u)\,du \right| |g'(x)|\,dx$$

$$\leq \|g'\|_\infty \frac{1}{b-a} \int_a^b \left| \int_a^x f(t)\,dt - \frac{x-a}{b-a} \int_a^b f(u)\,du \right| dx$$

$$\leq \frac{1}{2} \|g'\|_\infty \frac{\left(\frac{1}{b-a} \int_a^b f(x)\,dx - m \right)\left(M - \frac{1}{b-a} \int_a^b f(x)\,dx \right)}{M - m} (b-a),$$

where we have used the result (3.77).

To prove the sharpness of the constant $\frac{1}{2}$, assume that (3.86) holds with a constant $D > 0$. That is,

$$|T(f,g)|$$

$$\leq D \|g'\|_\infty \frac{\left(\frac{1}{b-a} \int_a^b f(x)\,dx - m \right)\left(M - \frac{1}{b-a} \int_a^b f(x)\,dx \right)}{M - m} (b-a). \qquad (3.87)$$

Consider the functions $g(x) = x - \frac{a+b}{2}$, $f : [a,b] \to \mathbb{R}$,

$$f(x) = \begin{cases} -1 \text{ if } x \in \left[a, \frac{a+b}{2}\right], \\ 1 \quad \text{if } x \in \left(\frac{a+b}{2}, b\right]. \end{cases}$$

Then

$$\int_a^b f(x)\,dx = \int_a^b g(x)\,dx = 0, \quad \|g'\|_\infty = 1, \quad m = -1, \quad M = 1$$

and

$$T(f,g) = \frac{1}{b-a} \int_a^b \left| x - \frac{a+b}{2} \right| dx = \frac{b-a}{4}.$$

Thus, from (3.87), we deduce

$$\frac{b-a}{4} \leq D \cdot \frac{b-a}{2},$$

giving $D \geq \frac{1}{2}$.

The last inequality and the sharpness of $\frac{1}{8}$ are obvious; and we omit the details. □

We note, by (3.83), that the equality is achieved in the last inequality of (3.86) if and only if $\alpha = \beta$ so that

$$\frac{1}{b-a} \int_a^b f(x)\,dx = \frac{m+M}{2}. \qquad (3.88)$$

Consequently, for any integrable function $f : [a, b] \rightarrow \mathbb{R}$, such that (3.88) does not hold, the last inequality in (3.86) is strict, showing that the above result indeed provides a refinement of Ostrowski's inequality. The same applies for the various applications of this inequality outlined below; and we omit the details.

Comments

(a) We may apply (3.86) to obtain the following result due to Agarwal and Dragomir [1]. Note that, to obtain (3.89), they used Hayashi's inequality and assumed differentiability of the function on the entire interval.

Let $f : [a, b] \rightarrow \mathbb{R}$ *be an absolutely continuous function so that there exist the real numbers* γ, Γ *with*

$$-\infty < \gamma \leq f'(x) \leq \Gamma < \infty \text{ for a.e. } x \in [a, b]. \tag{3.89}$$

If we denote by $[f; a, b] := \frac{f(b) - f(a)}{b - a}$ *the divided difference, then we have the inequality*

$$\left| \frac{f(a) + f(b)}{2} - \frac{1}{b - a} \int_a^b f(t)\, dt \right| \tag{3.90}$$

$$\leq \frac{1}{2} \cdot \frac{([f; a, b] - \gamma)(\Gamma - [f; a, b])}{\Gamma - \gamma} (b - a)$$

$$\leq \frac{1}{8} (\Gamma - \gamma)(b - a).$$

The constants $\frac{1}{2}$ and $\frac{1}{8}$ are best possible.

PROOF We start with the identity

$$\frac{f(a) + f(b)}{2} - \frac{1}{b - a} \int_a^b f(t)\, dt = \frac{1}{b - a} \int_a^b \left(t - \frac{a + b}{2} \right) f'(t)\, dt.$$

By using (3.86), we may state that

$$\left| \frac{1}{b - a} \int_a^b \left(t - \frac{a + b}{2} \right) f'(t)\, dt - \frac{1}{b - a} \int_a^b \left(t - \frac{a + b}{2} \right) dt \cdot \frac{1}{b - a} \int_a^b f'(t)\, dt \right|$$

$$\leq \frac{1}{2} \left\| \frac{d}{dt} \left(t - \frac{a + b}{2} \right) \right\|_\infty \cdot \frac{\left(\frac{1}{b-a} \int_a^b f'(t)\, dt - \gamma \right) \left(\Gamma - \frac{1}{b-a} \int_a^b f'(t)\, dt \right)}{\Gamma - \gamma} (b - a)$$

$$\leq \frac{1}{8} (\Gamma - \gamma)(b - a),$$

which is clearly equivalent to (3.90).

The case of sharpness for the constant $\frac{1}{2}$ may be proved in a similar way as above by choosing $f(t) = \left| t - \frac{a + b}{2} \right|$, $t \in [a, b]$.

We omit the details. ◻

(**b**) Let $I \subset \mathbb{R}$ be a closed interval, $a \in I$, and n be a positive integer. If $f : I \to \mathbb{R}$ is such that $f^{(n)}$ is absolutely continuous, then for each $x \in I$

$$f(x) = T_n(f; a, x) + R_n(f; a, x), \tag{3.91}$$

where $T_n(f; a, x)$ is Taylor's polynomial, namely,

$$T_n(f; a, x) = \sum_{k=0}^{n} \frac{(x-a)^k}{k!} f^{(k)}(a) \tag{3.92}$$

(note that $f^{(0)} = f$ and $0! = 1$), and the remainder is given by

$$R_n(f; a, x) = \frac{1}{n!} \int_a^x (x-t)^n f^{(n+1)}(t) \, dt. \tag{3.93}$$

By using (3.86), we may point out the following perturbation of the Taylor expansion [50]:

Let $f : I \to \mathbb{R}$ be as above and $a \in I$. Then we have Taylor's perturbed formula:

$$f(x) = T_n(f; a, x) + \frac{(x-a)^{n+1}}{(n+1)!} \left[f^{(n)}; a, x \right] + G_n(f; a, x); \tag{3.94}$$

and the remainder $G_n(f; a, x)$ satisfies the estimate

$$|G_n(f; a, x)| \leq \frac{1}{2(n-1)!}$$
$$\cdot \frac{\left[\left[f^{(n)}; a, x \right] - \gamma_{n+1}(x) \right] \left[\Gamma_{n+1}(x) - \left[f^{(n)}; a, x \right] \right]}{\Gamma_{n+1}(x) - \gamma_{n+1}(x)} |x-a|^{n+1} \tag{3.95}$$

where $x \in I$ and

$$\Gamma_{n+1}(x) = \sup \left\{ f^{(n+1)}(t), \quad t \in [a, x] \ ([x, a]) \right\},$$
$$\gamma_{n+1}(x) = \inf \left\{ f^{(n+1)}(t), \quad t \in [a, x] \ ([x, a]) \right\}.$$

PROOF Using (3.86), we may state the following inequality:

$$\left| \frac{1}{x-a} \int_a^x (x-t)^n f^{(n+1)}(t) \, dt \right.$$
$$\left. - \frac{1}{x-a} \int_a^x (x-t)^n \, dt \cdot \frac{1}{x-a} \int_a^x f^{(n+1)}(t) \, dt \right|$$
$$\leq \frac{1}{2} n \sup_{t \in [a,x]} |t-x|^{n-1} \frac{\left[\left[f^{(n)}; a, x \right] - \gamma_{n+1}(x) \right] \left[\Gamma_{n+1}(x) - \left[f^{(n)}; a, x \right] \right]}{\Gamma_{n+1}(x) - \gamma_{n+1}(x)} |x-a|,$$

which is clearly equivalent to

$$\left| \frac{1}{n!} \int_a^x (x-t)^n f^{(n+1)}(t)\, dt - \frac{1}{(n+1)!}(x-a)^{n+1}\left[f^{(n)}; a, x\right] \right|$$
$$\leq \frac{1}{2(n-1)!}|x-a|^{n+1} \frac{\left[\left[f^{(n)}; a, x\right] - \gamma_{n+1}(x)\right]\left[\Gamma_{n+1}(x) - \left[f^{(n)}; a, x\right]\right]}{\Gamma_{n+1}(x) - \gamma_{n+1}(x)}.$$

Using the representations (3.91) and (3.93), we deduce the desired result. □

(c) Let $f : [a, b] \to \mathbb{R}$ be a function such that the derivative $f^{(n-1)}$ $(n \geq 1)$ is absolutely continuous on $[a, b]$. Cerone, Dragomir, Roumeliotis, and Šunde [41] obtained the following generalisation of the trapezoidal rule:

$$\int_a^b f(t)\, dt$$
$$= \sum_{k=0}^{n-1} \frac{1}{(k+1)!}\left[(x-a)^{k+1} f^{(k)}(a) + (-1)^k (b-x)^{k+1} f^{(k)}(b)\right]$$
$$+ \frac{1}{n!} \int_a^b (x-t)^n f^{(n)}(t)\, dt. \quad (3.96)$$

By the use of (3.86), we may state the following perturbed version of (3.96):

With the above assumptions for $f : [a, b] \to \mathbb{R}$ and if there exist the constants $\gamma_n, \Gamma_n \in \mathbb{R}$ such that

$$-\infty < \gamma_n \leq f^{(n)}(t) \leq \Gamma_n < \infty \text{ for a.e. } t \in [a, b], \quad (3.97)$$

then we have the representation

$$\int_a^b f(t)\, dt$$
$$= \sum_{k=0}^{n-1} \frac{1}{(k+1)!}\left[(x-a)^{k+1} f^{(k)}(a) + (-1)^k (b-x)^{k+1} f^{(k)}(b)\right]$$
$$+ \frac{(x-a)^{n+1} + (-1)^n (b-x)^{n+1}}{(n+1)!}\left[f^{(n-1)}; b, a\right] + S_n(f, x) \quad (3.98)$$

and the remainder $S_n(f, x)$ satisfies the estimate

$$|S_n(f, x)| \leq \frac{1}{2(n-1)!}\left[\frac{1}{2}(b-a) + \left|x - \frac{a+b}{2}\right|\right]^{n-1} (b-a)^2$$
$$\times \frac{\left(\left[f^{(n-1)}; a, b\right] - \gamma_n\right)\left(\Gamma_n - \left[f^{(n-1)}; a, b\right]\right)}{\Gamma_n - \gamma_n}, \quad (3.99)$$

for any $x \in [a, b]$.

PROOF Applying (3.86), we may state that

$$\left| \frac{1}{b-a} \int_a^x (x-t)^n f^{(n)}(t)\, dt - \frac{1}{b-a} \int_a^x (x-t)^n\, dt \cdot \frac{1}{b-a} \int_a^x f^{(n)}(t)\, dt \right|$$

$$\leq \frac{1}{2} n \sup_{t \in [a,b]} |x-t|^{n-1} \cdot \frac{\left(\left[f^{(n-1)}; a, b \right] - \gamma_n \right) \left(\Gamma_n - \left[f^{(n-1)}; a, b \right] \right)}{\Gamma_n - \gamma_n} (b-a)$$

$$= \frac{1}{2} n \left[\frac{1}{2}(b-a) + \left| x - \frac{a+b}{2} \right| \right]^{n-1}$$

$$\times \frac{\left(\left[f^{(n-1)}; a, b \right] - \gamma_n \right) \left(\Gamma_n - \left[f^{(n-1)}; a, b \right] \right)}{\Gamma_n - \gamma_n} (b-a)$$

for any $x \in [a, b]$.

Further simplification gives

$$\left| \frac{1}{n!} \int_a^x (x-t)^n f^{(n)}(t)\, dt - \frac{(x-a)^{n+1} + (-1)^n (b-x)^{n+1}}{(n+1)!} \left[f^{(n-1)}; a, b \right] \right|$$

$$\leq \frac{1}{2(n-1)!} \left[\frac{1}{2}(b-a) + \left| x - \frac{a+b}{2} \right| \right]^{n-1} (b-a)^2$$

$$\times \frac{\left(\left[f^{(n-1)}; a, b \right] - \gamma_n \right) \left(\Gamma_n - \left[f^{(n-1)}; a, b \right] \right)}{\Gamma_n - \gamma_n}.$$

Using the representation (3.96) we deduce the desired result. □

It is natural to consider the following particular case.

With the above assumptions, we have

$$\int_a^b f(t)\, dt = \sum_{k=0}^{n-1} \frac{1}{(k+1)!} \left(\frac{b-a}{2} \right)^{k+1} \left[f^{(k)}(a) + (-1)^k f^{(k)}(b) \right]$$

$$+ \left(\frac{b-a}{2} \right)^n \frac{[1 + (-1)^n]}{(n+1)!} \left[f^{(n-1)}; b, a \right] + S_n(f) \quad (3.100)$$

and the remainder $S_n(f)$ satisfies the bound:

$$|S_n(f)| \leq \frac{1}{2^n (n-1)!} (b-a)^{n+1}$$

$$\times \frac{\left(\left[f^{(n-1)}; a, b \right] - \gamma_n \right) \left(\Gamma_n - \left[f^{(n-1)}; a, b \right] \right)}{\Gamma_n - \gamma_n}. \quad (3.101)$$

3.9   Ostrowski Type Inequality with End Interval Means

Let the functional $S(f; a, b)$ be defined by

$$S(f; a, b) = f(x) - \mathcal{M}(f; a, b), \tag{3.102}$$

where

$$\mathcal{M}(f; a, b) = \frac{1}{b-a} \int_a^b f(x)\, dx. \tag{3.103}$$

The functional $S(f; a, b)$ represents the deviation of $f(x)$ from its integral mean over $[a, b]$.

In 1938, Ostrowski proved the following integral inequality [147]:

Let $f : [a, b] \rightarrow \mathbb{R}$ be continuous on $[a, b]$ and differentiable on (a, b) and assume $|f'(x)| \leq M$ for all $x \in (a, b)$. Then the inequality

$$|S(f; a, b)| \leq \left[\left(\frac{b-a}{2} \right)^2 + \left(x - \frac{a+b}{2} \right)^2 \right] \frac{M}{b-a} \tag{3.104}$$

holds for all $x \in [a, b]$. The constant $\frac{1}{4}$ is best possible.

In a series of papers, Dragomir and Wang [111]–[114] proved (3.104) and other variants for $f' \in L_p[a, b]$ for $p \geq 1$, by making use of a Peano kernel approach and Montgomery's identity [142, p. 585]. Montgomery's identity states that for absolutely continuous mappings $f : [a, b] \rightarrow \mathbb{R}$

$$f(x) = \frac{1}{b-a} \int_a^b f(t)\, dt + \frac{1}{b-a} \int_a^b p(x, t) f'(t)\, dt, \tag{3.105}$$

where the kernel $p : [a, b]^2 \rightarrow \mathbb{R}$ is given by

$$p(x, t) = \begin{cases} t - a, & a \leq t \leq x \leq b, \\ t - b, & a \leq x < t \leq b. \end{cases}$$

If we assume that $f' \in L_\infty[a, b]$ and $\|f'\|_\infty := \operatorname*{ess\,sup}_{t \in [a,b]} |f'(t)|$, then M in (3.104) may be replaced by $\|f'\|_\infty$.

Dragomir and Wang [111]–[114] utilised an integration by parts argument, ostensibly Montgomery's identity (3.105), to obtain

$$|S(f; a, b)| \leq \begin{cases} \left[\left(\frac{b-a}{2} \right)^2 + \left(x - \frac{a+b}{2} \right)^2 \right] \frac{\|f'\|_\infty}{b-a}, & f' \in L_\infty[a, b]; \\[3mm] \left[\frac{(x-a)^{q+1} + (b-x)^{q+1}}{q+1} \right]^{\frac{1}{q}} \frac{\|f'\|_p}{b-a}, & f' \in L_p[a, b], \\ & p > 1, \frac{1}{p} + \frac{1}{q} = 1; \\[3mm] \left[\frac{b-a}{2} + \left| x - \frac{a+b}{2} \right| \right] \frac{\|f'\|_1}{b-a}, \end{cases} \tag{3.106}$$

where $f : [a, b] \to \mathbb{R}$ is absolutely continuous on $[a, b]$. The constants $\frac{1}{4}$, $\frac{1}{(q+1)^{\frac{1}{q}}}$ and $\frac{1}{2}$ are sharp.

The following identity was developed by Cerone [28] using integration by parts on separate intervals.

Let $f : [a, b] \to \mathbb{R}$ be an absolutely continuous mapping. Denote by $P(x, \cdot) :$ $[a, b] \to \mathbb{R}$ the kernel given by

$$P(x, t) = \begin{cases} \frac{\alpha}{\alpha+\beta} \left(\frac{t-a}{x-a} \right), & t \in [a, x] \\ \\ \frac{-\beta}{\alpha+\beta} \left(\frac{b-t}{b-x} \right), & t \in (x, b] \end{cases} \tag{3.107}$$

where $\alpha, \beta \in \mathbb{R}$ nonnegative and not both zero, then the identity

$$\int_a^b P(x, t) f'(t) \, dt$$

$$= f(x) - \frac{1}{\alpha + \beta} \left[\frac{\alpha}{x - a} \int_a^x f(t) \, dt + \frac{\beta}{b - x} \int_x^b f(t) \, dt \right] \tag{3.108}$$

holds.

The identity (3.108) was used to obtain the bounds by Cerone [28] given below.

Let $f : [a, b] \to \mathbb{R}$ be an absolutely continuous mapping and define

$$\mathcal{T}(x; \alpha, \beta) := f(x) - \frac{1}{\alpha + \beta} [\alpha \mathcal{M}(f; a, x) + \beta \mathcal{M}(f; x, b)], \tag{3.109}$$

where $\mathcal{M}(f; a, b)$ is the integral mean as defined by (3.103), then

$$|\mathcal{T}(x; \alpha, \beta)|$$

$$\leq \begin{cases} [\alpha(x-a) + \beta(b-x)] \frac{\|f'\|_\infty}{2(\alpha+\beta)}, & f' \in L_\infty[a, b]; \\ \\ [\alpha^q(x-a) + \beta^q(b-x)]^{\frac{1}{q}} \dfrac{\|f'\|_p}{(q+1)^{\frac{1}{q}}(\alpha+\beta)}, & f' \in L_p[a, b], \\ & p > 1, \frac{1}{p} + \frac{1}{q} = 1; \\ \\ \left[1 + \frac{|\alpha-\beta|}{\alpha+\beta} \right] \frac{\|f'\|_1}{2}, \end{cases} \tag{3.110}$$

where $\|h\|_p$ are the usual Lebesgue norms for $h \in L_p[a, b]$ with

$$\|h\|_\infty := \operatorname*{ess\ sup}_{t \in [a,b]} |h(t)| < \infty \quad and \quad \|h\|_p := \left(\int_a^b |h(t)|^p \, dt \right)^{\frac{1}{p}}, \quad 1 \leq p < \infty.$$

PROOF   By taking the modulus of (3.108) we have the following from (3.109) and (3.103):

$$|T(x;\alpha,\beta)| = \left| \int_a^b P(x,t) f'(t) dt \right| \leq \int_a^b |P(x,t)| |f'(t)| dt. \qquad (3.111)$$

Thus, for $f' \in L_\infty [a,b]$, (3.111) gives

$$|T(x;\alpha,\beta)| \leq \|f'\|_\infty \int_a^b |P(x,t)| dt.$$

A simple calculation using (3.107) gives

$$\int_a^b |P(x,t)| dt = \frac{\alpha}{\alpha+\beta} \int_a^x \frac{t-a}{x-a} dt + \frac{\beta}{\alpha+\beta} \int_x^b \frac{b-t}{b-x} dt$$

$$= \left[\frac{\alpha}{\alpha+\beta} (x-a) + \frac{\beta}{\alpha+\beta} (b-x) \right] \int_0^1 u \, du$$

and hence the first inequality results.

Further, using Hölder's integral inequality, we have for $f' \in L_p [a,b]$ from (3.111)

$$|T(x;\alpha,\beta)| \leq \|f'\|_p \left(\int_a^b |P(x,t)|^q dt \right)^{\frac{1}{q}},$$

where $\frac{1}{p} + \frac{1}{q} = 1$ with $p > 1$. Now

$$(\alpha+\beta) \left(\int_a^b |P(x,t)|^q dt \right)^{\frac{1}{q}} = \left[\alpha^q \int_a^x \left(\frac{t-a}{x-a} \right)^q dt + \beta^q \int_x^b \left(\frac{b-t}{b-x} \right)^q dt \right]^{\frac{1}{q}}$$

$$= [\alpha^q (x-a) + \beta^q (b-x)]^{\frac{1}{q}} \left(\int_0^1 u^q du \right)^{\frac{1}{q}}$$

and so the second inequality is obtained.

Finally, for $f' \in L_1 [a,b]$ we have the following from (3.111) and (3.107):

$$|T(x;\alpha,\beta)| \leq \sup_{t\in[a,b]} |P(x,t)| \|f'\|_1,$$

where

$$(\alpha+\beta) \sup_{t\in[a,b]} |P(x,t)| = \max\{\alpha,\beta\} = \frac{\alpha+\beta}{2} + \left| \frac{\alpha-\beta}{2} \right|.$$

This completes the proof.   ⬛

Comments
(a) It should be noted that from (3.109) and (3.102), we have

$$(\alpha + \beta) \, T \, (x; \alpha, \beta) = \alpha S \, (f; a, x) + \beta S \, (f; x, b) \, . \qquad (3.112)$$

It was shown by Cerone [28] that using the triangle inequality produces coarser bounds for $|(\alpha + \beta) \, T \, (x; \alpha, \beta)|$ than those in (3.104).

Cerone [28] further showed that (3.109) may be written in the equivalent form

$$T \, (x; \alpha, \beta) = f \, (x) - \left[\left(1 - \frac{\beta}{\alpha + \beta} \rho \, (x) \right) \mathcal{M} \, (f; a, x) + \frac{\beta}{\alpha + \beta} \rho \, (x) \, \mathcal{M} \, (f; a, b) \right],$$

where $\rho \, (x) = \frac{b-a}{b-x}$, so that for fixed $[a, b]$, $\mathcal{M} \, (f; a, b)$ is also fixed, which reduces the amount of work required for applications.

The result in (3.106) may be recaptured from specialisations of the above results by taking $\alpha = \beta$.

(b) Perturbed versions of the results may be obtained by using Grüss type results involving the Čebyšev functional

$$T \, (f, g) = \mathcal{M} \, (fg) - \mathcal{M} \, (f) \, \mathcal{M} \, (g) \qquad (3.113)$$

with $\mathcal{M} \, (f)$ being the integral mean of f over $[a, b]$, as defined in (3.103).

For $f, g : [a, b] \to \mathbb{R}$ and integrable on $[a, b]$, as is their product, then

$$
\begin{aligned}
|T \, (f, g)| &\leq T^{\frac{1}{2}} \, (f, f) \, T^{\frac{1}{2}} \, (g, g) \, , && \text{Dragomir [54]} \\
&&& \text{for } f, g \in L_2 \, [a, b] \, ; \\
&\leq \frac{\Gamma - \gamma}{2} T^{\frac{1}{2}} \, (f, f) \, , && \text{Matić et al. [135]} \\
&&& \text{for } \gamma \leq g \, (t) \leq \Gamma, \, t \in [a, b] \, , \qquad (3.114) \\
&\leq \frac{(\Gamma - \gamma)(\Phi - \phi)}{4} \, , && \text{Grüss (see Mitrinović, Pečarić,} \\
&&& \text{and Fink [141, pp. 295–310]),} \\
&&& \phi \leq f \leq \Phi, \, t \in [a, b] \, .
\end{aligned}
$$

Dragomir [54] obtains numerous results if either f, g or both are known, although the first inequality in (3.114) has a long history (see for example, Mitrinović, Pečarić, and Fink [141, pp. 295–310]. The inequalities in (3.114) when proceeding from top to bottom are in order of decreasing coarseness.

The following result is valid (see Cerone [28]).

Let $f : [a, b] \to \mathbb{R}$ *be an absolutely continuous mapping and* $\alpha \geq 0$, $\beta \geq 0$, $\alpha + \beta \neq 0$, *then*

$$
\begin{aligned}
\left| T \, (x, \alpha, \beta) - (x - \gamma) \frac{S}{2} \right| &\leq (b - a) \, \kappa \, (x) \left[\frac{1}{b-a} \|f'\|_2^2 - S^2 \right]^{\frac{1}{2}} \, , \\
&\qquad f' \in L_2 \, [a, b] \, ; \\
&\leq (b - a) \, \kappa \, (x) \, \frac{\Gamma - \gamma}{2} \, , \qquad (3.115) \\
&\qquad \gamma < f' \, (t) < \Gamma, \, t \in [a, b] \, ; \\
&\leq (b - a) \, \frac{\Gamma - \gamma}{4} \, ,
\end{aligned}
$$

where $T(x, \alpha, \beta)$ is as given by (3.109)

$$\gamma = \frac{\alpha a + \beta b}{\alpha + \beta}, \quad S = \frac{f(b) - f(a)}{b - a}, \tag{3.116}$$

$$\kappa^2(x) = \frac{1}{3} \left[\left(\frac{\alpha}{\alpha + \beta} \right)^2 (x - a) + \left(\frac{\beta}{\alpha + \beta} \right)^2 (b - x) \right] \tag{3.117}$$

$$- \left(\frac{x - \gamma}{2(b-a)} \right)^2.$$

PROOF (Sketch)   Associating $f(t)$ with $P(x, t)$ and $g(t)$ with $f'(t)$, then from (3.107) and (3.113) we obtain

$$T(P(x, \cdot), f'(\cdot)) = M(P(x, \cdot), f'(\cdot)) - M(P(x, \cdot)) M(f'(\cdot))$$

and so, on using identity (3.108),

$$(b - a) T(P(x, \cdot), f'(\cdot)) = T(x, \alpha, \beta) - (b - a) M(P(x, \cdot)) S \tag{3.118}$$

where S is the secant slope of f over $[a, b]$ as given in (3.116). The reader is referred to Cerone [28] for the details of the complete proof.   □

(**c**) The following is an application to the cumulative distribution function. Let X be a random variable taking values in the finite interval $[a, b]$ with cumulative distribution function $F(x) = \Pr(X \leq x) = \int_a^x f(u)\,du$, where f is a probability density function (p.d.f.). The following result holds [28].

Let X and F be as above. Then

$$|(\alpha(b - x) - \beta(x - a)) F(x) - (x - a)[(\alpha + \beta)(b - x) f(x) - \beta]| \tag{3.119}$$

$$\leq \begin{cases} (b - x)(x - a)[\alpha(x - a) + \beta(b - x)] \cdot \dfrac{\|f'\|_\infty}{2}, & f' \in L_\infty[a, b]; \\[2ex] (b - x)(x - a)[\alpha^q(x - a) + \beta^q(b - x)]^{\frac{1}{q}} \cdot \dfrac{\|f'\|_p}{(q+1)^{\frac{1}{q}}}, & f' \in L_p[a, b], \\ & p > 1, \frac{1}{p} + \frac{1}{q} = 1; \\[2ex] (b - x)(x - a)[\alpha + \beta + |\alpha - \beta|] \cdot \dfrac{\|f'\|_1}{2}, & f' \in L_1[a, b]. \end{cases}$$

The specialisation of $\alpha = \beta = \frac{1}{2}$ produces the following result.

Let X be a random variable, $F(x)$ a cumulative distribution function, and $f(x)$ the probability density function. Then

$$\left| \left(\frac{a+b}{2} - x \right) F(x) - (x - a) \left[(b - x) f(x) - \frac{1}{2} \right] \right|$$

$$\leq \begin{cases} (b-x)(x-a)(b-a) \cdot \dfrac{\|f'\|_\infty}{2}, & f' \in L_\infty[a,b]; \\[2ex] (b-x)(x-a)(b-a)^{\frac{1}{q}} \cdot \dfrac{\|f'\|_p}{2(q+1)^{\frac{1}{q}}}, & f' \in L_p[a,b], \\[1ex] & p>1, \ \frac{1}{p}+\frac{1}{q}=1; \\[2ex] (b-x)(x-a) \cdot \dfrac{\|f'\|_1}{2}, & f' \in L_1[a,b]. \end{cases} \qquad (3.120)$$

The above results allow the approximation of $F(x)$ in terms of $f(x)$. The approximation of $R(x) = 1 - F(x)$ could also be obtained by a simple substitution. $R(x)$ is of importance in reliability theory where $f(x)$ is the p.d.f. of failure.

We may take directly from (3.109) and (3.110) $\beta = 0$, by assuming that $\alpha \neq 0$, to give

$$|F(x) - (x-a)f(x)| \leq \begin{cases} \dfrac{(x-a)^2}{2} \|f'\|_\infty, & f' \in L_\infty[a,b]; \\[2ex] (x-a)^{1+\frac{1}{q}} \cdot \dfrac{\|f'\|_p}{(q+1)^{\frac{1}{q}}}, & f' \in L_p[a,b], \\[1ex] & p>1, \ \frac{1}{p}+\frac{1}{q}=1; \\[2ex] (x-a)\|f'\|_1, & f' \in L_1[a,b], \end{cases} \qquad (3.121)$$

which agrees with (3.106) for $|S(f;a,x)|$.

The perturbed results of Comment (b) could also be applied here; however, this will not be pursued further.

We may replace f by F in any of the Equations (3.119)–(3.121) so that the bounds are in terms of $\|f\|_p$, $p \geq 1$. Further, we note that

$$\int_a^b F(u)\, du = b - E[X].$$

3.10 Multidimensional Integration via Ostrowski Dimension Reduction

For $f : [a,b] \to \mathbb{R}$ we define the Ostrowski functional by

$$S(f;c,x,d) := f(x) - \mathcal{M}(f;c,d), \qquad (3.122)$$

where

$$\mathcal{M}(f;c,d) := \frac{1}{d-c} \int_c^d f(u)\, du, \quad \text{the integral mean.} \qquad (3.123)$$

The following identity may be easily shown to hold, for f of bounded variation, by an integration by parts argument of the Riemann-Stieltjes integrals, and so

$$S\left(f;c,x,d\right) = \int_{c}^{d} p\left(x,t,c,d\right) df\left(t\right),\qquad(3.124)$$

$$p\left(x,t,c,d\right) = \begin{cases} \dfrac{t-c}{d-c}, & t \in [c,x] \\ \dfrac{t-d}{d-c}, & t \in (x,d]. \end{cases}$$

Further, if $f\left(t\right)$ is assumed to be absolutely continuous for t over its respective interval, then $df\left(t\right) = f'\left(t\right)dt$ and the Riemann-Stieltjes integrals in (3.124) is equivalent to a Riemann integral. In this instance, the corresponding identity to (3.124) is known as Montgomery's identity (see Cerone and Dragomir [37]).

In Cerone [20], Ostrowski type results were procured for multidimensional integrals using an iterative approach from the one-dimensional result as a *seed* or *generator*.

The following result uses an iterative approach to extend the Ostrowski functional identity to multidimensions. First, we require some notation.

Let $I^n = \prod_{i=1}^{n} [a_i,b_i] = [a_1,b_1]\times[a_2,b_2]\times\cdots\times[a_n,b_n]$. Further, let $f : I^n \to \mathbb{R}$ and define operators $F_i\left(f\right)$ and $\lambda_i\left(f\right)$ by

$$F_i\left(f\right) := f\left(t_1,\ldots,t_{i-1},x_i,t_{i+1},\ldots,t_n\right) \text{ where } x_i \in [a_i,b_i]\qquad(3.125)$$

and

$$\lambda_i\left(f\right) := \frac{1}{d_i}\int_{a_i}^{b_i} f\left(t_1,\ldots,t_{i-1},t_i,t_{i+1},\ldots,t_n\right)dt_i.\qquad(3.126)$$

This means that $F_i\left(f\right)$ evaluates $f\left(\cdot\right)$ in the ith variable at $x_i \in [a_i,b_i]$ and $\lambda_i\left(f\right)$ is the integral mean of $f\left(\cdot\right)$ in the ith variable. Assuming that $f\left(\cdot\right)$ is absolutely continuous in the ith variable $t_i \in [a_i,b_i]$, we have

$$\mathcal{L}_i\left(f\right) := \frac{1}{d_i}\int_{a_i}^{b_i} p_i\left(x_i,t_i\right)\frac{\partial f}{\partial t_i}dt_i = \left(F_i - \lambda_i\right)\left(f\right),\qquad(3.127)$$

for $i = 1,2,\ldots,n$, where

$$\frac{p_i\left(x_i,t_i\right)}{d_i} = \begin{cases} \dfrac{t_i-a_i}{b_i-a_i}, & t_i \in [a_i,x_i] \\ \dfrac{t_i-b_i}{b_i-a_i}, & t_i \in (x_i,b_i], \end{cases}\qquad(3.128)$$

and $d_i = b_i - a_i$.

Thus (3.127) and (3.128) are ostensibly equivalent to identity (3.124) for absolutely continuous $f(t_1, \ldots, t_{i-1}, t_i, t_{i+1}, \ldots, t_n)$ where $t_i \in [a_i, b_i]$.

The following result was obtained by Cerone [20]:

Let $f : I^n \to \mathbb{R}$ be absolutely continuous in such a manner that the partial derivatives of order one with respect to every variable exist. Then

$$E_n(f)$$

$$= f(x_1, x_2, \ldots, x_n) - \sum_{i=1}^{n} \frac{1}{d_i} \int_{a_i}^{b_i} f(x_1, x_2, \ldots, x_{i-1}, t_i, x_{i+1}, \ldots, x_n) \, dt_i$$

$$+ \sum_{i<j}^{n} \frac{1}{d_i d_j} \int_{a_j}^{b_j} \int_{a_i}^{b_i} f(x_1, \ldots, x_{i-1}, t_i, x_{i+1}, \ldots, t_j, \ldots, x_n) \, dt_i dt_j$$

$$- \cdots\cdots - \frac{(-1)^n}{D_n} \int_{a_n}^{b_n} \cdots \int_{a_i}^{b_i} f(t_1, \ldots, t_n) \, dt_1 \ldots dt_n \quad (3.129)$$

where

$$E_n(f) := \frac{1}{D_n} \int_{a_n}^{b_n} \cdots \int_{a_1}^{b_1} \prod_{i=1}^{n} p_i(x_i, t_i) \frac{\partial^n f}{\partial t_n \ldots \partial t_1} \, dt_1 \ldots dt_n, \quad (3.130)$$

$$D_n = \prod_{i=1}^{n} d_i, \quad d_i = b_i - a_i, \quad (3.131)$$

and $p_i(x_i, t_i)$ is given by (3.128).

PROOF Define $E_r(f)$ by [20]

$$E_r(f) = \left(\prod_{i=1}^{r} \mathcal{L}_i \right)(f). \quad (3.132)$$

Then, from the left identity in (3.127), $E_n(f)$ is as given by (3.130). Further,

$$E_r(f) = \mathcal{L}_r(E_{r-1}(f)), \quad \text{for } r = 1, 2, \ldots, n \quad (3.133)$$

where $E_0(f) = f$.

Now, from (3.132),

$$E_1(f) = \mathcal{L}_1(f) = (F_1 - \lambda_1)(f),$$

which is the Montgomery identity for $t_1, x_1 \in [a_1, b_1]$:

$$E_1(f) = \frac{1}{d_1} \int_{a_1}^{b_1} p_1(x_1, t_1) \frac{\partial f}{\partial t_1}(t_1, t_2, \ldots, t_n) \, dt_1 \quad (3.134)$$

$$= f(x_1, t_2, \ldots, t_n) - \frac{1}{d_1} \int_{a_1}^{b_1} f(t_1, t_2, \ldots, t_n) \, dt_1.$$

Further,

$$E_2 (f)$$
$$= \mathcal{L}_2 (E_1 (f)) = (F_2 - \lambda_2) (E_1 (f))$$
$$= F_2 (E_1 (f)) - \lambda_2 (E_1 (f))$$
$$= f (x_1, x_2, t_3, \ldots, t_n) - \frac{1}{d_1} \int_{a_1}^{b_1} f (t_1, x_2, t_3, \ldots, t_n) \, dt_1$$
$$- \frac{1}{d_2} \int_{a_2}^{b_2} \left[f (x_1, t_2, \ldots, t_n) - \frac{1}{d_1} \int_{a_1}^{b_1} f (t_1, t_2, \ldots, t_n) \, dt_1 \right] dt_2$$
$$= f (x_1, x_2, t_3, \ldots, t_n) - \frac{1}{d_1} \int_{a_1}^{b_1} f (t_1, x_2, t_3, \ldots, t_n) \, dt_1$$
$$- \frac{1}{d_2} \int_{a_2}^{b_2} f (t_1, t_2, t_3, \ldots, t_n) \, dt_2 + \frac{1}{d_1 d_2} \int_{a_2}^{b_2} \int_{a_1}^{b_1} f (t_1, t_2, \ldots, t_n) \, dt_1 dt_2.$$

Continuing in this manner until $r = n$ gives the result as stated in (3.129). □

The result given by (3.129) may be utilised to approximate the n-dimensional integral in terms of lower dimensional integrals and a function evaluation $f (x_1, x_2, \ldots, x_n)$ where $x_i \in [b_i, a_i]$, $i = 1, 2, \ldots, n$. Specifically, there are $\binom{n}{0}$ function evaluations, $\binom{n}{1}$ single integral evaluations, and so forth, in each of the axes, $\binom{n}{2}$ double integral evaluations and so on, and, of course, $\binom{n}{n}$ n-dimensional integral evaluations. This results from the fact that

$$E_n (f) = \left(\prod_{i=1}^{n} \mathcal{L}_i \right) (f) = \left(\prod_{i=1}^{n} (F_i - \lambda_i) \right) (f), \tag{3.135}$$

on using (3.132) and (3.125)–(3.127).

It is subsequently demonstrated that the above procedure of utilising a one-dimensional identity as the *seed* or *generator* to recursively obtain a multidimensional identity may be extended to other seed identities.

In the following result, bounds for $T_n \left(\underset{\sim}{a}, \underset{\sim}{x}, \underset{\sim}{b} \right)$ are obtained where

$$T_n \left(\underset{\sim}{a}, \underset{\sim}{x}, \underset{\sim}{b} \right)$$
$$= f (x_1, x_2, \ldots, x_n) - \sum_{i=1}^{n} \frac{1}{d_i} \int_{a_i}^{b_i} f (x_1, x_2, \ldots, x_{i-1}, t_i, x_{i+1}, \ldots, x_n) \, dt_i$$
$$+ \sum_{i<j}^{n} \frac{1}{d_j d_i} \int_{a_j}^{b_j} \int_{a_i}^{b_i} f (x_1, \ldots, x_{i-1}, t_i, x_{i+1}, \ldots, x_{j-1}, t_j, x_{j+1}, \ldots, x_n) \, dt_i dt_j$$
$$- \cdots - \frac{(-1)^n}{D_n} \int_{a_n}^{b_n} \cdots \int_{a_1}^{b_1} f (t_1, \ldots, t_n) \, dt_1 \ldots dt_n \tag{3.136}$$

and $\underset{\sim}{z} = (z_1, z_2, \ldots, z_n)$ [20].

Let $f : I^n \rightarrow \mathbb{R}$ be absolutely continuous in such a manner that the partial derivatives of order one with respect to every variable exist. Then

$$\left| T_n \left(\underset{\sim}{a}, \underset{\sim}{x}, \underset{\sim}{b} \right) \right| \tag{3.137}$$

$$\leq \begin{cases} \prod_{i=1}^{n} P_i(1) \left\| \dfrac{\partial^n f}{\partial t_n \ldots \partial t_1} \right\|_{\infty}, & \dfrac{\partial^n f}{\partial t_n \ldots \partial t_1} \in L_{\infty}[I^n]; \\[2em] \left(\prod_{i=1}^{n} P_i(q) \right)^{\frac{1}{q}} \left\| \dfrac{\partial^n f}{\partial t_n \ldots \partial t_1} \right\|_p, & \dfrac{\partial^n f}{\partial t_n \ldots \partial t_1} \in L_p[I^n], \\ & p > 1, \ \frac{1}{p} + \frac{1}{q} = 1; \\[2em] \prod_{i=1}^{n} \theta_i \left\| \dfrac{\partial^n f}{\partial t_n \ldots \partial t_1} \right\|_1, & \dfrac{\partial^n f}{\partial t_n \ldots \partial t_1} \in L_1[I^n], \end{cases}$$

where $T_n \left(\underset{\sim}{a}, \underset{\sim}{x}, \underset{\sim}{b} \right)$ is as defined by (3.136),

$$(q+1) P_i(q) = (x_i - a_i)^{q+1} + (b_i - x_i)^{q+1}, \tag{3.138}$$

$$\theta_i = \frac{b_i - a_i}{2} + \left| x_i - \frac{a_i + b_i}{2} \right|. \tag{3.139}$$

PROOF (Follows Cerone [20]; see also Cerone [17]). From (3.130) and (3.136), we obtain

$$\left| T_n \left(\underset{\sim}{a}, \underset{\sim}{x}, \underset{\sim}{b} \right) \right|$$
$$= |E_n(f)| \leq \frac{1}{D_n} \int_{a_n}^{b_n} \cdots \int_{a_1}^{b_1} \left| \prod_{i=1}^{n} p_i(x_i, t_i) \frac{\partial^n f}{\partial t_n \ldots \partial t_1} \right| dt_1 \ldots dt_n. \tag{3.140}$$

Now, for $\frac{\partial^n f}{\partial t_n \ldots \partial t_1} \in L_{\infty}[I^n]$, we have

$$D_n |E_n(f)| \tag{3.141}$$

$$\leq \int_{a_n}^{b_n} \cdots \int_{a_1}^{b_1} \left| \prod_{i=1}^{n} p_i(x_i, t_i) \right| dt_1 \ldots dt_n \left\| \frac{\partial^n f}{\partial t_n \ldots \partial t_1} \right\|_{\infty}$$

$$= \prod_{i=1}^{n} \int_{a_i}^{b_i} |p_i(x_i, t_i)| \, dt_i \left\| \frac{\partial^n f}{\partial t_n \ldots \partial t_1} \right\|_{\infty}$$

$$= \prod_{i=1}^{n} \left[\int_{a_i}^{x_i} (t_i - a_i)\, dt_i + \int_{x_i}^{b_i} (b_i - t_i)\, dt_i \right] \left\| \frac{\partial^n f}{\partial t_n \ldots \partial t_1} \right\|_{\infty}$$

$$= \frac{1}{2^n} \prod_{i=1}^{n} \left[(x_i - a_i)^2 + (b_i - x_i)^2 \right] \left\| \frac{\partial^n f}{\partial t_n \ldots \partial t_1} \right\|_{\infty}.$$

Hence combining (3.140) and (3.141) gives the first inequality of (3.137).

Further, using the Hölder inequality, we have for $\frac{\partial^n f}{\partial t_n \ldots \partial t_1} \in L_p[I^n]$, $1 \le p < \infty$,

$$D_n \left| E_n(f) \right| \le \left(\int_{a_n}^{b_n} \cdots \int_{a_1}^{b_1} \left| \prod_{i=1}^{n} p_i(x_i, t_i) \right|^q dt_1 \ldots dt_n \right)^{\frac{1}{q}} \left\| \frac{\partial^n f}{\partial t_n \ldots \partial t_1} \right\|_{p},$$

where

$$\int_{a_n}^{b_n} \cdots \int_{a_1}^{b_1} \left| \prod_{i=1}^{n} p_i(x_i, t_i) \right|^q dt_1 \ldots dt_n$$

$$= \prod_{i=1}^{n} \int_{a_i}^{b_i} |p_i(x_i, t_i)|^q dt_i$$

$$= \prod_{i=1}^{n} \left[\int_{a_i}^{x_i} (t_i - a_i)^q dt_i + \int_{x_i}^{b_i} (b_i - t_i)^q dt_i \right]$$

$$= \frac{1}{(q+1)^n} \prod_{i=1}^{n} \left[(x_i - a_i)^{q+1} + (b_i - x_i)^{q+1} \right]$$

and so the second inequality is valid on noting (3.138).

The final inequality in (3.137) is obtained from (3.140) for $\frac{\partial^n f}{\partial t_n \ldots \partial t_1} \in L_1[I^n]$, giving

$$D_n \left| E_n(f) \right| \le \sup_{\underset{\sim}{t} \in \left[\underset{\sim}{a}, \underset{\sim}{b}\right]} \left| \prod_{i=1}^{n} p_i(x_i, t_i) \right| \int_{a_n}^{b_n} \cdots \int_{a_1}^{b_1} \left| \frac{\partial^n f}{\partial t_n \ldots \partial t_1} \right| dt_1 \ldots dt_n$$

$$= \prod_{i=1}^{n} \sup_{t_i \in [a_i, b_i]} |p_i(x_i, t_i)| \left\| \frac{\partial f^n}{\partial t_n \ldots \partial t_1} \right\|_{1}$$

$$= \prod_{i=1}^{n} \max\{x_i - a_i, b_i - x_i\} \left\| \frac{\partial f^n}{\partial t_n \ldots \partial t_1} \right\|_{1}.$$

On noting that $\max\{X, Y\} = \frac{X+Y}{2} + \frac{|X-Y|}{2}$ readily produces the stated result.

☐

Comments

(a) The expression for $T_n \left(\underset{\sim}{a}, \underset{\sim}{x}, \underset{\sim}{b} \right)$ may be written in a less explicit form which is perhaps more appealing. Namely,

$$T_n \left(\underset{\sim}{a}, \underset{\sim}{x}, \underset{\sim}{b} \right) = f(x_1, x_2, \dots, x_n) + \sum_{k=1}^{n-1} (-1)^k \sum_k \mathcal{M}_k + (-1)^n \mathcal{M}_n, \quad (3.142)$$

where $\mathcal{M}_k$ represents the integral means in k variables with the remainder being evaluated at their respective interior point and $\sum_k \mathcal{M}_k$ is a sum over all $\binom{n}{k}$, k-dimensional integral means. Here

$$\mathcal{M}_n = \frac{1}{D_n} \int_{a_n}^{b_n} \cdots \int_{a_1}^{b_1} f(t_1, \dots, t_n) \, dt_1 \dots dt_n$$

and

$$\sum_1 \mathcal{M}_1 = \frac{1}{d_1} \int_{a_1}^{b_1} f(t_1, x_2, \dots, x_n) \, dt_1 + \frac{1}{d_2} \int_{a_2}^{b_2} f(x_1, t_2, x_3, \dots, x_n) \, dt_2$$

$$+ \cdots + \frac{1}{d_n} \int_{a_n}^{b_n} f(x_1, x_2, \dots, x_{n-1}, t_n) \, dt_n.$$

(b) Taking $\alpha_i = \frac{a_i + b_i}{2}$ above gives a midpoint type rule.
(c) Perturbed rules via the Čebyšev functional were obtained by Cerone [20].
(d) Other rules may be utilised as the seed or generator of multidimensional integrals in an iterative fashion as shown by Cerone [20].

3.11 Multidimensional Integration via Trapezoid and Three Point Generators with Dimension Reduction

For $f : [a, b] \to \mathbb{R}$ we define the Ostrowski and trapezoidal functionals by

$$S(f; c, x, d) := f(x) - \mathcal{M}(f; c, d) \quad (3.143)$$

and

$$T(f; c, x, d) := \left(\frac{x - c}{d - c} \right) f(c) + \left(\frac{d - x}{d - c} \right) f(d) - \mathcal{M}(f; c, d), \quad (3.144)$$

respectively, where

$$\mathcal{M}(f; c, d) := \frac{1}{d - c} \int_c^d f(u) \, du, \text{ the integral mean.} \quad (3.145)$$

The following identities may be easily shown to hold for f of bounded variation, by an integration by parts argument of the Riemann-Stieltjes integrals and so

$$S(f; c, x, d) = \int_c^d p(x, t, c, d) \, df(t), \quad p(x, t, c, d) = \begin{cases} \frac{t-c}{d-c}, & t \in [c, x] \\ \frac{t-d}{d-c}, & t \in (x, d] \end{cases} \tag{3.146}$$

and

$$T(f; c, x, d) = \int_c^d q(x, t, c, d) \, df(t), \quad q(x, t, c, d) = \frac{t-x}{d-c}, \quad x, t \in [c, d]. \tag{3.147}$$

Dragomir and Rassias's book *Ostrowski Type Inequalities and Applications in Numerical Integration* [109] is devoted to Ostrowski type results involving (3.143) and numerous generalisations (see also Cerone [23]).

Further, define the three point functional $\mathfrak{T}(f; a, \alpha, x, \beta, b)$ which involves the difference between the integral mean and a weighted combination of a function evaluated at the end points and an interior point. Namely, for $a \leq \alpha < x < \beta \leq b$,

$$\mathfrak{T}(f; a, \alpha, x, \beta, b)$$
$$:= \left(\frac{\alpha - a}{b - a} \right) f(a) + \left(\frac{\beta - \alpha}{b - a} \right) f(x) + \left(\frac{b - \beta}{b - a} \right) f(b) - \mathcal{M}(f; a, b). \tag{3.148}$$

Cerone and Dragomir [37] showed that for f of bounded variation, the identity

$$\mathfrak{T}(f; a, \alpha, x, \beta, b) = \int_a^b r(x, t) \, df(t), \quad r(x, t) = \begin{cases} \frac{t-\alpha}{b-a}, & t \in [a, x] \\ \frac{t-\beta}{b-a}, & t \in (x, b] \end{cases} \tag{3.149}$$

is valid. They effectively demonstrated that the Ostrowski functional and the trapezoid functional could be recaptured as particular instances. Specifically, from (3.148) and (3.149), we have

$$S(f; a, x, b) = \mathfrak{T}(f; a, a, x, b, b)$$

and

$$T(f; a, x, b) = \mathfrak{T}(f; a, x, x, x, b),$$

where $S(f; a, x, b)$ and $T(f; a, x, b)$ are defined by (3.143) and (3.144) and satisfy identities (3.146) and (3.147), respectively.

It should be noted at this stage that

$$(b - a) \mathfrak{T}\left(f; a, \frac{5a + b}{6}, \frac{a + b}{2}, \frac{a + 5b}{6}, b \right)$$
$$= \frac{b - a}{6} \left[f(a) + 4f\left(\frac{a + b}{2} \right) + f(b) \right] - \int_a^b f(x) \, dx$$

is the Simpson functional.

Further, if $f(t)$ is assumed to be absolutely continuous for t over its respective interval, then $df(t) = f'(t)\, dt$ and the Riemann-Stieltjes integrals in (3.148) and (3.149) are equivalent to Riemann integrals.

In the current work, the generalised trapezoidal (3.146) and (3.147), and three point identities and (3.148) and (3.149) for absolutely continuous functions are used as generators to produce identities involving multidimensional integrals in terms of lower dimensional integrals and function evaluations. These are used to procure bounds for $\frac{\partial^n f}{\partial t_n \cdots \partial t_1} \in L_p[I^n]$, $1 \le p \le \infty$, where $I^n = [a_1, b_1] \times \cdots \times [a_n, b_n]$. Here for $h : I^n \to \mathbb{R}$ we mean by $h \in L_p[I^n]$, the Lebesgue norms, that is,

$$\|h\|_p := \left(\int_{a_n}^{b_n} \cdots \int_{a_1}^{b_1} |h(t_1, t_2, \ldots, t_n)|^p \, dt_1 \ldots dt_n \right)^{\frac{1}{p}},$$

$$1 \le p < \infty, \text{ for } h \in L_p[I^n], \quad (3.150)$$

and

$$\|h\|_\infty := ess \sup_{\underset{\sim}{t} \in I^n} |h(t_1, t_2, \ldots, t_n)|, \text{ for } h \in L_\infty[I^n], \quad (3.151)$$

where $\underset{\sim}{t} = (t_1, t_2, \ldots, t_n)$ and $t_i \in [a_i, b_i]$, $i = 1, 2, \ldots, n$.

The methodology of Cerone [20] (see also Section 3.10) is used for the current work. That work turns out to be a particular case of the three point development starting from Equation (3.171). The generalised trapezoidal results below are also specialisations of the three point results.

We follow the presentation by Cerone [22]. The generalised trapezoidal identity (3.147) with (3.144) is used as the *generator* of a higher dimensional result. We restrict the current work to absolutely continuous functions so that the Riemann integral identity corresponding to (3.147) is used. Let $f : I^n \to \mathbb{R}$ and define the operator

$$G_i(f) := \frac{A_i}{d_i} f(t_1, \ldots, t_{i-1}, a_i, t_{i+1}, \ldots, t_n)$$

$$+ \frac{B_i}{d_i} f(t_1, \ldots, t_{i-1}, b_i, t_{i+1}, \ldots, t_n) \quad (3.152)$$

and

$$\lambda_i(f) := \frac{1}{d_i} \int_{a_i}^{b_i} f(t_1, \ldots, t_{i-1}, t_i, t_{i+1}, \ldots, t_n) \, dt_i \quad (3.153)$$

where

$$A_i = x_i - a_i, \quad B_i = b_i - x_i, \quad d_i = b_i - a_i. \quad (3.154)$$

Here $d_i G_i(f)$ represents the generalised trapezoid in the ith variable, giving the standard trapezoid when $x_i = \frac{a_i + b_i}{2}$.

We note that $A_i + B_i = d_i$ and if we extend the notation to

$$\tilde{A}_i(x_i) = x_i - a_i, \text{ and } \tilde{B}_i(x_i) = b_i - x_i,$$

then we see that

$$\tilde{A}_i(x_i) = \begin{cases} 0, & x_i = a_i \\ d_i, & x_i = b_i \end{cases} \text{ and } \tilde{B}_i(x_i) = \begin{cases} d_i, & x_i = a_i \\ 0, & x_i = b_i \end{cases}.$$

Now, for $f(\cdot)$ absolutely continuous in the ith variable $t_i \in [a_i, b_i]$ we have

$$\mathfrak{M}_i(f) = \frac{1}{d_i} \int_{a_i}^{b_i} q_i(x_i, t_i) \frac{\partial f}{\partial t_i} dt_i = (G_i - \lambda_i)(f), \quad i = 1, 2, \ldots, n, \quad (3.155)$$

where $G_i(f)$ and $\lambda_i(f)$ are as given by (3.152)–(3.154), and

$$\frac{q_i(x_i, t_i)}{d_i} = \frac{t_i - x_i}{b_i - a_i}, \quad x_i, t_i \in [a_i, b_i]. \tag{3.156}$$

Let $c^{(0)} = (c_1, c_2, \ldots, c_n)$, where $c_i = a_i$ or b_i in the ith position for $i = 1, 2, \ldots, n$. Also, let $\sigma_0(c^{(0)})$ be the set of all such vectors which consists of 2^n possibilities. Further, let

$$\chi_k = \prod_{j=1}^{n}{}^{(k)} \frac{C_j}{d_j}, \quad k = 0, 1, \ldots, n, \tag{3.157}$$

where $C_j = A_j$ or B_j with the exception that k of the $C_j = d_j$ and so $\frac{C_j}{d_j} = 1$.

In a similar fashion, let $c^{(k)}$ be a vector taking on the fixed values a_i or b_i in the ith position except for k of the positions which are variable, $t_\bullet$. Let $\mathcal{M}_k$ be k-dimensional integral means for $f(c^{(k)})$. Here $c^{(k)} \in \sigma_k(c^{(k)})$, the set of all such elements, of which there are $\binom{n}{k} 2^{n-k}$.

With the above notation in place, the following identity holds [22].

Let $f : I^n \to \mathbb{R}$ be absolutely continuous and be such that all partial derivatives of order one in each of the variables exist. Then

$$R_n(f) = \sum_0 \chi_0 f\left(c^{(0)}\right) - \sum_1 \chi_1 \mathcal{M}_1 + \sum_2 \chi_2 \mathcal{M}_2 \tag{3.158}$$

$$- \cdots - (-1)^{n-1} \sum_{n-1} \chi_{n-1} \mathcal{M}_{n-1} + (-1)^n \mathcal{M}_n$$

$$:= \rho_n\left(\underset{\sim}{a}, \underset{\sim}{x}, \underset{\sim}{b}\right),$$

where, χ_k is as defined in (3.157), $\mathcal{M}_k$ is the k-dimensional integral mean for $f(c^{(k)})$, specifically,

$$\mathcal{M}_n = \frac{1}{D_n} \int_{a_n}^{b_n} \cdots \int_{a_1}^{b_1} f(t_1, t_2, \ldots, t_n) \, dt_1 \cdots dt_n,$$

where

$$D_n = \prod_{i=1}^{n} d_i, \tag{3.159}$$

and $\sum_k$ is a sum involving each of the elements of $\sigma_k\left(c^{(k)}\right)$ of which there are $\binom{n}{k} 2^{n-k}$ terms. Further,

$$R_n(f) = \frac{1}{D_n} \int_{a_n}^{b_n} \cdots \int_{a_1}^{b_1} \prod_{i=1}^{n} q_i(x_i, t_i) \frac{\partial^n f}{\partial t_n \cdots \partial t_1} dt_1 \cdots dt_n, \tag{3.160}$$

and $q_i(x_i, t_i)$ is given by (3.156), D_n by (3.159). Here, c_i is equal to either a_i or b_i in which case $C_i = A_i$ or B_i.

PROOF (Follows from Cerone [22]). Let $R_r(f)$ be defined by

$$R_r(f) := \left(\prod_{i=1}^{r} \mathfrak{M}_i \right)(f), \tag{3.161}$$

then from the left identity in (3.155), $R_n(f)$ is as given by (3.160). Now,

$$R_r(f) = \mathfrak{M}_r(R_{r-1}(f)), \text{ for } r = 1, 2, \ldots, n, \tag{3.162}$$

where $R_0(f) = f$.
  Thus, from (3.161)

$$R_1(f) = \mathfrak{M}_1(f) = (G_1 - \lambda_1)(f),$$

which is the generalised trapezoidal identity for $t_1, x_1 \in [a_1, b_1]$,

$$R_1(f) = \frac{1}{d_1} \int_{a_1}^{b_1} q_1(x_1, t_1) \frac{\partial f}{\partial t_1} (t_1, t_2, \ldots, t_n) dt_1 \tag{3.163}$$

$$= \frac{A_1}{d_1} f(a_1, t_2, \ldots, t_n) + \frac{B_1}{d_1} f(b_1, t_2, \ldots, t_n)$$

$$- \frac{1}{d_1} \int_{a_1}^{b_1} f(t_1, t_2, \ldots, t_n) dt_1$$

contains three entities; two function evaluations and one integral. Further,

$$R_2(f) = \mathfrak{M}_2(R_1(f)) = (G_2 - \lambda_2)(R_1(f))$$
$$= G_2(R_1(f)) - \lambda_2(R_1(f))$$
$$= \frac{A_2}{d_2} R_1(f) \Big|_{t_2=a_2} + \frac{B_2}{d_2} R_1(f) \Big|_{t_2=b_2} - \frac{1}{d_2} \int_{a_2}^{b_2} R_1(f) dt_2$$

contains nine entities. Thus,

$$R_2 (f) \tag{3.164}$$

$$= \frac{A_2}{d_2} \left\{ \frac{A_1}{d_1} f(a_1, a_2, t_3, \ldots, t_n) + \frac{B_1}{d_1} f(b_1, a_2, t_3, \ldots, t_n) \right.$$

$$\left. - \frac{1}{d_1} \int_{a_1}^{b_1} f(t_1, a_2, t_3, \ldots, t_n)\, dt_1 \right\}$$

$$+ \frac{B_2}{d_2} \left\{ \frac{A_1}{d_1} f(a_1, b_2, t_3, \ldots, t_n) + \frac{B_1}{d_1} f(b_1, b_2, t_3, \ldots, t_n) \right.$$

$$\left. - \frac{1}{d_1} \int_{a_1}^{b_1} f(t_1, b_2, t_3, \ldots, t_n)\, dt_1 \right\}$$

$$- \frac{1}{d_2} \int_{a_2}^{b_2} \left\{ \frac{A_1}{d_1} f(a_1, t_2, \ldots, t_n) + \frac{B_1}{d_1} f(b_1, t_2, t_3, \ldots, t_n) \right.$$

$$\left. - \frac{1}{d_1} \int_{a_1}^{b_1} f(t_1, t_2, \ldots, t_n)\, dt_1 \right\} dt_2$$

$$= \frac{A_2}{d_2} \frac{A_1}{d_1} f(a_1, a_2, t_3, \ldots, t_n) + \frac{A_2}{d_2} \frac{B_1}{d_1} f(b_1, a_2, t_3, \ldots, t_n)$$

$$+ \frac{B_2}{d_2} \frac{A_1}{d_1} f(a_1, b_2, t_3, \ldots, t_n) + \frac{B_2}{d_2} \frac{B_1}{d_1} f(b_1, b_2, t_3, \ldots, t_n)$$

$$- \frac{A_2}{d_2} \cdot \frac{1}{d_1} \int_{a_1}^{b_1} f(t_1, a_2, t_3, \ldots, t_n)\, dt_1$$

$$- \frac{B_2}{d_2} \cdot \frac{1}{d_1} \int_{a_1}^{b_1} f(t_1, b_2, t_3, \ldots, t_n)\, dt_1$$

$$- \frac{A_1}{d_1} \cdot \frac{1}{d_2} \int_{a_2}^{b_2} f(a_1, t_2, \ldots, t_n)\, dt_2$$

$$- \frac{B_1}{d_1} \cdot \frac{1}{d_2} \int_{a_2}^{b_2} f(b_1, t_2, t_3, \ldots, t_n)\, dt_2$$

$$+ \frac{1}{d_2 d_1} \int_{a_2}^{b_2} \int_{a_1}^{b_1} f(t_1, t_2, \ldots, t_n)\, dt_1 dt_2.$$

From the 3^2 entities of $R_2(f)$ there are 2^2 function evaluations, 2×2 single integrals, and one double integral:

$$R_3(f) = \mathfrak{M}_3(R_2(f)) = (G_3 - \lambda_3)(R_2(f)) = G_3(R_2(f)) - \lambda_3(R_2(f))$$

$$= \frac{A_3}{d_3} R_2(f) \Big|_{t_3 = a_3} + \frac{B_3}{d_3} R_2(f) \Big|_{t_3 = b_3} - \frac{1}{d_3} \int_{a_3}^{b_3} R_2(f)\, dt_3.$$

This will produce 3^2 entities with $\binom{3}{0} 2^3$ function evaluations, $\binom{3}{1} 2^2$ single integrals, $\binom{3}{2} 2^1$ double integrals, and $\binom{3}{0} 2^0$ triple integrals. The 2 occurs since evaluation is at either the a_i or the b_i.

Continuing in this manner we obtain the result as stated where there are 3^n entities for $R_n(f)$, with $\binom{n}{0} 2^n$ function evaluations only, $\binom{n}{1} 2^{n-1}$ single integrals, $\binom{n}{2} 2^{n-2}$ double integrals, ... , $\binom{n}{n-1} 2$, $(n-1)$th integrals, and one n-dimensional integral. $\qquad\qquad$ ▯

We now obtain bounds for $\rho_n\left(\underset{\sim}{a}, \underset{\sim}{x}, \underset{\sim}{b}\right)$ as defined in (3.158) for $\frac{\partial^n f}{\partial t_n \cdots \partial t_1} \in L_p[I^n]$, $1 \le p \le \infty$ with the usual Lebesgue norms [22].

Let $f : I^n \to \mathbb{R}$ be absolutely continuous such that all partial derivatives of order one in each of the variables exist. Then

$$D_n \left| \rho_n \left(\underset{\sim}{a}, \underset{\sim}{x}, \underset{\sim}{b} \right) \right|$$

$$\le \begin{cases} \prod_{i=1}^n Q_i(1) \left\| \frac{\partial^n f}{\partial t_n \cdots \partial t_1} \right\|_\infty, & \frac{\partial^n f}{\partial t_n \cdots \partial t_1} \in L_\infty[I^n]; \\[2mm] \left(\prod_{i=1}^n Q_i(q) \right)^{\frac{1}{q}} \left\| \frac{\partial^n f}{\partial t_n \cdots \partial t_1} \right\|_p, & \frac{\partial^n f}{\partial t_n \cdots \partial t_1} \in L_p[I^n], \\ & p > 1, \ \frac{1}{p} + \frac{1}{q} = 1 \\[2mm] \prod_{i=1}^n \phi_i \left\| \frac{\partial^n f}{\partial t_n \cdots \partial t_1} \right\|_1, & \frac{\partial^n f}{\partial t_n \cdots \partial t_1} \in L_1[I^n]; \end{cases} \quad (3.165)$$

where $\rho_n\left(\underset{\sim}{a}, \underset{\sim}{x}, \underset{\sim}{b}\right)$ is as defined by (3.158),

$$(q+1) Q_i(q) = A_i^{q+1} + B_i^{q+1}, \qquad (3.166)$$

$$\phi_i = \frac{d_i}{2} + |x_i - \gamma_i|. \qquad (3.167)$$

PROOF (Follows Cerone [22].) $\quad$ From (3.158) and (3.160) we have

$$\left| \rho_n \left(\underset{\sim}{a}, \underset{\sim}{x}, \underset{\sim}{b} \right) \right| = |R_n(f)| \qquad (3.168)$$

$$\le \frac{1}{D_n} \int_{a_n}^{b_n} \cdots \int_{a_1}^{b_1} \left| \prod_{i=1}^n q_i(x_i, t_i) \frac{\partial^n f}{\partial t_n \cdots \partial t_1} \right| dt_1 \cdots dt_n.$$

Now, for $\frac{\partial^n f}{\partial t_n \cdots \partial t_1} \in L_\infty[I^n]$, we obtain

$$|R_n(f)| \le \left\| \frac{\partial^n f}{\partial t_n \cdots \partial t_1} \right\|_\infty \int_{a_n}^{b_n} \cdots \int_{a_1}^{b_1} \left| \prod_{i=1}^n q_i(x_i, t_i) \right| dt_1 \cdots dt_n \qquad (3.169)$$

$$= \left\| \frac{\partial^n f}{\partial t_n \cdots \partial t_1} \right\|_\infty \prod_{i=1}^n \int_{a_i}^{b_i} |q_i(x_i, t_i)| \, dt_i$$

$$= \left\| \frac{\partial^n f}{\partial t_n \cdots \partial t_1} \right\|_\infty \prod_{i=1}^{n} \left[\int_{a_i}^{x_i} (t_i - a_i) \, dt_i + \int_{x_i}^{b_i} (b_i - t_i) \, dt_i \right]$$

$$= \frac{1}{2^n} \prod_{i=1}^{n} [A_i^2 + B_i^2] \left\| \frac{\partial^n f}{\partial t_n \cdots \partial t_1} \right\|_\infty .$$

Thus, combining (3.168) and (3.169) gives the first inequality in (3.165).

Further, using the Hölder inequality for multiple integrals, we have from (3.168) the following for $\frac{\partial^n f}{\partial t_n \dots \partial t_1} \in L_p [I^n]$, $1 < p < \infty$:

$$D_n \left| R_n (f) \right|$$

$$\leq \left\| \frac{\partial^n f}{\partial t_n \cdots \partial t_1} \right\|_p \left(\int_{a_n}^{b_n} \cdots \int_{a_1}^{b_1} \left| \prod_{i=1}^{n} q_i (x_i, t_i) \right|^q dt_1 \cdots dt_n \right)^{\frac{1}{q}} . \quad (3.170)$$

Here, on using (3.156), we have

$$\int_{a_n}^{b_n} \cdots \int_{a_1}^{b_1} \left| \prod_{i=1}^{n} q_i (x_i, t_i) \right|^q dt_1 \cdots dt_n$$

$$= \prod_{i=1}^{n} \int_{a_i}^{b_i} |p_i (x_i, t_i)|^q \, dt_i = \prod_{i=1}^{n} \left[\int_{a_i}^{x_i} (t_i - a_i)^q \, dt_i + \int_{x_i}^{b_i} (b_i - t_i)^q \, dt_i \right]$$

$$= \frac{1}{(q+1)^n} \prod_{i=1}^{n} \left[A_i^{q+1} + B_i^{q+1} \right] .$$

The second inequality in (3.165) holds on utilising (3.168) and noting (3.166).

The final inequality in (3.165) is obtained from (3.169) for $\frac{\partial^n f}{\partial t_n \dots \partial t_1} \in L_1 [I^n]$, giving from (3.168)

$$D_n \left| R_n (f) \right| \leq \sup_{\underset{\sim}{t} \in \left[\underset{\sim}{a}, \underset{\sim}{b} \right]} \left| \prod_{i=1}^{n} q_i (x_i, t_i) \right| \left\| \frac{\partial^n f}{\partial t_n \cdots \partial t_1} \right\|_1$$

$$= \prod_{i=1}^{n} \sup_{t_i \in [a_i, b_i]} |q_i (x_i, t_i)| \left\| \frac{\partial^n f}{\partial t_n \cdots \partial t_1} \right\|_1$$

$$= \prod_{i=1}^{n} \max \{ A_i, B_i \} \left\| \frac{\partial^n f}{\partial t_n \cdots \partial t_1} \right\|_1 .$$

Noting that

$$\max \{ X, Y \} = \frac{X + Y}{2} + \frac{|X - Y|}{2}$$

gives the final result. $\qquad \Box$

We follow the development by Cerone [22].

For $f(\cdot)$ absolutely continuous, then from (3.148) and (3.149)

$$\mathfrak{T}(f;a,\alpha,x,\beta,b) = \int_a^b r(x,t) f'(t)\, dt, \quad r(x,t) = \begin{cases} \frac{t-\alpha}{b-a}, & t \in [a,x] \\ \\ \frac{t-\beta}{b-a}, & t \in (x,b] \end{cases} . \quad (3.171)$$

However, it may be noticed that

$$\mathfrak{T}(f;a,\alpha,x,\beta,b) = T(f;a,\alpha,x) + T(f;x,\beta,b) \quad (3.172)$$

and

$$r(x,t) = \begin{cases} q(\alpha,t), & t \in [a,x] \\ \\ q(\beta,t), & t \in (x,b] \end{cases} \quad (3.173)$$

where $q(x,t) = \frac{t-x}{b-a}$, $t, x \in [a,b]$.

Thus we have the identity

$$\Psi_n\left(\underset{\sim}{a}, \underset{\sim}{\alpha}, \underset{\sim}{x}, \underset{\sim}{\beta}, \underset{\sim}{b}\right) \quad (3.174)$$

$$:= \rho_n\left(\underset{\sim}{a}, \underset{\sim}{\alpha}, \underset{\sim}{x}\right) + \rho_n\left(\underset{\sim}{x}, \underset{\sim}{\beta}, \underset{\sim}{b}\right)$$

$$= \frac{1}{D_n} \int_{a_n}^{x_n} \cdots \int_{a_1}^{x_1} \prod_{i=1}^{n} (t_i - \alpha_i) \frac{\partial^n f}{\partial t_n \cdots \partial t_1} dt_1 \cdots dt_n$$

$$+ \frac{1}{D_n} \int_{x_n}^{b_n} \cdots \int_{x_1}^{b_1} \prod_{i=1}^{n} (t_i - \beta_i) \frac{\partial^n f}{\partial t_n \cdots \partial t_1} dt_1 \cdots dt_n$$

$$= \int_{a_n}^{b_n} \cdots \int_{a_1}^{b_1} \prod_{i=1}^{n} r_i(x_i,t_i) \frac{\partial^n f}{\partial t_n \cdots \partial t_1} dt_1 \cdots dt_n$$

where

$$\frac{r_i(x_i,t_i)}{d_i} = \begin{cases} \dfrac{t_i - \alpha_i}{d_i}, & t_i \in [a_i,x_i] \\ \\ \dfrac{t_i - \beta_i}{d_i}, & t_i \in (x_i,b_i] \end{cases} . \quad (3.175)$$

It is important to obtain an identity for the three point rule since the bounds are tighter than using the bounds of the two trapezoidal rules, as this would entail using the triangle inequality. We notice that $\Psi_n\left(\underset{\sim}{a}, \underset{\sim}{\alpha}, \underset{\sim}{x}, \underset{\sim}{\beta}, \underset{\sim}{b}\right)$ in (3.174) is not expressed explicitly. This may be accomplished by returning to (3.158), or else we may use the *generator* methodology which was utilised to obtain the results for the Ostrowski and trapezoidal functionals.

Let $f : I^n \to \mathbb{R}$ and define the operator

$$H_i(f) := \frac{\nu_i^{(a)}}{d_i} f(t_1, \ldots, t_{i-1}, a_i, t_{i+1}, \ldots, t_n)$$

$$+ \frac{\nu_i^{(x)}}{d_i} f(t_1, \ldots, t_{i-1}, x_i, t_{i+1}, \ldots, t_n)$$

$$+ \frac{\nu_i^{(b)}}{d_i} f(t_1, \ldots, t_{i-1}, b_i, t_{i+1}, \ldots, t_n), \quad (3.176)$$

where

$$\nu_i^{(a)} = \alpha_i - a_i, \quad \nu_i^{(x)} = \beta_i - \alpha_i, \quad \nu_i^{(b)} = b_i - \beta_i, \quad d_i = b_i - a_i. \quad (3.177)$$

Then, from (3.153)–(3.171) for $f(\cdot)$ absolutely continuous in the ith variable $t_i \in [a_i, b_i]$ we have

$$\mathfrak{N}_i(f) = \frac{1}{d_i} \int_{a_i}^{b_i} r_i(x_i, t_i) \frac{\partial f}{\partial t_i} dt_i = (H_i - \lambda_i)(f), \quad i = 1, 2, \ldots, n. \quad (3.178)$$

where $H_i(f)$ and $\lambda_i(f)$ are as given by (3.176)–(3.177), and (3.153) respectively.

If we now follow the work of the previous subsection and let $c^{(0)} = (c_1, c_2, \ldots, c_n)$, where now $c_i = a_i, x_i$, or b_i in the ith partition for $i = 1, 2, \ldots, n$, then $\sigma_0(c^{(0)})$ is the set of all such vectors consists of 3^n possibilities. Let χ_n be as in (3.157) where now, $C_j = \nu_j^{(a)}$ or $\nu_j^{(x)}$ or $\nu_j^{(b)}$ are as defined by (3.177) with the exception that k of the $C_j = d_j$ and so $\frac{C_j}{d_j} = 1$. Further, $c^{(k)}$ is a vector taking on fixed values of either a_i, x_i or b_i in the ith position except for k of the positions which are variable, $t_\bullet$. Let $\mathcal{M}_k$ be k-dimensional integral means for $f(c^{(k)})$, then the following result holds [22].

Let $f : I^n \to \mathbb{R}$ be absolutely continuous and be such that all partial derivatives of order one in each of the variables exist. Then

$$B_n(f) = \sum_0 \chi_0 f\left(c^{(0)}\right) - \sum_1 \chi_1 \mathcal{M}_1 + \sum_2 \chi_2 \mathcal{M}_2 \quad (3.179)$$

$$- \cdots - (-1)^{n-1} \sum_{n-1} \chi_{n-1} \mathcal{M}_{n-1} + (-1)^n \mathcal{M}_n$$

$$:= \Psi_n\left(\underset{\sim}{a}, \underset{\sim}{\alpha}, \underset{\sim}{x}, \underset{\sim}{\beta}, \underset{\sim}{b}\right),$$

where

$$\chi_k = \prod_{j=1}^n {}^{(k)} \frac{C_j}{d_j}, \quad k = 0, 1, \ldots, n, \quad (3.180)$$

with $C_j = \nu_j^{(a)}, \nu_j^{(x)}$, or $\nu_j^{(b)}$ as defined in (3.177) except for k of the $C_j = d_j$, giving $\frac{C_j}{d_j} = 1$, $\mathcal{M}_k$ is the k-dimensional integral mean of $f(c^{(k)})$, and *specifically,*

$$\mathcal{M}_n = \frac{1}{D_n} \int_{a_n}^{b_n} \cdots \int_{a_1}^{b_1} f\left(c^{(n)}\right) dt_1 \ldots dt_n, \ c^{(n)} = (t_1, t_2, \ldots, t_n).$$

Finally, $\sum_k$ is a sum over all $\binom{n}{k} 3^{n-k}$ terms and $B_n(f)$ is as defined in (3.174).

The proof is similar to the result in (3.158), except that for each variable t_i there are now three possible choices for evaluation of either $a_i, x_i,$ or b_i.

The following result gives bounds for the $\Psi_n\left(\underset{\sim}{a}, \underset{\sim}{\alpha}, \underset{\sim}{x}, \underset{\sim}{\beta}, \underset{\sim}{b}\right)$ as given in either (3.179) or (3.174) [22].

Let $f : I^n \to \mathbb{R}$ be absolutely continuous and be such that all partial derivatives of order one in each of the variables exist. Then

$$D_n \left| \Psi_n\left(\underset{\sim}{a}, \underset{\sim}{\alpha}, \underset{\sim}{x}, \underset{\sim}{\beta}, \underset{\sim}{b}\right) \right|$$

$$\leq \begin{cases} \prod_{i=1}^{n} S_i(1) \left\| \dfrac{\partial^n f}{\partial t_n \cdots \partial t_1} \right\|_{\infty}, & \dfrac{\partial^n f}{\partial t_n \cdots \partial t_1} \in L_\infty[I^n]; \\[12pt] \left(\prod_{i=1}^{n} S_i(q)\right)^{\frac{1}{q}} \left\| \dfrac{\partial^n f}{\partial t_n \cdots \partial t_1} \right\|_{p}, & \dfrac{\partial^n f}{\partial t_n \cdots \partial t_1} \in L_\infty[I^n], \quad (3.181) \\ & p > 1, \ \frac{1}{p} + \frac{1}{q} = 1 \\[12pt] \prod_{i=1}^{n} \zeta_i \left\| \dfrac{\partial^n f}{\partial t_n \cdots \partial t_1} \right\|_{1}, & \dfrac{\partial^n f}{\partial t_n \cdots \partial t_1} \in L_1[I^n]; \end{cases}$$

where $\|h\|_p$, $1 \leq p < \infty$, and $\|h\|_\infty$ are defined by (3.150) and (3.151), $\Psi_n\left(\underset{\sim}{a}, \underset{\sim}{\alpha}, \underset{\sim}{x}, \underset{\sim}{\beta}, \underset{\sim}{b}\right)$ is defined by (3.174) or, explicitly, by (3.179),

$$(q+1) S_i(q)$$
$$= (\alpha_i - a_i)^{q+1} + (x_i - \alpha_i)^{q+1} + (\beta_i - x_i)^{q+1} + (b_i - \beta_i)^{q+1}, \quad (3.182)$$

$$\zeta_i = \frac{1}{2} \left\{ \frac{b_i - a_i}{2} + \left| \alpha_i - \frac{a_i + x_i}{2} \right| + \left| \beta_i - \frac{x_i + b_i}{2} \right| \right.$$
$$\left. + \left| x_i - \frac{a_i + b_i}{2} + \left| \alpha_i - \frac{a_i + x_i}{2} \right| - \left| \beta_i - \frac{x_i + b_i}{2} \right| \right| \right\}. \quad (3.183)$$

Comments

(a) It should be noted that the bounds given on $\left|\mathcal{T}_n\left(\underset{\sim}{a}, \underset{\sim}{x}, \underset{\sim}{b}\right)\right|$ by (3.137) in Section 3.10 are exactly the same bounds as those given by (3.165) for $\left|\mathcal{P}_n\left(\underset{\sim}{a}, \underset{\sim}{x}, \underset{\sim}{b}\right)\right|$. It was shown by Cerone [25] that the bounds in terms of the Lebesgue norms on $|S\left(f; c, x, d\right)|$ and $|T\left(f; c, x, d\right)|$, as defined in (3.143) and (3.144), are the same. Since the identities for these two functionals were used as **generators** for the multidimensional extension, the equality of the two sets of bounds (3.137) and (3.165) should not come as a great surprise. Finally we notice, due to convexity of the bounds in (3.165), that the sharpest bounds occur at $x_i = \gamma_i = \frac{a_i + b_i}{2}$.

(b) The bounds obtained above in the result (3.181) are the product of the bounds for the one dimensional integral results. It should further be noted that the three point results recapture the generalised trapezoidal results of the previous section if we take $\alpha_i = \beta_i = x_i$. In addition, the Ostrowski type results of Cerone [20] are recaptured if we take $\alpha_i = a_i$ and $\beta_i = b_i$.

3.12   Relationships between Ostrowski, Trapezoidal, and Čebyšev Functionals

In this section, we consider the relationship between the three functionals, for $f : [a, b] \to \mathbb{R}$,

$$S\left(f; a, x, b\right) := f\left(x\right) - \mathcal{M}\left(f; a, b\right), \tag{3.184}$$

$$T\left(f; a, x, b\right) := \left(\frac{x - a}{b - a}\right) f\left(a\right) + \left(\frac{b - x}{b - a}\right) f\left(b\right) - \mathcal{M}\left(f; a, b\right), \tag{3.185}$$

and

$$\mathfrak{T}\left(f, g; a, b\right) := \mathcal{M}\left(fg; a, b\right) - \mathcal{M}\left(f; a, b\right) \mathcal{M}\left(g; a, b\right), \tag{3.186}$$

where

$$\mathcal{M}\left(f; a, b\right) := \frac{1}{b - a} \int_a^b f\left(u\right) du, \qquad \text{the integral mean.} \tag{3.187}$$

The above functionals will be termed as the Ostrowski, trapezoidal, and Čebyšev functionals, respectively.

It may easily be demonstrated through an integration by parts argument of the Riemann-Stieltjes integral on the intervals $[a, x]$ and $(x, b]$ that the functional $S\left(f; a, x, b\right)$ satisfies the identity

$$S\left(f; a, x, b\right) = \int_a^b p\left(x, t\right) df\left(t\right), \quad p\left(x, t\right) = \begin{cases} \frac{t - a}{b - a}, & t \in [a, x], \\ \frac{t - b}{b - a}, & t \in (x, b]. \end{cases} \tag{3.188}$$

For f absolutely continuous, $df(t) \equiv f'(t) dt$. Thus, (3.188) becomes the Montgomery identity.

It may further be demonstrated by integration by parts of the Riemann-Stieltjes integral that the identity [39]

$$T(f; a, x, b) = \int_a^b q(x, t) \, df(t), \quad q(x, t) = \frac{t - x}{b - a}, \quad x, t \in [a, b] \quad (3.189)$$

holds.

More recently, Cerone [23] showed that

$$\mathfrak{T}(f, g; a, b) = \frac{1}{(b - a)^2} \int_a^b \psi(t) \, df(t), \quad (3.190)$$

where

$$\psi(t) = (t - a) G(t, b) - (b - t) G(a, t), \quad G(a, b) = \int_a^b g(u) \, du. \quad (3.191)$$

In the current section, relationships between the functionals (3.184), (3.185), and (3.186) will be investigated through their respective identities (3.188), (3.189), and (3.190). It will be shown that one identity may be obtained from the other through some transformation. An investigation of the effect on their respective bounds is also undertaken.

It may be noticed that if $H(u)$ is the Heaviside unit function defined by

$$H(u) := \begin{cases} 1, & u > 0 \\ 0, & u < 0 \end{cases} \quad (3.192)$$

then $p(x, t)$ from (3.188) may be written in the following form:

$$p(x, t) = \left(\frac{t - a}{b - a}\right) H(x - t) + \left(\frac{t - b}{b - a}\right) H(t - x), \quad x, t \in [a, b], \quad (3.193)$$

and from (3.189) let

$$q(x, t) = \hat{q}(x, t; a, b) = \left(\frac{t - x}{b - x}\right) H(t - a) H(b - t), \quad x \in [a, b]. \quad (3.194)$$

The following result holds [25]:

For $p(x, t)$ defined by (3.193) and $q(x, t)$ by (3.194), then the following relationships are valid for $x, t \in [a, b]$:

$$q(x, t) = \frac{(x - a) p(a, t) + (b - x) p(b, t)}{b - a} \quad (3.195)$$

and

$$p(x,t) = \frac{(x-a)\,\hat{q}(a,t;a,x) + (b-x)\,\hat{q}(b,t;x,b)}{b-a}. \tag{3.196}$$

PROOF From (3.193) and using (3.192), we have

$$p(a,t) = \frac{t-b}{b-a} \quad \text{and} \quad p(b,t) = \frac{t-a}{b-a}$$

and so

$$\frac{(x-a)\,p(a,t) + (b-x)\,p(b,t)}{b-a} = \frac{(x-a)\,(t-b) + (b-x)\,(t-a)}{(b-a)^2}$$

$$= \frac{t-x}{b-a} = q(x,t),$$

producing identity (3.195).

Now to prove (3.196) use the more explicit representation of $q(x,t)$ as given by (3.194), then

$$\hat{q}(a,t;a,x) = \left(\frac{t-a}{x-a}\right) H(x-t) \quad \text{and} \quad \hat{q}(b,t;x,b) = \left(\frac{t-b}{b-x}\right) H(t-x)$$

and so

$$\frac{(x-a)\,\hat{q}(a,t;a,x) + (b-x)\,\hat{q}(b,t;x,b)}{b-a}$$

$$= \left(\frac{t-a}{b-a}\right) H(x-t) + \left(\frac{t-b}{b-a}\right) H(t-x)$$

$$= p(x,t).$$

☐

It should be emphasized that there is a strong relationship between the Ostrowski and trapezoidal functionals. This is highlighted by the symmetric transformations amongst their kernels that provide the identities (3.188) and (3.189).

The following result is from Cerone [25].

The Lebesgue norms for the generalised trapezoidal rule and those for the Ostrowski functional are equal. That is,

$$\|p(x,\cdot)\|_\gamma = \|q(x,\cdot)\|_\gamma, \quad \gamma \in [1,\infty], \tag{3.197}$$

where $p(x,t)$ and $q(x,t)$ are as given in (3.188) and (3.189) or (3.193) and (3.194), respectively.

PROOF   Let $\lambda = \frac{x-a}{b-a}$ and $1-\lambda = \frac{b-x}{b-a}$, then $|p(x,t)| = \frac{t-a}{b-a}$ varies linearly from 0 to λ, for $a \leq t \leq x$ and $|q(x,t)| = \frac{x-t}{b-a}$ varies linearly from λ to 0 for $a \leq t \leq x$. Further, $|p(x,t)|$ and $|q(x,t)|$ are symmetric about $t = \frac{a+x}{2}$ within the interval $t \in [a,x]$. Similarly, over the interval $t \in (x,b]$, $|p(x,t)| = \frac{b-t}{b-a}$ which varies linearly from 0 to $1-\lambda$ for $x < t \leq b$, and $|q(x,t)| = \frac{t-x}{b-a}$ varies linearly from 0 to $1-\lambda$ for $t \in (x,b]$. Again $|p(x,t)|$ and $|q(x,t)|$ are symmetric about $t = \frac{x+b}{2}$, the midpoint of the interval $t \in (x,b]$. We thus conclude that (3.197) is true.   □

All the bounds obtained for the Ostrowski functional are valid for the trapezoidal functional, due to the symmetric transformations amongst their kernels.

We now illustrate the process for acquiring the identity (3.190) for the Čebyšev functional from the identity (3.188) for the Ostrowski functional.

We start with (3.188), multiplying by $\frac{g(x)}{b-a}$ and then integrating over $[a,b]$ to give, on assuming that the interchange of order is permissible,

$$
\begin{aligned}
\mathfrak{T}(f,g;a,b) &= \frac{1}{b-a} \int_a^b \int_a^b g(x)\,p(x,t)\,df(t)\,dx \\
&= \frac{1}{b-a} \int_a^b \left[\int_a^x g(x)\left(\frac{t-a}{b-a}\right) + \int_x^b g(x)\left(\frac{t-b}{b-a}\right) \right] df(t)\,dx \\
&= \frac{1}{b-a} \int_a^b \left[\left(\frac{t-a}{b-a}\right) \int_t^b g(x)\,dx + \left(\frac{t-b}{b-a}\right) \int_a^t g(x)\,dx \right] df(t) \\
&= \frac{1}{(b-a)^2} \int_a^b \psi(t)\,df(t),
\end{aligned}
$$

which is identity (3.190) where $\psi(t)$ is as given by (3.191).

Thus, we have noticed how multiplication and integration transform the Ostrowski functional identity to the Čebyšev functional identity.

The reverse path is also possible.

From (3.190) and (3.191) let

$$g(t) = H(t-\alpha)\,H(\beta-t), \quad a \leq \alpha < \beta \leq b, \tag{3.198}$$

where $H(\cdot)$ is the Heaviside unit function defined by (3.192), to give

$$\left(\frac{b-a}{\beta-\alpha}\right) \mathfrak{T}(f,g;a,b) = D(f;a,\alpha,\beta,b) := \mathcal{M}(f;\alpha,\beta) - \mathcal{M}(f;a,b) \tag{3.199}$$

$$= \frac{1}{(b-a)(\beta-\alpha)} \int_a^b \tilde{\psi}(t)\,df(t)$$

where $\tilde{\psi}(t) = \psi(t)$ with $g(t)$ as given by (3.198), and so from (3.191)

$$\tilde{\psi}(t) = \begin{cases} (\beta - \alpha)(t - a), & a \leq t \leq \alpha; \\ \alpha(b-a) - a(\beta - \alpha) - [(b-a) - (\beta - \alpha)]t, & \alpha < t < \beta; \\ (\beta - \alpha)(t - b), & \beta \leq t \leq b. \end{cases} \quad (3.200)$$

Now, if we take $\alpha = x$ and $\beta = x+h$, then from the left-hand side of (3.199)

$$D(f; a, x, x+h, b) = \mathcal{M}(f; x, x+h) - \mathcal{M}(f; a, b),$$

giving

$$S(f; a, x, b) = \lim_{h \to 0} D(f; a, x, x+h, b).$$

The right-hand side of (3.199) and using (3.200) produces, for f continuous,

$$\lim_{h \to 0} \left\{ \frac{1}{b-a} \left[\int_a^x (t-a)\, df(t) \right. \right.$$

$$+ \frac{1}{h} \int_x^{x+h} [x(b-a) - ah - (b-a-h)t]\, df(t)$$

$$\left. \left. + \int_{x+h}^b (t-b)\, df(t) \right] \right\} = \int_a^b p(x,t)\, df(t)$$

and so (3.188) is obtained.

For further work on various bounds for $D(f; a, \alpha, \beta, b)$, the difference between two means, the reader is referred to Barnett et al. [5] and Cerone and Dragomir [32].

If in (3.199) and (3.200) we take $\alpha = a$ and $\beta = a + h$ and allow $h \to 0$, then after some simplification, we get

$$f(a) - \mathcal{M}(f; a, b) = \int_a^b \left(\frac{t-b}{b-a} \right) df(t).$$

Further, taking $\alpha = b - h$ and $\beta = b$ gives

$$f(b) - \mathcal{M}(f; a, b) = \int_a^b \left(\frac{t-a}{b-a} \right) df(t)$$

from (3.199) and (3.200) after allowing $h \to 0$. Combining the above results gives

$$\left(\frac{x-a}{b-a} \right) f(a) + \left(\frac{b-x}{b-a} \right) f(b)$$

$$= \frac{1}{(b-a)^2} \int_a^b [(x-a)(t-b) + (b-x)(t-a)]\, df(t)$$

$$= \frac{1}{b-a} \int_a^b (t-x)\, df(t).$$

The identity for the trapezoidal functional (3.189) is thus recaptured.

Comments

(a) The functional $\mathfrak{T}(f; a, \alpha, x, \beta, b)$ involves the difference between the integral mean, a weighted combination of a function evaluated at the end points and an interior point. Namely, for $a \le \alpha < x < \beta \le b$,

$$\mathfrak{T}(f; a, \alpha, x, \beta, b) := \left(\frac{\alpha - a}{b - a}\right) f(a) + \left(\frac{\beta - \alpha}{b - a}\right) f(x)$$

$$+ \left(\frac{b - \beta}{b - a}\right) f(b) - M(f; a, b). \quad (3.201)$$

Cerone and Dragomir [34] showed that the three point functional (3.201) satisfies the identity

$$\mathfrak{T}(f; a, \alpha, x, \beta, b) = \int_a^b r(x, t)\, df(t), \quad r(x, t) = \begin{cases} \frac{t - \alpha}{b - a}, & t \in [a, x] \\ \\ \frac{t - \beta}{b - a}, & t \in (x, b]. \end{cases} \quad (3.202)$$

It was shown by Cerone [19] that (3.202) may be obtained from (3.188) and (3.189) and vice versa.

(b) The following results hold:

Let $\alpha, \beta : [c, d] \to \mathbb{R}$ *be continuous on* $[c, d]$ *and* $|\alpha(t)| = |\beta(c + d - t)|$, *then the Lebesgue norms*

$$\|\alpha\|_\gamma = \|\beta\|_\gamma, \quad \gamma \in [1, \infty]. \quad (3.203)$$

PROOF First, note that

$$\|\alpha\|_\infty := \operatorname*{ess\,sup}_{t \in [c,d]} |\alpha(t)| = \operatorname*{ess\,sup}_{t \in [c,d]} |\beta(c + d - t)|$$

$$= \operatorname*{ess\,sup}_{u \in [d,c]} |\beta(u)| = \|\beta\|_\infty.$$

Further, for $1 \le p < \infty$ we have

$$\|\alpha\|_p^p := \int_c^d |\alpha(t)|^p\, dt = \int_c^d |\beta(c + d - t)|^p\, dt = \int_c^d |\beta(u)|^p\, du = \|\beta\|_p^p.$$

Thus, the result is proved. ∎

Let $\alpha_i, \beta_i : [c_i, d_i] \to \mathbb{R}$ *be continuous on* $[c_i, d_i]$, *where* $I_i = [c_i, d_i] \subset [a, b]$ *and* $\bigcup_{i=1}^n I_i = [a, b]$. *If*

$$|\alpha_i(t)| = |\beta_i(c_i + d_i - t)|,$$

then

$$\|\alpha\|_\gamma = \|\beta\|_\gamma, \quad \gamma \in [0, \infty],$$

where $\alpha, \beta : [a, b] \to \mathbb{R}$ *and* $\alpha(t) = \alpha_i(t)$, $t \in [c_i, d_i]$ *and* $\beta(t) = \beta_i(t)$, $t \in [c_i, d_i]$.

The proof follows directly from the result (3.203) applied to each subinterval $[c_i, d_i]$, $i = 1, 2, \ldots, n$ with $c_1 = a$ and $c_n = b$.

The generalised Peano kernels involving $f^{(n)}(\cdot)$ for Ostrowski and trapezoidal type inequalities are defined as follows:

$$P_n(x, t) = \begin{cases} \dfrac{(t - a)^n}{n!}, & t \in [a, x] \\[2mm] \dfrac{(t - b)^n}{n!}, & t \in (x, b] \end{cases} \tag{3.204}$$

and

$$Q_n(x, t) = \frac{(t - x)^n}{n!}, \quad t, x \in [a, b], \tag{3.205}$$

respectively.

It may be noticed that over each of the intervals $[a, x]$ and $(x, b]$, the moduli of the two kernels are equivalent, that is,

$$|P_n(x, t)| = \begin{cases} |Q_n(x, a + x - t)|, & t \in [a, x] \\[2mm] |Q_n(x, x + b - t)|, & t \in (x, b]. \end{cases} \tag{3.206}$$

It was shown by Cerone [25] that the bounds are the same for these kernels. The result was proved from first principles by Cerone, Dragomir, and Roumeliotis [40] and Cerone, Dragomir, Roumeliotis, and Šunde [41]. Bounds for the trapezoidal and Ostrowski rules were found independently by Cerone and Dragomir [34, 35].

3.13   Perturbed Trapezoidal and Midpoint Rules

Let $f : [a, b] \to \mathbb{R}$ and define the functionals

$$I(f) := \int_a^b f(t) \, dt \tag{3.207}$$

$$I^{(T)}(f) := \frac{b - a}{2} [f(a) + f(b)] \tag{3.208}$$

and

$$I^{(M)}(f) := (b - a) f \left(\frac{a+b}{2} \right). \tag{3.209}$$

Here $I^{(T)}(f)$ and $I^{(M)}(f)$ are the well-known trapezoidal and midpoint rules used to approximate the functional $I(f)$.

Atkinson [3] defined the corrected (or perturbed) trapezoidal and midpoint rules by

$$PI^{(T)}(f) := I^{(T)}(f) - \frac{c^2}{3} [f'(b) - f'(a)] \tag{3.210}$$

and

$$PI^{(M)}(f) := I^{(M)}(f) + \frac{c^2}{6} [f'(b) - f'(a)] \tag{3.211}$$

respectively, where $c = \frac{b-a}{2}$.

Let the trapezoidal functional $T(f; a, b)$ be defined by

$$T(f; a, b) := I(f) - I^{(T)}(f) = \frac{1}{b-a} \int_a^b f(t) \, dt - \frac{b-a}{2} [f(a) + f(b)]. \tag{3.212}$$

Then it is well known that the identity

$$T(f; a, b) = -\frac{1}{2} \int_a^b (t - a)(b - t) f''(t) \, dt \tag{3.213}$$

holds. The following result was obtained by Dragomir, Cerone, and Sofo [89] using identity (3.213).

Let $f : [a, b] \to \mathbb{R}$ be a twice differentiable function on (a, b). Then we have the estimate

$$|T(f; a, b)|$$
$$\leq \begin{cases} \frac{\|f''\|_\infty}{12} (b - a)^3 & \text{if } f'' \in L_\infty[a, b]; \\ \frac{1}{2} \|f''\|_q [B(q + 1, q + 1)]^{\frac{1}{p}} (b - a)^{2 + \frac{1}{p}}, & \text{if } f'' \in L_p[a, b], \\ \hspace{5.5cm} \frac{1}{p} + \frac{1}{q} = 1, \, p > 1 \\ \frac{\|f''\|_1}{8} (b - a)^2 & \text{if } f'' \in L_1[a, b], \end{cases} \tag{3.214}$$

where B is the beta function, that is,

$$B(r, s) := \int_0^1 t^{r-1} (1 - t)^{s-1} \, dt, \, r, s > 0.$$

Let

$$PT(f; a, b) := I(f) - PI^{(T)}(f) = T(f; a, b) + \frac{c^2}{3} [f'(b) - f'(a)], \tag{3.215}$$

where $c = \frac{b-a}{2}$, then the following results hold (see Cerone [19]):

Let $f : [a, b] \to \mathbb{R}$ be such that f' is absolutely continuous on $[a, b]$, then

$$PT(f; a, b) = \frac{1}{2} \int_a^b \kappa(t) f''(t) \, dt \qquad (3.216)$$

is valid with

$$\kappa(t) = \left(t - \frac{a+b}{2}\right)^2 - \frac{c^2}{3}, \quad c = \frac{b-a}{2}. \qquad (3.217)$$

PROOF From (3.215) and (3.217), we have

$$PT(f; a, b) = -\frac{1}{2} \int_a^b (t-a)(b-t) f''(t) \, dt + \frac{c^2}{3} [f'(b) - f'(a)]$$

$$= \int_a^b \left[\frac{c^2}{3} - \frac{1}{2} (t-a)(b-t)\right] f''(t) \, dt$$

$$= \frac{1}{2} \int_a^b \left[t^2 - (a+b) t + ab + 2\frac{c^2}{3}\right] f''(t) \, dt$$

$$= \frac{1}{2} \int_a^b \left[\left(t - \frac{a+b}{2}\right)^2 - \frac{c^2}{3}\right] f''(t) \, dt.$$

Thus, (3.216) holds with $\kappa(t)$ as given by (3.217). $\qquad \Box$

Let $f : [a, b] \to \mathbb{R}$ be such that f' is absolutely continuous on $[a, b]$, then

$$|PT(f; a, b)| \qquad (3.218)$$

$$\leq \begin{cases} \frac{4c^3}{9\sqrt{3}} \|f''\|_\infty, & \text{if } f'' \in L_\infty[a, b]; \\[2mm] \frac{c^2}{6} \left\{\frac{c}{\sqrt{3}} B\left(\frac{1}{2}, q+1\right) + 2 \int_1^{\sqrt{3}} (u^2-1)^q \, du\right\}^{\frac{1}{q}} \|f''\|_p, & \text{if } f'' \in L_p[a, b], \\ & \qquad \frac{1}{p} + \frac{1}{q} = 1, \, p > 1 \\[2mm] \frac{c^2}{6} \|f''\|_1 & \text{if } f'' \in L_1[a, b], \end{cases}$$

where B is the beta function and $c = \frac{b-a}{2}$.

PROOF From identity (3.216) and using (3.217) we have

$$|PT(f; a, b)| \leq \frac{1}{2} \int_a^b |\kappa(t)| |f''(t)| \, dt \qquad (3.219)$$

$$\leq \frac{\|f''\|_p}{2} \left(\int_a^b |\kappa(t)|^q \, dt^{\frac{1}{q}}\right), \quad p > 1.$$

Now we need to examine the behaviour of $\kappa(t)$ in order to proceed further. We notice from (3.217) that $\kappa(a) = \kappa(b) = \frac{2}{3}c^2$ and $\kappa(t) = 0$ where $t = \frac{a+b}{2} \pm \frac{c}{\sqrt{3}}$.

Further,

$$\kappa'(t) = 2\left(t - \frac{a+b}{2}\right) \begin{cases} < 0, t < \frac{a+b}{2}; \\ = 0, t = \frac{a+b}{2}; \\ > 0, t > \frac{a+b}{2}. \end{cases}$$

Also, $\kappa(t)$ is a symmetric function about $\frac{a+b}{2}$ since $\kappa\left(\frac{a+b}{2} + x\right) = \kappa\left(\frac{a+b}{2} - x\right)$, so that from (3.219)

$$\|\kappa\|_q^q = \int_{\frac{a+b}{2}}^{\frac{a+b}{2}+\frac{c}{\sqrt{3}}} [-\kappa(t)]^q \, dt + \int_{\frac{a+b}{2}+\frac{c}{\sqrt{3}}}^{b} \kappa^q(t) \, dt \qquad (3.220)$$

$$:= 2\left[I_1(q) + I_2(q)\right].$$

Now, from (3.217)

$$I_1(q) = \int_{\frac{a+b}{2}}^{\frac{a+b}{2}+\frac{c}{\sqrt{3}}} \left[\frac{c^2}{3} - \left(t - \frac{a+b}{2}\right)^2\right]^q \, dt.$$

Let $\frac{c}{\sqrt{3}}u = t - \frac{a+b}{2}$, then

$$I_1(q) = \frac{c}{\sqrt{3}} \int_0^1 \left(\frac{c^2}{3}\right)^q (1 - u^2)^q \, du = \frac{c^{2q+1}}{3^{q+\frac{1}{2}}} \cdot \frac{1}{2} B\left(\frac{1}{2}, q+1\right) \qquad (3.221)$$

since $\int_0^1 (1 - u^2)^q \, du = \frac{1}{2}B\left(\frac{1}{2}, q+1\right)$.

Also,

$$I_2(q) \int_{\frac{a+b}{2}+\frac{c}{\sqrt{3}}}^{b} \left[\left(t - \frac{a+b}{2}\right)^2 - \frac{c^2}{3}\right]^q \, dt$$

and substituting $\frac{c}{\sqrt{3}}v = t - \frac{a+b}{2}$ gives

$$I_2(q) = \frac{c}{\sqrt{3}} \left(\frac{c^2}{3}\right)^q \int_0^1 [v^2 - 1]^q \, dv. \qquad (3.222)$$

Combining (3.221) and (3.222) into (3.220) gives from (3.219) the second inequality in (3.218).

The first inequality is obtained by taking $q = 1$ in the second inequality of (3.218) as may be noticed from (3.219).

Thus

$$\frac{1}{2} \int_a^b |\kappa(t)| \, dt = \frac{c^3}{6\sqrt{3}} \left[B\left(\frac{1}{2}, 2\right) + 2\int_1^{\sqrt{3}} (u^2 - 1) \, du\right]$$

$$= \frac{c^3}{6\sqrt{3}} \left[\frac{4}{3} + \frac{4}{3}\right] = \frac{4c^3}{3^{\frac{5}{2}}}.$$

Now, for the final inequality, from (3.219) we obtain

$$|PT(f;a,b)| \leq \frac{1}{2} \sup_{t \in [a,b]} |\kappa(t)| \, \|f''\|_1$$

and so, from the behaviour of $\kappa(t)$ discussed earlier,

$$\sup_{t \in [a,b]} |\kappa(t)| = \max \left\{ \frac{2}{3}c^2, \frac{1}{3}c^2 \right\} = \frac{2}{3}c^2.$$

This completes the proof. □

The following result is a particular case of the preceding ones which involve the Euclidean norm (see Cerone [19]):

Let $f; [a,b] \rightarrow \mathbb{R}$ be such that $f'' \in L_2[a,b]$. Then we have the inequality

$$|PT(f;a,b)| \leq \frac{\sqrt{2}c^{\frac{5}{2}}}{3\sqrt{5}} \|f''\|_2 = \frac{(b-a)^{\frac{5}{2}}}{6\sqrt{5}} \|f''\|_2, \qquad (3.223)$$

where $c = \frac{b-a}{2}$. The proof follows by letting $p = 2$.

Let, from (3.207) and (3.209), the midpoint functional, $M(f;a,b)$, be defined by

$$M(f;a,b) := I(f) - I^{(M)}(f) \qquad (3.224)$$

$$= \frac{1}{b-a} \int_a^b f(t)\, dt - (b-a) f\left(\frac{a+b}{2}\right).$$

Then the identity

$$M(f;a,b) = \int_a^b \phi(t) f''(t)\, dt \qquad (3.225)$$

is well known, where

$$\phi(t) = \begin{cases} \dfrac{(t-a)^2}{2}, & t \in \left[a, \dfrac{a+b}{2}\right], \\[2mm] \dfrac{(b-t)^2}{2}, & t \in \left(\dfrac{a+b}{2}, b\right]. \end{cases} \qquad (3.226)$$

The following result concerning the classical midpoint functional (3.224) with bounds involving the $L_p[a,b]$ norms of the second derivative is known (see Atkinson [3]).

Let $f : [a, b] \to \mathbb{R}$ be such that f' is absolutely continuous on $[a, b]$. Then

$$
|M(f; a, b)| \leq \begin{cases} \frac{(b-a)^3}{24} \|f''\|_{\infty} & \text{if } f'' \in L_{\infty}[a, b]; \\[2mm] \frac{(b-a)^{2+\frac{1}{q}}}{8(2q+1)^{\frac{1}{q}}} \|f''\|_p, & \text{if } f'' \in L_p[a, b], \\[1mm] \qquad\qquad \frac{1}{p} + \frac{1}{q} = 1, \ p > 1 \\[2mm] \frac{(b-a)^2}{8} \|f''\|_1 & \text{if } f'' \in L_1[a, b]. \end{cases} \tag{3.227}
$$

The first inequality in (3.227) is the one that is traditionally most well known. Further, from (3.211) and (3.224) define the perturbed or corrected midpoint functional as

$$
PM(f; a, b) := I(f) - PI^{(M)}(f) = M(f; a, b) - \frac{c^2}{6}[f'(b) - f'(a)], \tag{3.228}
$$

where $c = \frac{b-a}{2}$.

The following identity concerning $PM(f; a, b)$ holds (see Cerone [19]):

Let $f : [a, b] \to \mathbb{R}$ be such that f' is absolutely continuous on $[a, b]$. Then

$$
PM(f; a, b) = \frac{1}{2} \int_a^b \chi(t) f''(t) \, dt, \tag{3.229}
$$

where

$$
\chi(t) = \begin{cases} (t-a)^2 - \frac{1}{3}c^2, \ t \in \left[a, \dfrac{a+b}{2}\right], \\[4mm] (b-t)^2 - \frac{1}{3}c^2, \ t \in \left(\dfrac{a+b}{2}, b\right] \end{cases} \tag{3.230}
$$

with $c = \frac{b-a}{2}$.

Further, the following result is valid [19].

The Lebesgue norms for the perturbed midpoint functional $PM(f; a, b)$ as given by (3.229) are the same as those for the perturbed trapezoid function $PT(f; a, b)$ given by (3.218).

PROOF   To prove the result, it suffices to demonstrate that

$$
\|\chi\|_p = \|\kappa\|_p, \quad p \geq 1. \tag{3.231}
$$

The properties of κ were investigated in the proof of (3.218). Now for $\chi(t)$, we note that $\chi(a) = \chi(b) = -\frac{c^2}{3}$ and $\chi(t) = 0$ when $t = a + \frac{c}{\sqrt{3}}, b - \frac{c}{\sqrt{3}}$ for

$t \in [a, b]$. Further, $\chi(t)$ is continuous at $t = \frac{a+b}{2}$ and $\chi\left(\frac{a+b}{2}\right) = \frac{2}{3}c^2$. Also

$$\chi'(t) = \begin{cases} 2(t-a) > 0, & t \in \left[a, \dfrac{a+b}{2}\right], \\[2mm] -2(b-t) < 0, & t \in \left(\dfrac{a+b}{2}, b\right]. \end{cases}$$

In fact, for $t \in \left[a, \frac{a+b}{2}\right]$, $\chi(t) = \kappa\left(a + \frac{a+b}{2} - t\right)$ and for $t \in \left(\frac{a+b}{2}, b\right]$, $\chi(t) = \kappa\left(b + \frac{a+b}{2} - t\right)$. It implies that $\chi(t)$ and $\kappa(t)$ are symmetric about $\frac{3a+b}{4}$ (the midpoint of $\left[a, \frac{a+b}{2}\right]$) and $\frac{a+3b}{4}$ (the midpoint of $\left(\frac{a+b}{2}, b\right]$). Thus, (3.231) holds; and the result is valid as stated. □

The bound given in (3.223) also holds for $PM(f; a, b)$ given the above results.

Comments

(a) *Perturbed Rules from the Čebyšev Functional.* For $g, h : [a, b] \to \mathbb{R}$ the following $\mathfrak{T}(g, h)$ is well known as the Čebyšev functional. Namely,

$$\mathfrak{T}(g, h) = \mathcal{M}(gh) - \mathcal{M}(g)\mathcal{M}(h), \qquad (3.232)$$

where $\mathcal{M}(g) = \frac{1}{b-a}\int_a^b g(t)\,dt$ is the integral mean.

The Čebyšev functional (3.232) is known to satisfy a number of identities including

$$\mathfrak{T}(g, h) = \frac{1}{b-a}\int_a^b h(t)[g(t) - \mathcal{M}(g)]\,dt. \qquad (3.233)$$

Further, a number of sharp bounds for $|\mathfrak{T}(g, h)|$ exist, under various assumptions about g and h, including (see Cerone [25], for example):

$$|\mathfrak{T}(g, h)|$$

$$\leq \begin{cases} [\mathfrak{T}(g, g)]^{\frac{1}{2}} [\mathfrak{T}(h, h)]^{\frac{1}{2}}, & g, h \in L_2[a, b] \\[2mm] \frac{A_u - A_l}{2} [\mathfrak{T}(h, h)]^{\frac{1}{2}}, & A_l \leq g(t) \leq A_u, \ t \in [a, b] \\[2mm] \left(\frac{A_u - A_l}{2}\right)\left(\frac{B_u - B_l}{2}\right), & B_l \leq h(t) \leq B_u, \ t \in [a, b] \ (\text{Grüss}). \end{cases} \qquad (3.234)$$

The following result holds [19]:

Let $f : [a, b] \to \mathbb{R}$ be such that f' is absolutely continuous, then

$$|PT(f; a, b)| \qquad (3.235)$$

$$\leq \frac{(b-a)^3}{12\sqrt{5}}\left[\frac{1}{b-a}\|f''\|_2^2 - [f'; a, b]^2\right]^{\frac{1}{2}}, \qquad f'' \in L_2[a, b]$$

$$\leq \frac{(b-a)^3}{24\sqrt{5}}(B_u - B_l), \quad B_l \leq f''(t) \leq B_u, \ t \in [a, b],$$

where $PT(f; a, b)$ is the perturbed trapezoidal rule defined by (3.215).

PROOF Let $g(t) = -\frac{1}{2}(t-a)(b-t)$, the trapezoidal kernel, and $h(t) = f''(t)$, then from (3.232) [19]

$$(b-a)\mathfrak{T}(g(t), f''(t)) = \int_a^b g(t) f''(t) \, dt - \mathcal{M}(g) \int_a^b f''(t) \, dt \qquad (3.236)$$

$$= T(f; a, b) + \frac{c^2}{3}[f'(b) - f'(a)] = PT(f; a, b),$$

where $\mathcal{M}(g) = -\frac{c^2}{3}$.

Now, from (3.233),

$$(b-a)\mathfrak{T}(g(t), f''(t)) = \int_a^b f''(t) \left[g(t) + \frac{c^2}{3} \right] dt \qquad (3.237)$$

$$= \frac{1}{2} \int_a^b \kappa(t) f''(t) \, dt$$

and so (3.235) and (3.236) produce identities (3.216) and (3.217). Using (3.234) gives (3.235) from (3.236) and (3.237). ⬜

(b) Even though $A_l = -\frac{c^2}{2} \le g(t) \le 0 = A_u$, it is not worthwhile using this in the second and third inequalities of (3.236) as this would produce a coarser bound than those stated in (3.235). For a different proof of the sharpness of (3.235) see Barnett, Cerone, and Dragomir [4].

3.14 A Čebyšev Functional and Some Ramifications

For two measurable functions $f, g : [a, b] \to \mathbb{R}$, define the functional, which is known in the literature as Čebyšev's functional, by

$$T(f, g) := \mathcal{M}(fg) - \mathcal{M}(f)\mathcal{M}(g), \qquad (3.238)$$

where the integral mean is given by

$$\mathcal{M}(f) = \frac{1}{b-a} \int_a^b f(x) \, dx. \qquad (3.239)$$

The integrals in (3.238) are assumed to exist.

Further, the weighted Čebyšev functional is defined by

$$\mathfrak{T}(f, g; p) := \mathfrak{M}(f, g; p) - \mathfrak{M}(f; p)\mathfrak{M}(g; p), \qquad (3.240)$$

where the weighted integral mean is given by

$$\mathfrak{M}(f;p) = \frac{\int_a^b p(x) f(x) \, dx}{\int_a^b p(x) \, dx}. \tag{3.241}$$

We note that,

$$\mathfrak{T}(f,g;1) \equiv T(f,g) \quad \text{and} \quad \mathfrak{M}(f;1) \equiv \mathcal{M}(f).$$

It is worthwhile noting that a number of identities relating to the Čebyšev functional already exist. The reader is referred to Mitrinović, Pečarić, and Fink [141, Chapters IX and X]. Korkine's identity is well known (see Mitrinović, Pečarić, and Fink [141, p. 296]) and is given by

$$T(f,g) = \frac{1}{2(b-a)^2} \int_a^b \int_a^b (f(x) - f(y))(g(x) - g(y)) \, dx dy. \tag{3.242}$$

It is identity (3.242) that is often used to prove an inequality of Grüss for functions which are bounded above and below [141].

The Grüss inequality is given by

$$|T(f,g)| \leq \frac{1}{4}(\Phi_f - \phi_f)(\Phi_g - \phi_g) \tag{3.243}$$

where $\phi_f \leq f(x) \leq \Phi_f$ for $x \in [a,b]$.

If we let $S(f)$ be an operator defined by

$$S(f)(x) := f(x) - \mathcal{M}(f), \tag{3.244}$$

which shifts a function by its integral mean, then the following identity holds. Namely,

$$T(f,g) = T(S(f),g) = T(f,S(g)) = T(S(f),S(g)), \tag{3.245}$$

and so

$$T(f,g) = \mathcal{M}(S(f)g) = \mathcal{M}(fS(g)) = \mathcal{M}(S(f)S(g)) \tag{3.246}$$

since $\mathcal{M}(S(f)) = \mathcal{M}(S(g)) = 0$.

For the last term in (3.245) (or 3.246) only one of the functions needs to be shifted by its integral mean. If the other were to be shifted by any other quantity, the identities would still hold. A weighted version of (3.246) related to $\mathfrak{T}(f,g) = \mathcal{M}((f(x) - \kappa) S(g))$ for κ arbitrary was given by Sonin [161] (see Mitrinović, Pečarić, and Fink [141, p. 246]).

The following result presents an identity for the Čebyšev functional that involves a Riemann-Stieltjes integral and provides a Peano kernel representation [23].

A. *Let $f, g : [a, b] \to \mathbb{R}$, where f is of bounded variation and g is continuous on $[a, b]$, then*

$$T(f, g) = \frac{1}{(b-a)^2} \int_a^b \psi(t)\, df(t),$$ (3.247)

where

$$\psi(t) = (t-a) A(t, b) - (b-t) A(a, t)$$ (3.248)

with

$$A(a, b) = \int_a^b g(x)\, dx.$$ (3.249)

PROOF From (3.247), integrating the Riemann-Stieltjes integral by parts produces

$$\frac{1}{(b-a)^2} \int_a^b \psi(t)\, df(t)$$

$$= \frac{1}{(b-a)^2} \left\{ \psi(t) f(t) \Big]_a^b - \int_a^b f(t)\, d\psi(t) \right\}$$

$$= \frac{1}{(b-a)^2} \left\{ \psi(b) f(b) - \psi(a) f(a) - \int_a^b f(t) \psi'(t)\, dt \right\}$$

since $\psi(t)$ is differentiable. Thus, from (3.248), $\psi(a) = \psi(b) = 0$ and so

$$\frac{1}{(b-a)^2} \int_a^b \psi(t)\, df(t) = \frac{1}{(b-a)^2} \int_a^b [(b-a) g(t) - A(a, b)] f(t)\, dt$$

$$= \frac{1}{b-a} \int_a^b [g(t) - \mathcal{M}(g)] f(t)\, dt$$

$$= \mathcal{M}(fS(g))$$

from which the result (3.247) is obtained on noting identity (3.246). ◻

The following well-known results will prove useful and are stated here for lucidity.

B. *Let $g, v : [a, b] \to \mathbb{R}$ be such that g is continuous and v is of bounded variation on $[a, b]$. Then the Riemann-Stieltjes integral $\int_a^b g(t)\, dv(t)$ exists and is such that*

$$\left| \int_a^b g(t)\, dv(t) \right| \leq \sup_{t \in [a,b]} |g(t)| \bigvee_a^b (v),$$ (3.250)

where $\bigvee_a^b (v)$ is the total variation of v on $[a, b]$.

C. *Let $g, v : [a, b] \to \mathbb{R}$ be such that g is Riemann integrable on $[a, b]$ and v is L-Lipschitzian on $[a, b]$. Then*

$$\left| \int_a^b g(t)\, dv(t) \right| \le L \int_a^b |g(t)|\, dt \tag{3.251}$$

with v is L-Lipschitzian if it satisfies

$$|v(x) - v(y)| \le L\,|x - y|$$

for all $x, y \in [a, b]$.

D. *Let $g, v : [a, b] \to \mathbb{R}$ be such that g is continuous on $[a, b]$ and v is monotonic nondecreasing on $[a, b]$. Then*

$$\left| \int_a^b g(t)\, dv(t) \right| \le \int_a^b |g(t)|\, dv(t). \tag{3.252}$$

It should be noted that if v is nonincreasing then $-v$ is nondecreasing. The following provides bounds for the Čebyšev function [23].

Let $f, g : [a, b] \to \mathbb{R}$, where f is of bounded variation and g is continuous on $[a, b]$. Then

$$(b - a)^2\, |T(f, g)|$$

$$\le \begin{cases} \sup_{t \in [a,b]} |\psi(t)| \bigvee_a^b (f), & \\ L \int_a^b |\psi(t)|\, dt, & \text{for } f \text{ } L\text{-Lipschitzian,} \\ \int_a^b |\psi(t)|\, df(t), & \text{for } f \text{ monotonic nondecreasing,} \end{cases} \tag{3.253}$$

where $\bigvee_a^b (f)$ is the total variation of f on $[a, b]$.

The proof follows directly from the above results A through D. That is, from the identity (3.247) and (3.250)–(3.252).

The following result gives an identity for the weighted Čebyšev functional that involves a Riemann-Stieltjes integral [23].

Let $f, g, p : [a, b] \to \mathbb{R}$, where f is of bounded variation and g, p are continuous on $[a, b]$. Further, let $P(b) = \int_a^b p(x)\, dx > 0$, then

$$\mathfrak{T}(f, g; p) = \frac{1}{P^2(b)} \int_a^b \Psi(t)\, df(t), \tag{3.254}$$

where $\mathfrak{T}(f,g;p)$ is as given in (3.240),

$$\Psi(t) = P(t)\bar{G}(t) - \bar{P}(t)G(t) \tag{3.255}$$

with

$$\begin{cases} P(t) = \int_a^t p(x)\,dx, & \bar{P}(t) = P(b) - P(t) \\ and \\ G(t) = \int_a^t p(x)g(x)\,dx, & \bar{G}(t) = G(b) - G(t). \end{cases} \tag{3.256}$$

The following bounds were obtained by Cerone [23] from identity (3.254).

Let the conditions of (3.254) *on f, g, and p continue to hold. Then*

$$P^2(b)\,|\mathfrak{T}(f,g;p)|$$

$$\leq \begin{cases} \sup_{t\in[a,b]} |\Psi(t)| \bigvee_a^b (f), \\ L\int_a^b |\Psi(t)|\,dt, & \text{for } f \text{ L-Lipschitzian,} \\ \int_a^b |\Psi(t)|\,df(t), & \text{for } f \text{ monotonic nondecreasing.} \end{cases} \tag{3.257}$$

where $\mathfrak{T}(f,g;p)$ is as given by (3.240) *and* $\Psi(t) = P(t)G(b) - P(b)G(t)$, *with* $P(t) = \int_a^t p(x)\,dx$, $G(t) = \int_a^t p(x)g(x)\,dx$.

PROOF The proof uses results A through D and follows closely the proof in procuring the bounds in (3.253). $\square$

Grüss type inequalities obtained from bounds on the Čebyšev functional have been applied in a variety of areas including in obtaining perturbed rules in numerical integration (see for example, Cerone and Dragomir [34]). In the following, the above work will be applied to the approximation of moments. For other related results, see also Cerone and Dragomir [31].

If f is differentiable, then the identity (3.247) would become

$$T(f,g) = \frac{1}{(b-a)^2}\int_a^b \psi(t)f'(t)\,dt \tag{3.258}$$

and so

$$(b-a)^2\,|T(f,g)| \leq \begin{cases} \|\psi\|_1\|f'\|_\infty, & f' \in L_\infty[a,b]; \\ \|\psi\|_q\|f'\|_p, & f' \in L_p[a,b], \\ & p>1, \frac{1}{p}+\frac{1}{q}=1; \\ \|\psi\|_\infty\|f'\|_1, & f' \in L_1[a,b]; \end{cases}$$

where the Lebesgue norms $\|\cdot\|$ are defined in the usual way as

$$\|g\|_p := \left(\int_a^b |g(t)|^p\, dt\right)^{\frac{1}{p}}, \quad \text{for } g \in L_p[a,b],\, p \geq 1,\, \frac{1}{p}+\frac{1}{q}=1$$

and

$$\|g\|_\infty := \operatorname*{ess\,sup}_{t\in[a,b]} |g(t)|, \quad \text{for } g \in L_\infty[a,b].$$

We note from (3.248) and (3.259) that in order to obtain bounds on the Čebyšev functional, norms of $\psi(\cdot)$ are required. To this end, since

$$\psi(t) = (t-a)(b-t)D(g;a,t,b),$$

where

$$D(g;a,t,b) := M(g;t,b) - M(g;a,t), \tag{3.259}$$

then the following result provides bounds for $D(g;a,t,b)$. We further note that (3.259) is an expression for the means over the end intervals in $[a,b]$ for all $t \in [a,b]$ (see Cerone [23, 28]).

Let $g : [a,b] \to \mathbb{R}$ be *absolutely continuous on* $[a,b]$, *then for* $D(g;a,t,b)$ *given by* (3.259):

$$|D(g;a,t,b)| \leq \begin{cases} \left(\frac{b-a}{2}\right)\|g'\|_\infty, & g' \in L_\infty[a,b]; \\[2ex] \left[\frac{(t-a)^q+(b-t)^q}{q+1}\right]^{\frac{1}{q}}\|g'\|_p, & g' \in L_p[a,b], \\ & p>1,\, \frac{1}{p}+\frac{1}{q}=1; \\ & g' \in L_1[a,b]; \\[2ex] \|g'\|_1, & \\[1ex] V_a^b(g), & g \text{ of bounded variation;} \\[2ex] \left(\frac{b-a}{2}\right)L, & g \text{ is } L\text{-Lipschitzian.} \end{cases} \tag{3.260}$$

PROOF Let the kernel $r(t,u)$ be defined by

$$r(t,u) := \begin{cases} \frac{u-a}{t-a}, & u \in [a,t], \\[1ex] \frac{b-u}{b-t}, & u \in (t,b], \end{cases} \tag{3.261}$$

then a straightforward integration by parts argument of the Riemann-Stieltjes integral over each of the intervals $[a,t]$ and $(t,b]$ gives the identity

$$\int_a^b r(t,u)\,dg(u) = D(g;a,t,b). \tag{3.262}$$

Now for g an absolutely continuous function, we have

$$D\left(g; a, t, b\right) = \int_a^b r\left(t, u\right) g'\left(u\right) du, \tag{3.263}$$

and so

$$\left|D\left(g; a, t, b\right)\right| \leq \operatorname*{ess\,sup}_{u \in [a,b]} \left|r\left(t, u\right)\right| \int_a^b \left|g'\left(u\right)\right| du, \quad \text{for } g' \in L_1\left[a, b\right],$$

where from (3.261)

$$\operatorname*{ess\,sup}_{u \in [a,b]} \left|r\left(t, u\right)\right| = 1; \tag{3.264}$$

and so the third inequality in (3.260) results. Further, the Hölder inequality gives

$$\left|D\left(g; a, t, b\right)\right| \leq \left(\int_a^b \left|r\left(t, u\right)\right|^q du\right)^{\frac{1}{q}} \left(\int_a^b \left|g'\left(t\right)\right|^p dt\right)^{\frac{1}{p}} \tag{3.265}$$

$$\text{for } p > 1, \ \frac{1}{p} + \frac{1}{q} = 1;$$

where explicitly from (3.261), we get

$$\left(\int_a^b \left|r\left(t, u\right)\right|^q du\right)^{\frac{1}{q}} = \left[\frac{\left(t - a\right)^q + \left(b - t\right)^q}{q + 1}\right]^{\frac{1}{q}}. \tag{3.266}$$

Also

$$\left|D\left(g; a, t, b\right)\right| \leq \operatorname*{ess\,sup}_{u \in [a,b]} \left|g'\left(u\right)\right| \int_a^b \left|r\left(t, u\right)\right| du, \tag{3.267}$$

and so from (3.266) with $q = 1$ gives the first inequality in (3.260).

Now, for $g\left(u\right)$ of bounded variation on $[a, b]$, then from result (3.250), equation (3.250), and identity (3.262), we get

$$\left|D\left(g; a, t, b\right)\right| \leq \operatorname*{ess\,sup}_{u \in [a,b]} \left|r\left(t, u\right)\right| \bigvee_a^b \left(g\right)$$

producing the fourth inequality in (3.260) on using (3.264). From (3.251) and (3.262) we have, by associating g with v and $r\left(t, \cdot\right)$ with $g\left(\cdot\right)$,

$$\left|D\left(g; a, t, b\right)\right| \leq L \int_a^b \left|r\left(t, u\right)\right| du$$

and so from (3.266) with $q = 1$ gives the final inequality in (3.260). $\quad\square$

Comments

(a) *Results Involving Moments*

Bounds on the nth moment about a point γ are investigated from the above results. Define for n a nonnegative integer,

$$M_n(\gamma) := \int_a^b (x - \gamma)^n h(x)\, dx, \quad \gamma \in \mathbb{R}. \tag{3.268}$$

If $\gamma = 0$ then $M_n(0)$ are the moments about the origin, while taking $\gamma = M_1(0)$ gives the central moments. Further, the expectation of a continuous random variable is given by

$$E(X) = \int_a^b h(x)\, dx, \tag{3.269}$$

where $h(x)$ is the probability density function of the random variable X and so $E(X) = M_1(0)$. Also, the variance of the random variable X, $\sigma^2(X)$ is given by

$$\sigma^2(X) = E\left[(X - E(X))^2\right] = \int_a^b (x - E(X))^2 h(x)\, dx, \tag{3.270}$$

which may be seen to be the second moment about the mean, namely

$$\sigma^2(X) = M_2(M_1(0)).$$

The following result obtained by Cerone [23] provides bounds on moments:

Let $f : [a, b] \to \mathbb{R}$ be integrable on $[a, b]$, then

$$\left| M_n(\gamma) - \frac{B^{n+1} - A^{n+1}}{n+1} \mathcal{M}(f) \right|$$

$$\leq \begin{cases} \displaystyle \sup_{t \in [a,b]} |\phi(t)| \cdot \frac{1}{n+1} \overset{b}{\underset{a}{\mathrm{V}}}(f), & \text{for } f \text{ of bounded variation on } [a, b], \\[2ex] \displaystyle \frac{L}{n+1} \int_a^b |\phi(t)|\, dt, & \text{for } f \text{ L-Lipschitzian}, \\[2ex] \displaystyle \frac{1}{n+1} \int_a^b |\phi(t)|\, df(t), & \text{for } f \text{ monotonic nondecreasing}, \end{cases} \tag{3.271}$$

where $M_n(\gamma)$ is as given by (3.268), $\mathcal{M}(f)$ is the integral mean of f as defined in (3.239),

$$B = b - \gamma, \quad A = a - \gamma,$$

and

$$\phi(t) = (t - \gamma)^n - \left[\left(\frac{t - a}{b - a}\right)(b - \gamma)^{n+1} + \left(\frac{b - t}{b - a}\right)(a - \gamma)^{n+1}\right]. \tag{3.272}$$

PROOF   Taking $g(t) = (t - \gamma)^n$ in (3.253), then using (3.238) and (3.239) gives

$$(b - a) |T(f, (t - \gamma)^n)| = \left| M_n(\gamma) - \frac{B^{n+1} - A^{n+1}}{n+1} \mathcal{M}(f) \right|.$$

The right-hand side is obtained on noting that for $g(t) = (t - \gamma)^n$, $\phi(t) = -\frac{\psi(t)}{b-a}$. $\square$

(b) *Approximations for the Moment Generating Function.*

Let X be a random variable on $[a, b]$ with probability density function $h(x)$, then the moment generating function $M_X(p)$ is given by

$$M_X(p) = E\left[e^{pX}\right] = \int_a^b e^{px} h(x)\, dx. \tag{3.273}$$

The following result will prove useful later as it examines the behaviour of the function $\theta(t)$:

$$(b - a)\theta(t) = tA_p(a, b) - [aA_p(t, b) + bA_p(a, t)], \tag{3.274}$$

where

$$A_p(a, b) = \frac{e^{bp} - e^{ap}}{p}. \tag{3.275}$$

The following result was obtained by Cerone [23]:

Let $\theta(t)$ be as defined by (3.274) and (3.275) then for any $a, b \in \mathbb{R}$, $\theta(t)$ has the following characteristics:

(i) $\theta(a) = \theta(b) = 0$,

(ii) $\theta(t)$ is convex for $p < 0$ and concave for $p > 0$,

(iii) there is one turning point at $t^* = \frac{1}{p} \ln\left(\frac{A_p(a,b)}{b-a}\right)$ and $a \le t^* \le b$.

The following bounds were developed for approximants to the moment generating function (Cerone [23]) based on the above result.

Let $f : [a, b] \to \mathbb{R}$ be of bounded variation on $[a, b]$, then

$$\left| \int_a^b e^{pt} f(t)\, dt - A_p(a, b)\, \mathcal{M}(f) \right|$$

$$\leq \begin{cases} \left(m \left(\ln(m) - 1 \right) + \frac{be^{ap} - ae^{bp}}{b-a} \right) \frac{\bigvee_a^b(f)}{|p|}, \\[2mm] (b-a)\, m \left[\left(\frac{b-a}{2} \right) p - 1 \right] \frac{L}{|p|} \quad \text{for } f \text{ L-Lipschitzian on } [a, b], \\[2mm] \frac{p}{|p|} (b-a)\, m \left[f(b) - f(a) \right], \quad f \text{ monotonic nondecreasing,} \end{cases}$$

$$\text{(3.276)}$$

where

$$m = \frac{A_p(a, b)}{b - a} = \frac{e^{bp} - e^{ap}}{p(b-a)}. \qquad \text{(3.277)}$$

3.15   Weighted Three Point Quadrature Rules

In this section, weighted (or product) inequalities are developed. The weight function is assumed to be nonnegative and integrable over its entire domain. In order to simplify the working, some notation needs to be presented.

Let $w : (a, b) \to [0, \infty)$ be an integrable function so that $\int_a^b w(t)\, dt < \infty$. Define the zeroth and first moments of $w(\cdot)$ by

$$m(a, b) = \int_a^b w(t)\, dt \qquad \text{(3.278)}$$

and

$$M(a, b) = \int_a^b tw(t)\, dt, \qquad \text{(3.279)}$$

respectively. Both are assumed to exist over the entire domain of $w(\cdot)$. The weight function may be zero at the end points.

The following result involving the supremum norm of the first derivative was developed by Cerone, Roumeliotis, and Hanna [42]:

Let $f : [a, b] \to \mathbb{R}$ be a differentiable mapping on (a, b) whose derivative is bounded on (a, b), and denote $\|f'\|_\infty = \sup_{t \in (a, b)} |f'(t)| < \infty$. Further, let a nonnegative weight function $w(\cdot)$ have the properties as outlined above. Then *for $x \in [a,b]$, $\alpha \in [a,x]$, $\beta \in (x,b]$, the following inequality holds:*

$$\left| \int_a^b w(t) f(t) \, dt - [m(\alpha,\beta) f(x) + m(a,\alpha) f(a) + m(\beta,b) f(b)] \right|$$

$$\leq I(\alpha,x,\beta) \|f'\|_\infty, \quad (3.280)$$

where

$$I(\alpha,x,\beta) = \int_a^b k(x,t) w(t) \, dt, \quad (3.281)$$

$$k(x,t) = \begin{cases} t - a, & t \in [a,\alpha] \\ |x - t|, & t \in (\alpha,\beta] \\ b - t, & t \in (\beta,b]. \end{cases} \quad (3.282)$$

PROOF   Define the mapping $K(\cdot,\cdot) : [a,b]^2 \to \mathbb{R}$ by

$$K(x,t) = \begin{cases} m(\alpha,t), & t \in [a,x] \\ m(\beta,t), & t \in (x,b], \end{cases} \quad (3.283)$$

where $m(a,b)$ is the zeroth moment of $w(\cdot)$ over the interval $[a,b]$ and is given by (3.278).

It should be noted that $m(c,d)$ is nonnegative for $d \geq c$.

Integration by parts gives, on using (3.283),

$$\int_a^b K(x,t) f'(t) \, dt$$

$$= \int_a^x m(\alpha,t) f'(t) \, dt + \int_x^b m(\beta,t) f'(t) \, dt$$

$$= m(\alpha,t) f(t) \Big]_{t=a}^{x} - \int_a^x w(t) f(t) \, dt + m(\beta,t) f(t) \Big]_{t=x}^{b} - \int_x^b w(t) f(t) \, dt,$$

producing the identity

$$\int_a^b K(x,t) f'(t) \, dt$$

$$= m(\alpha,\beta) f(x) + m(a,\alpha) f(a) + m(\beta,b) f(b) - \int_a^b w(t) f(t) \, dt, \quad (3.284)$$

which is valid for all $x \in [a,b]$.

Taking the modulus of (3.284) gives

$$\left| \int_a^b w(t) f(t) \, dt - [m(\alpha, \beta) f(x) + m(a, \alpha) f(a) + m(\beta, b) f(b)] \right|$$

$$= \left| \int_a^b K(x,t) f'(t) \, dt \right| \leq \|f'\|_\infty \int_a^b |K(x,t)| \, dt. \quad (3.285)$$

Now, we wish to determine $\int_a^b |K(x,t)| \, dt$. To this end notice that, from (3.283), $K(x,t)$ is a monotonically nondecreasing function of t over each of its branches. Thus, there are points $\alpha \in [a,x]$ and $\beta \in [x,b]$ such that

$$K(x, \alpha) = K(x, \beta) = 0.$$

Thus,

$$\int_a^b |K(x,t)| \, dt = - \int_a^\alpha m(\alpha, t) \, dt + \int_\alpha^x m(\alpha, t) \, dt$$

$$+ \int_x^\beta m(\beta, t) \, dt - \int_\beta^b m(\beta, t) \, dt. \quad (3.286)$$

Integration by parts gives, for example,

$$-\int_a^\alpha m(\alpha, t) \, dt = -(t-a) m(\alpha, t) \Big]_{t=a}^\alpha + \int_a^\alpha (t-a) w(t) \, dt$$

$$= \int_a^\alpha (t-a) w(t) \, dt.$$

A similar development for the remainder of the three integrals on the right-hand side of (3.286) produces the result

$$\int_a^b |K(x,t)| \, dt = I(\alpha, x, \beta), \quad (3.287)$$

where $I(\alpha, x, \beta)$ is as given by (3.281) and (3.282). Combining (3.285) and (3.287) produces the result (3.280); and hence the result is proved. ▯

The following results were also proven by Cerone, Roumeliotis, and Hanna [42]:

A. *Inequality* (3.280) *is minimised at* $x = x^*$ *where* x^* *satisfies*

$$m(\alpha^*, x^*) = m(x^*, \beta^*) \quad (3.288)$$

and

$$\alpha^* = \frac{a + x^*}{2} \quad \text{and} \quad \beta^* = \frac{x^* + b}{2}. \quad (3.289)$$

B. *Let the conditions imposed on result (3.280) be maintained here. Then the following inequalities hold:*

$$\left| \int_a^b w(t) f(t) \, dt - [m(\alpha, \beta) f(x) + m(a, \alpha) f(a) + m(\beta, b) f(b)] \right|$$

$$\leq \|f'\|_\infty \times \begin{cases} \|w\|_\infty \cdot K_1(x) \\ \\ \|w\|_1 \cdot K_\infty(x) \end{cases}, \quad (3.290)$$

where

$$K_1(x) = \frac{1}{2} \left[\left(\frac{b-a}{2} \right)^2 + \left(x - \frac{a+b}{2} \right)^2 \right]$$

$$+ \left(\alpha - \frac{a+x}{2} \right)^2 + \left(\beta - \frac{x+b}{2} \right)^2 \quad (3.291)$$

and

$$K_\infty(x) = \frac{1}{2} \left[\frac{b-a}{2} + \left| \alpha - \frac{a+x}{2} \right| + \left| \beta - \frac{x+b}{2} \right| \right.$$

$$\left. + \left| x - \frac{a+b}{2} + \left| \alpha - \frac{a+x}{2} \right| - \left| \beta - \frac{x+b}{2} \right| \right| \right] \quad (3.292)$$

with $\|g\|_1 := \int_a^b |g(s)| \, ds$ *meaning* $g \in L_1[a,b]$, *the linear space of absolutely integrable functions, and* $\|g\|_\infty := \sup_{t \in [a,b]} |g(t)| < \infty$.

The following result involving the one-norm of the first derivative was obtained by Cerone, Roumeliotis, and Hanna [42]:

Let $f : I \subseteq \mathbb{R} \to \mathbb{R}$ *be a differentiable mapping on* $\mathring{I}$ *(the interior of I) and* $a, b \in \mathring{I}$ *are such that* $b > a$. *If* $f' \in L_1[a,b]$, *then* $\|f'\|_1 = \int_a^b |f'(t)| \, dt < \infty$. *In addition, let a nonnegative weight function* $w(\cdot)$ *have the properties as outlined on 190. Then for* $x \in [a,b]$, $\alpha \in [a,x]$, *and* $\beta \in (x,b]$ *the following inequality holds:*

$$\left| \int_a^b w(t) f(t) \, dt - [m(\alpha, \beta) f(x) + m(a, \alpha) f(a) + m(\beta, b) f(b)] \right|$$

$$\leq \theta(\alpha, x, \beta) \|f'\|_1, \quad (3.293)$$

where

$$\theta(\alpha, x, \beta) \quad (3.294)$$

$$= \frac{1}{4} \{ m(a, b) + |m(\alpha, x) - m(a, \alpha)| + |m(\beta, b) - m(x, \beta)|$$

$$+ |m(a, x) - m(x, b) + |m(\alpha, x) - m(a, \alpha)| - |m(\beta, b) - m(x, \beta)|| \}$$

and $m(a, b)$ *is the zeroth moment of* $w(\cdot)$ *over* $[a, b]$ *as defined by* (3.278).

PROOF   From identity (3.284) we obtain, from taking the modulus

$$\theta(\alpha, x, \beta) = \sup_{t \in [a,b]} |K(x, t)| \,,$$

where $K(x, t)$ is as given by (3.283). As discussed in the proof of (3.280), $K(x, t)$ is a monotonic nondecreasing function of t in each of its two branches so that

$$\theta(\alpha, x, \beta) = \max\{m(a, \alpha), m(\alpha, x), m(x, \beta), m(\beta, b)\}.$$

Now, by using the fact that

$$\max\{X, Y\} = \frac{X + Y}{2} + \frac{|Y - X|}{2},$$

we have

$$m_1 = \max\{m(a, \alpha), m(\alpha, x)\} = \frac{1}{2}[m(a, x) + |m(\alpha, x) - m(a, \alpha)|]$$

$$\text{and } m_2 = \max\{m(x, \beta), m(\beta, b)\} = \frac{1}{2}[m(x, b) + |m(\beta, b) - m(x, \beta)|],$$

to give

$$\theta(\alpha, x, \beta) = \max\{m_1, m_2\} = \frac{m_1 + m_2}{2} + \left|\frac{m_1 - m_2}{2}\right|.$$

The result (3.294) is obtained after some simplification. This completes the proof. ⬚

It should be noted that the tightest bound in (3.294) is obtained when $\alpha, x,$ and β are taken as their respective medians. Thus, the best quadrature rule in the above sense is given by

$$\left|\int_a^b w(t) f(t) \, dt - \left[m(a, \tilde{\alpha}) f(a) + m(\tilde{\alpha}, \tilde{\beta}) f(\tilde{x}) + m(\tilde{\beta}, b) f(b)\right]\right|$$

$$\leq \frac{m(a, b)}{4} \|f'\|_1, \quad (3.295)$$

where

$$m(a, \tilde{x}) = m(\tilde{x}, b), \; m(a, \tilde{\alpha}) = m(\tilde{\alpha}, \tilde{x}) \text{ and } m(\tilde{\beta}, b) = m(\tilde{x}, \tilde{\beta}).$$

Chapter 4

Grüss Type Inequalities and Related Results

A number of inequalities related to the famous Grüss result for approximating the integral mean of the product of two functions by the product of the integral means are investigated in this chapter. The Grüss-Čebyšev integral inequality and the multiplicative version of Karamata are provided. Various recent generalisations of the Grüss inequality for Riemann-Stieltjes integrals with Lipschitzian, monotonic, and of bounded variation integrators and integrands are presented. The famous Steffensen and Young inequalities, complemented by evocative new remarks and comments, are given as well. Generalisations of Steffensen's inequality over subintervals are also considered.

4.1 The Grüss Integral Inequality

Let $f, g : [a, b] \to \mathbb{R}$ be integrable on $[a, b]$ and satisfy

$$\phi \le f(x) \le \Phi, \quad \gamma \le g(x) \le \Gamma \text{ for all } x \in [a, b]. \tag{4.1}$$

Then we have the inequality

$$
\begin{aligned}
|T(f, g)| \quad &\tag{4.2}\\
:= \left| \frac{1}{b-a} \int_a^b f(x) g(x)\, dx - \frac{1}{b-a} \int_a^b f(x)\, dx \cdot \frac{1}{b-a} \int_a^b g(x)\, dx \right| \\
\le \frac{1}{4} (\Phi - \phi)(\Gamma - \gamma),
\end{aligned}
$$

where $T(f, g)$ is known as the Čebyšev functional.

The constant $\frac{1}{4}$ is the best possible.

PROOF Start with the well-known Korkine identity:

$$(b-a) \int_a^b f(x) g(x) \, dx - \int_a^b f(x) \, dx \int_a^b g(x) \, dx$$

$$= \frac{1}{2} \int_a^b \int_a^b (f(x) - f(y))(g(x) - g(y)) \, dx dy. \quad (4.3)$$

Applying the Cauchy-Bunyakovsky-Schwarz integral inequality for double integrals, we have

$$\left| \int_a^b \int_a^b (f(x) - f(y))(g(x) - g(y)) \, dx dy \right|$$

$$\leq \left[\int_a^b \int_a^b (f(x) - f(y))^2 \, dx dy \right]^{\frac{1}{2}} \left[\int_a^b \int_a^b (g(x) - g(y))^2 \, dx dy \right]^{\frac{1}{2}}. \quad (4.4)$$

Now, observe from (4.3) that we have

$$\frac{1}{2} \int_a^b \int_a^b (f(x) - f(y))^2 \, dx dy$$

$$= (b-a) \int_a^b f^2(x) \, dx - \left(\int_a^b f(x) \, dx \right)^2, \quad (4.5)$$

and a similar identity for g.

A simple calculation shows that

$$\frac{1}{b-a} \int_a^b f^2(x) \, dx - \left(\frac{1}{b-a} \int_a^b f(x) \, dx \right)^2$$

$$= \left(\Phi - \frac{1}{b-a} \int_a^b f(x) \, dx \right) \left(\frac{1}{b-a} \int_a^b f(x) \, dx - \phi \right)$$

$$- \frac{1}{b-a} \int_a^b (f(x) - \phi)(\Phi - f(x)) \, dx \quad (4.6)$$

and a similar identity for g.

By the assumption (4.1) we have $(f(x) - \phi)(\Phi - f(x)) \geq 0$ for all $x \in [a, b]$, and so

$$\int_a^b (f(x) - \phi)(\Phi - f(x)) \, dx \geq 0$$

which, from (4.6), implies that

$$\frac{1}{b-a} \int_a^b f^2(x)\,dx - \left(\frac{1}{b-a} \int_a^b f(x)\,dx\right)^2 \tag{4.7}$$

$$\leq \left(\Phi - \frac{1}{b-a} \int_a^b f(x)\,dx\right) \left(\frac{1}{b-a} \int_a^b f(x)\,dx - \phi\right)$$

$$\leq \frac{1}{4} \left[\left(\Phi - \frac{1}{b-a} \int_a^b f(x)\,dx\right) + \left(\frac{1}{b-a} \int_a^b f(x)\,dx - \phi\right)\right]^2$$

$$= \frac{1}{4}(\Phi - \phi)^2,$$

where we have used the fact that $\left(\frac{A+B}{2}\right)^2 \geq AB$.

A similar argument gives

$$\int_a^b \int_a^b (g(x) - g(y))^2 \,dx\,dy \leq \frac{1}{4}(\Gamma - \gamma)^2. \tag{4.8}$$

Using the inequality (4.4) via (4.5) and the estimations (4.7) and (4.8), we get

$$\left|\frac{1}{2} \int_a^b \int_a^b (f(x) - f(y))(g(x) - g(y))\,dx\,dy\right| \leq \frac{1}{4}(\Phi - \phi)(\Gamma - \gamma)(b-a)^2$$

and then, by (4.3), we deduce the desired inequality (4.2).

To prove the sharpness of the constant $\frac{1}{4}$, let us choose $f, g : [a, b] \to \mathbb{R}$, $f(x) = \mathrm{sgn}\left(x - \frac{a+b}{2}\right)$, $g(x) = \mathrm{sgn}\left(x - \frac{a+b}{2}\right)$. Then

$$\int_a^b f(x)g(x)\,dx = 1, \quad \int_a^b f(x)\,dx = \int_a^b g(x)\,dx = 0$$

and

$$\Phi - \phi = \Gamma - \gamma = 2$$

and the equality is realised in (4.2). $\qquad\qquad\qquad\qquad\qquad\square$

Comments

(a) The condition (4.1) can be relaxed by assuming the weaker condition

$$\int_a^b (f(x) - \phi)(\Phi - f(x))\,dx \geq 0, \quad \int_a^b (\Gamma - g(x))(g(x) - \gamma)\,dx \geq 0. \tag{4.9}$$

(b) If we assume that $f(x) = (x - a)^p$, $p > 0$, then we get the following inequality for the moments of g:

$$\left|\int_a^b (x-a)^p g(x)\,dx - \frac{(b-a)^p}{p+1} \int_a^b g(x)\,dx\right| \leq \frac{1}{4}(b-a)^{p+1}(M-m),$$

where we presume that $m \leq g(x) \leq M$ for $x \in [a, b]$.

(c) If g, for example, is known explicitly, then tighter bounds may be obtained from (4.4) by calculating $\left[\int_a^b \int_a^b (g(x) - g(y))^2 \, dx dy \right]^{\frac{1}{2}}$ rather than utilising the bound $\frac{\Gamma - \gamma}{2}$.

4.2 The Grüss-Čebyšev Integral Inequality

Let $f, g : [a, b] \rightarrow \mathbb{R}$ be L_1, L_2-Lipschitzian mappings on $[a, b]$, so that

$$|f(x) - f(y)| \leq L_1 |x - y|, \quad |g(x) - g(y)| \leq L_2 |x - y| \qquad (4.10)$$

for all $x, y \in [a, b]$. We then have the inequality [83]:

$$\left| \frac{1}{b-a} \int_a^b f(x) g(x) \, dx - \frac{1}{b-a} \int_a^b f(x) \, dx \cdot \frac{1}{b-a} \int_a^b g(x) \, dx \right|$$
$$\leq \frac{1}{12} L_1 L_2 (b-a)^2. \qquad (4.11)$$

The constant $\frac{1}{12}$ is the best possible.

PROOF   We have the Korkine identity

$$(b-a) \int_a^b f(x) g(x) \, dx - \int_a^b f(x) \, dx \cdot \int_a^b g(x) \, dx$$
$$= \frac{1}{2} \int_a^b \int_a^b (f(x) - f(y)) (g(x) - g(y)) \, dx dy. \qquad (4.12)$$

From condition (4.10), we have

$$|(f(x) - f(y)) (g(x) - g(y))| \leq L_1 L_2 (x - y)^2 \quad \text{for all } x, y \in [a, b]. \qquad (4.13)$$

Integrating (4.13) on $[a, b]^2$, we get

$$\int_a^b \int_a^b |(f(x) - f(y))(g(x) - g(y))| \, dx dy$$

$$\leq L_1 L_2 \int_a^b \int_a^b (x - y)^2 \, dx dy$$

$$= L_1 L_2 \int_a^b \int_a^b (x^2 - 2xy + y^2) \, dx dy$$

$$= L_1 L_2 \left[2(b - a) \int_a^b x^2 dx - 2 \left(\int_a^b x dx \right)^2 \right]$$

$$= 2L_1 L_2 \left[(b - a) \frac{b^3 - a^3}{3} - \left(\frac{b^2 - a^2}{2} \right)^2 \right] = \frac{L_1 L_2 (b - a)^4}{6}.$$

Using (4.12) we get (4.11).

Choose $f(x) = g(x) = x$. Then, $L_1 = L_2 = 1$ and

$$\frac{1}{b - a} \int_a^b f(x) g(x) \, dx - \frac{1}{b - a} \int_a^b f(x) \, dx \cdot \frac{1}{b - a} \int_a^b g(x) \, dx = \frac{(b - a)^2}{12}$$

and the identity (4.11) is realised. $\qquad\qquad\square$

Comments

(a) If we assume that f and g are differentiable on (a, b) and whose derivatives are bounded on (a, b), so that $\|f'\|_\infty := \sup_{t \in (a,b)} |f'(t)| < \infty$, then Čebyšev's result holds, that is,

$$\left| \frac{1}{b - a} \int_a^b f(x) g(x) \, dx - \frac{1}{b - a} \int_a^b f(x) \, dx \cdot \frac{1}{b - a} \int_a^b g(x) \, dx \right|$$

$$\leq \frac{1}{12} \|f'\|_\infty \|g'\|_\infty (b - a)^2. \quad (4.14)$$

The constant $\frac{1}{12}$ is the best possible.

(b) The inequality (4.11) can be generalised for Hölder type mappings. Assume that $f : [a, b] \to \mathbb{R}$ is of s-Hölder type, that is,

$$|f(x) - f(y)| \leq H_1 |x - y|^s, \quad H_1 > 0 \qquad (4.15)$$

for all $x, y \in [a, b]$, where $s \in (0, 1]$ is fixed. Further, let $g : [a, b] \to \mathbb{R}$ be of r-Hölder type, with $r \in (0, 1]$ and the constant $H_2 > 0$. Then, we have the inequality [57]:

$$\left| \frac{1}{b-a} \int_a^b f(x) g(x) \, dx - \frac{1}{b-a} \int_a^b f(x) \, dx \cdot \frac{1}{b-a} \int_a^b g(x) \, dx \right|$$
$$\leq \frac{H_1 H_2}{(r+s+1)(r+s+2)} (b-a)^{r+s}. \quad (4.16)$$

Indeed, as above, we have

$$\int_a^b \int_a^b |(f(x) - f(y))(g(x) - g(y))| \, dx dy$$
$$\leq H_1 H_2 \int_a^b \int_a^b |x-y|^{r+s} \, dx dy = H_1 H_2 \int_a^b \left[\frac{(b-y)^{r+s+1} + (y-a)^{r+s+1}}{r+s+1} \right] dy$$
$$= \frac{2(b-a)^{r+s+2} H_1 H_2}{(r+s+1)(r+s+2)}.$$

Using the identity (4.12) we deduce (4.16).

4.3 Karamata's Inequality

Let $f, g : [0,1] \rightarrow \mathbb{R}$ be integrable and satisfy the conditions $0 < a \leq f(t) \leq A$ and $0 < b \leq g(t) \leq B$ for $t \in [0,1]$, then [125]

$$K^{-2} \leq \frac{\int_0^1 f(x) g(x) \, dx}{\int_0^1 f(x) \, dx \int_0^1 g(x) \, dx} \leq K^2, \quad (4.17)$$

where

$$K = \frac{\sqrt{ab} + \sqrt{AB}}{\sqrt{aB} + \sqrt{Ab}} \geq 1. \quad (4.18)$$

PROOF Let $F = \int_0^1 f(t) \, dt$, $G = \int_0^1 g(t) \, dt$, and $V = \int_0^1 f(t) g(t) \, dt$. Then, by the mean value theorem, there are a y_1 and a y_2 such that

$$V - bF = \int_0^1 f(t)(g(t) - b) \, dt = (G - b) y_1, \quad a \leq y_1 \leq A$$

and

$$BF - V = \int_0^1 f(t)(B - g(t)) \, dt = (B - G) y_2, \quad a \leq y_2 \leq A.$$

Now,

$$\frac{a}{A} \le \frac{y_1}{y_2} \le \frac{A}{a}$$

and so

$$\left(\frac{G-b}{B-G}\right)\frac{a}{A} \le \frac{V-bF}{BF-V} \le \left(\frac{G-b}{B-G}\right)\frac{A}{a}. \tag{4.19}$$

Some algebraic manipulation of (4.19) gives

$$h\left(G\right) \le \frac{V}{F} \le H\left(G\right), \tag{4.20}$$

where

$$H\left(x\right) = \frac{ab\left(B-x\right) + AB\left(x-b\right)}{a\left(B-x\right) + A\left(x-b\right)}$$

and

$$h\left(x\right) = \frac{Ab\left(B-x\right) + aB\left(x-b\right)}{A\left(B-x\right) + a\left(x-b\right)}.$$

Comparing (4.17) with (4.20), it may be seen that $\frac{H(x)}{x}$ and $\frac{h(x)}{x}$ need to be investigated. It may be further noticed that $H\left(x\right)$ and $h\left(x\right)$ are related by an interchange of a and A so that only one needs to be analysed in detail.

Let

$$\theta\left(x\right) := \frac{H\left(x\right)}{x} = 1 + \frac{\left(A-a\right)\left(B-x\right)\left(x-b\right)}{x\left[a\left(B-x\right) + A\left(x-b\right)\right]}, \quad x \in [b, B]$$

then $\theta\left(b\right) = \theta\left(B\right) = 1$. Further,

$$\theta'\left(x\right) = \frac{\left(A-a\right)}{x^2} \cdot \frac{ab\left(B-x\right)^2 - AB\left(x-b\right)^2}{\left[a\left(B-x\right) + A\left(x-b\right)\right]^2}$$

and $\theta'\left(b\right) = \frac{(A-a)}{ab} > 0$, $\theta'\left(B\right) = -\frac{(A-a)}{AB} < 0$, so that a maximum exists at x^* where

$$\theta'\left(x^*\right) = 0 \quad \text{and} \quad x^* = \frac{\sqrt{aB} + \sqrt{Ab}}{\sqrt{ab} + \sqrt{AB}} \cdot \sqrt{bB}$$

with $b < x^* < \sqrt{bB} < B$.

Here $\theta\left(x^*\right) = K$ as given by (4.18).

Similarly, $\frac{h(x)}{x}$ attains its minimum at $x_* = \frac{\sqrt{AB}+\sqrt{ab}}{\sqrt{Ab}+\sqrt{aB}} \cdot \sqrt{bB} = \frac{bB}{x^*}$, with $b < \sqrt{bB} < x_* < B$. Also, $\frac{h(x_*)}{x_*} = \frac{1}{\theta(x^*)} = \frac{1}{K} < 1$. ◻

Comments
(a) The inequality (4.17) may easily be transferred to a general interval $[\alpha, \beta]$ to give

$$K^{-2} \le \frac{\int_\alpha^\beta f\left(x\right)dx \int_\alpha^\beta g\left(x\right)dx}{\left(\beta - \alpha\right)\int_\alpha^\beta f\left(x\right)g\left(x\right)dx} \le K^2, \tag{4.21}$$

where $0 < a \leq f(t) \leq A$ and $0 < b \leq g(t) \leq B$ for $t \in [\alpha, \beta]$.

Lupaş [134] extends Karamata's inequality for positive linear functionals. See Cerone [21] for applications to approximating and bounding the Gini mean difference used as a measure in business and social sciences.

(b) If we let $g(t) = \frac{1}{f(t)}$, then we obtain

$$0 < k^{-2}(\beta - \alpha)^2 \leq \int_\alpha^\beta f(x)\, dx \int_\alpha^\beta \frac{dx}{f(x)} \leq k^2 (\beta - \alpha)^2 \tag{4.22}$$

where

$$k^2 = \left(\frac{\sqrt{\frac{a}{A}} + \sqrt{\frac{A}{a}}}{2}\right)^2 = \frac{(a + A)^2}{4aA} \geq 1.$$

The right inequality is due to Schweitzer [159], and it may also be obtained from the Grüss inequality (4.2). Using Karamata's inequality (4.21) gives a tighter lower bound rather than zero.

4.4 Steffensen's Inequality

Let $f, g : [a, b] \to \mathbb{R}$ be integrable mappings on $[a, b]$ such that f is nonincreasing and $0 \leq g(t) \leq 1$ for $t \in [a, b]$. Then

$$\int_{b-\lambda}^b f(t)\, dt \leq \int_a^b f(t) g(t)\, dt \leq \int_a^{a+\lambda} f(t)\, dt, \tag{4.23}$$

where $\lambda = \int_a^b g(t)\, dt$.

PROOF First, notice that $\lambda \leq b - a$ so that $a + \lambda, b - \lambda \in [a, b]$. By direct calculation the identity

$$\int_a^{a+\lambda} f(t)\, dt - \int_a^b f(t) g(t)\, dt$$

$$= \int_a^{a+\lambda} (f(t) - f(a + \lambda))(1 - g(t))\, dt$$

$$+ \int_{a+\lambda}^b (f(a + \lambda) - f(t)) g(t)\, dt \tag{4.24}$$

may be shown to hold. Now, since f is nonincreasing and $0 \leq g(t) \leq 1$, then

$$\int_a^{a+\lambda} f(t)\, dt - \int_a^b f(t) g(t)\, dt \geq 0$$

and thus the second inequality in (4.23) is valid.

Replacing $g(t)$ by $1 - g(t)$ in (4.24) gives

$$\int_a^b f(t)g(t)\,dt - \int_{b-\lambda}^b f(t)\,dt$$

$$= \int_a^{b-\lambda} (f(t) - f(b-\lambda))g(t)\,dt$$

$$+ \int_{b-\lambda}^b (f(b-\lambda) - f(t))(1 - g(t))\,dt \quad (4.25)$$

which is again nonnegative by the postulates and thus proves the first inequality in (4.23). $\Box$

Comments

(a) The assumption that $0 \le g(t) \le 1$ may be relaxed in a number of ways. Hayashi showed a similar result to (4.23) for $0 \le g(t) \le A$, which may be obtained by replacing $g(t)$ by $\frac{g(t)}{A}$ in (4.23).

In his original paper, Steffensen [162] obtains a generalisation of the better known (4.23), which includes Hayashi's result as a special case. He shows that for $\phi \le g(x) \le \Phi$, $x \in [a, b]$ then

$$\phi \int_a^{b-\lambda} f(t)\,dt + \Phi \int_{b-\lambda}^b f(t)\,dt \le \int_a^b f(t)g(t)\,dt$$

$$\le \Phi \int_a^{a+\lambda} f(t)\,dt + \phi \int_{a+\lambda}^b f(t)\,dt,$$

where $\lambda = \int_a^b G(t)\,dt$, $G(t) = \frac{g(t) - \phi}{\Phi - \phi}$, $\Phi \ne \phi$.

Vasić and Pečarić [164] show that (4.23) holds if and only if

$$0 \le \int_x^b g(t)\,dt \le b - x \text{ and } 0 \le \int_a^x g(t)\,dt \le x - a \text{ for all } x \in [a, b].$$

(b) For $m \le h'(t) \le M$, Cerone and Dragomir [30] obtained the trapezoid inequality

$$\left| \int_a^b h(x)\,dx - \frac{b-a}{2}[h(a) + h(b)] \right| \le \frac{(b-a)^2}{2(M-m)}(S - m)(M - S),$$

where $S = \frac{h(b) - h(a)}{b-a}$ by taking $f(x) = \theta - x$, $\theta \in [a, b]$ and $(M - m)g(x) = h'(x) - m$ in (4.23). An inequality due to Iyengar [124] is recaptured by taking $-m = M$.

(c) For $[c,d] \subseteq [a,b]$ and $\lambda = d - c = \int_a^b g(t)\,dt$ the identity

$$\int_c^d f(t)\,dt - \int_a^b f(t)\,g(t)\,dt$$

$$= \int_a^c (f(d) - f(t))\,g(t)\,dt + \int_c^d (f(t) - f(d))(1 - g(t))\,dt$$

$$+ \int_d^b (f(d) - f(t))\,g(t)\,dt \quad (4.26)$$

holds. Taking $c = a$ and $d = a + \lambda$ reproduces (4.24). A comparable identity to (4.25) may be obtained from (4.26) by replacing $g(t)$ by $1 - g(t)$.

These identities were used by Cerone [26] to obtain

$$\int_{c_2}^{d_2} f(t)\,dt - r(c_2, d_2) \le \int_a^b f(t)\,g(t)\,dt \le \int_{c_1}^{d_1} f(t)\,dt - R(c_1, d_1)$$

where

$$r(c_2, d_2) = \int_{d_2}^b (f(c_2) - f(t))\,g(t)\,dt \ge 0$$

and

$$R(c_1, d_1) = \int_a^{c_1} (f(t) - f(d_1))\,g(t)\,dt \ge 0,$$

with $[c_i, d_i] \subset [a,b]$, $d_1 < d_2$ and $\lambda = d_i - c_i$, $i = 1, 2$.

Taking $c_1 = a$ and $d_2 = b$ recaptures the Steffensen inequality (4.23).

(d) Steffensen's inequality was first derived for actuarial applications but has since found application in many directions, including in special functions, bounding the Gini mean difference which arises in business and social sciences (see Cerone [21, 27]).

4.5 Young's Inequality

Let f be a continuous and increasing function on $[0, c]$ with $c > 0$. If $f(0) = 0$, $a \in [0, c]$ and $b \in [0, f(c)]$, then for f^{-1}, the inverse function of f, we have [141]

$$\int_0^a f(x)\,dx + \int_0^b f^{-1}(x)\,dx \ge ab. \quad (4.27)$$

Equality holds if and only if $b = f(a)$.

PROOF We start with the expression

$$g(a) = ba - \int_0^a f(x)\,dx \quad (4.28)$$

with a parameter $b > 0$. Then, $g(a)$ is an increasing function. Further, since $g'(a) = b - f(a)$, we have

$$g'(a) \begin{cases} > 0, \ 0 < a < f^{-1}(b) \\ = 0, \ a = f^{-1}(b) \\ < 0, \ a > f^{-1}(b). \end{cases}$$

Thus, $g(a)$ is the maximum value of g attained at $a = f^{-1}(b)$, that is,

$$g(a) \le \max g(x) = g\left(f^{-1}(b)\right). \tag{4.29}$$

Integration by parts shows that, from (4.28),

$$g(a) = ba - xf(x)\big]_0^a + \int_0^a xf'(x)\,dx$$

and so

$$g(a) = ba - af(a) + \int_0^a xf'(x)\,dx.$$

The substitution $y = f(x)$ produces

$$g(a) = ba - af(a) + \int_0^{f(a)} f^{-1}(y)\,dy \tag{4.30}$$

and so

$$g\left(f^{-1}(b)\right) = \int_0^b f^{-1}(y)\,dy. \tag{4.31}$$

Substitution of (4.31) into (4.29) and using (4.28) gives (4.27).

It may readily be seen from (4.30) that equality is obtained in (4.27) if and only if $b = f(a)$. ☐

Comments

(a) If $f(x) = x^{p-1}$, $p > 1$ then $f^{-1}(x) = x^{q-1}$ where $p \ge 1$, $\frac{1}{p} + \frac{1}{q} = 1$. Now from (4.27) we have

$$\frac{a^p}{p} + \frac{b^q}{q} \ge ab \ \text{ for } \ a, b \ge 0,$$

which is the inequality between the arithmetic and geometric means (see Equation 1.13).

(b) If $f(x) = \ln(x+1)$ and a is replaced by $a - 1$, then we obtain

$$ab \le a(\ln a - 1) + e^b.$$

(c) If $f : \mathbb{R}_+ \to \mathbb{R}_+$ is an increasing and continuous function on $\mathbb{R}_+$, then f^{-1} exists. Let $f(0) = 0$ and $f(x) \to +\infty$ as $x \to +\infty$. Further, let these properties hold for f^{-1}. The functions

$$F(x) = \int_0^x f(s)\,ds \ \text{ and } \ F_*(y) = \int_0^y f^{-1}(t)\,dt$$

are then convex on $\mathbb{R}^+$ and the following statements hold:

(i) $xy \le F(x) + F_*(y)$, $x \ge 0$, $y \ge 0$ (Young's inequality 4.27);

(ii) $xy = F(x) + F_*(y)$, $y = f(x) = F'(x)$;

(iii) $F'_*(y) = (F'(x))^{-1}$;

(iv) $F_{**} = F$;

(v) $F_*(y) = \sup\limits_{x \ge 0}(xy - f(x))$.

If $f^* : I^* \to \mathbb{R}$ and f is convex on I, then the conjugate function

$$f^*(y) = \sup_{x \in I}(xy - f(x))$$

is important in optimisation theory (see Mitrinović, Pečarić, and Fink [141, Chapter XIV]).

(d) Witkowski [166] gave a reverse for Young's inequality in 2007. Namely,

$$\int_0^a f(x)dx + \int_0^b f^{-1}(x)dx \le ab + (f^{-1}(b) - a)(b - f(a)).$$

4.6 Grüss Type Inequalities for the Stieltjes Integral of Bounded Integrands

Consider the *weighted Čebyšev functional*

$$T_w(f, g) := \frac{1}{\int_a^b w(t)\,dt} \int_a^b w(t)f(t)g(t)\,dt$$

$$- \frac{1}{\int_a^b w(t)\,dt} \int_a^b w(t)f(t)\,dt \cdot \frac{1}{\int_a^b w(t)\,dt} \int_a^b w(t)g(t)\,dt \quad (4.32)$$

where $f, g, w : [a, b] \to \mathbb{R}$ and $w(t) \ge 0$ for almost every (a.e.) $t \in [a, b]$ are measurable functions such that the involved integrals exist and $\int_a^b w(t)\,dt > 0$.

Cerone and Dragomir [29] obtained, among others, the following inequalities:

$$|T_w(f,g)| \tag{4.33}$$

$$\leq \frac{1}{2}(M-m)\frac{1}{\int_a^b w(t)\,dt}\int_a^b w(t)\left|g(t) - \frac{1}{\int_a^b w(s)\,ds}\int_a^b w(s)\,g(s)\,ds\right|dt$$

$$\leq \frac{1}{2}(M-m)\left[\frac{1}{\int_a^b w(t)\,dt}\int_a^b w(t)\right.$$

$$\left. \times \left|g(t) - \frac{1}{\int_a^b w(s)\,ds}\int_a^b w(s)\,g(s)\,ds\right|^p dt\right]^{\frac{1}{p}} \quad (p>1)$$

$$\leq \frac{1}{2}(M-m)\,\operatorname*{ess\,sup}_{t\in[a,b]}\left|g(t) - \frac{1}{\int_a^b w(s)\,ds}\int_a^b w(s)\,g(s)\,ds\right|$$

provided

$$-\infty < m \leq f(t) \leq M < \infty \quad \text{for a.e. } t \in [a,b] \tag{4.34}$$

and the corresponding integrals are finite. The constant $\frac{1}{2}$ is sharp in all the inequalities in (4.33) in the sense that it cannot be replaced by a smaller constant.

In addition, if

$$-\infty < n \leq g(t) \leq N < \infty \quad \text{for a.e. } t \in [a,b], \tag{4.35}$$

then the following refinement of the celebrated Grüss inequality is obtained:

$$|T_w(f,g)| \tag{4.36}$$

$$\leq \frac{1}{2}(M-m)\frac{1}{\int_a^b w(t)\,dt}\int_a^b w(t)\left|g(t) - \frac{1}{\int_a^b w(s)\,ds}\int_a^b w(s)\,g(s)\,ds\right|dt$$

$$\leq \frac{1}{2}(M-m)\left[\frac{1}{\int_a^b w(t)\,dt}\int_a^b w(t)\right.$$

$$\left. \times \left|g(t) - \frac{1}{\int_a^b w(s)\,ds}\int_a^b w(s)\,g(s)\,ds\right|^2 dt\right]^{\frac{1}{2}}$$

$$\leq \frac{1}{4}(M-m)(N-n).$$

Here, the constants $\frac{1}{2}$ and $\frac{1}{4}$ are also sharp in the sense mentioned above.

Before stating the next result, let us denote $C[a,b]$ to be the set of all continuous functions on $[a,b]$, and $BV[a,b]$ to be the set of all functions of bounded variation on $[a,b]$.

In this section, we extend the above results for Riemann-Stieltjes integrals.

For this purpose, we introduce the following Čebyšev functional for the Stieltjes integral:

$$T\left(f,g;u\right) := \frac{1}{u\left(b\right)-u\left(a\right)}\int_a^b f\left(t\right)g\left(t\right)du\left(t\right)$$

$$-\frac{1}{u\left(b\right)-u\left(a\right)}\int_a^b f\left(t\right)du\left(t\right)\cdot\frac{1}{u\left(b\right)-u\left(a\right)}\int_a^b g\left(t\right)du\left(t\right),\quad(4.37)$$

where $f,g\in C\left[a,b\right]$ and $u\in BV\left[a,b\right]$ with $u\left(b\right)\neq u\left(a\right)$.

For some recent inequalities for the Stieltjes integral, see Dragomir [68, 69]. The following result holds [78]:

Let $f,g:\left[a,b\right]\to\mathbb{R}$ be continuous on $\left[a,b\right]$ and $u:\left[a,b\right]\to\mathbb{R}$ with $u\left(a\right)\neq u\left(b\right)$. Assume also that there exist the real constants m,M such that

$$m\leq f\left(t\right)\leq M\quad\text{for each }t\in\left[a,b\right].\qquad(4.38)$$

If u is of bounded variation on $\left[a,b\right]$, then we have the inequality

$$\left|T\left(f,g;u\right)\right|\leq\frac{1}{2}\left(M-m\right)\frac{1}{\left|u\left(b\right)-u\left(a\right)\right|}$$

$$\times\left\|g-\frac{1}{u\left(b\right)-u\left(a\right)}\int_a^b g\left(s\right)du\left(s\right)\right\|_\infty\bigvee_a^b\left(u\right),\quad(4.39)$$

where $\bigvee_a^b\left(u\right)$ denotes the total variation of u in $\left[a,b\right]$. The constant $\frac{1}{2}$ is sharp, in the sense that it cannot be replaced by a smaller constant.

PROOF It is easy to see, by simple computation from (4.37) with the Stieltjes integral, that the following identity,

$$T\left(f,g;u\right)=\frac{1}{u\left(b\right)-u\left(a\right)}\int_a^b\left[f\left(t\right)-\frac{m+M}{2}\right]$$

$$\times\left[g\left(t\right)-\frac{1}{u\left(b\right)-u\left(a\right)}\int_a^b g\left(s\right)du\left(s\right)\right]du\left(t\right),\quad(4.40)$$

holds.

Using the known inequality

$$\left|\int_a^b p\left(t\right)dv\left(t\right)\right|\leq\sup_{t\in\left[a,b\right]}\left|p\left(t\right)\right|\bigvee_a^b\left(v\right),\qquad(4.41)$$

provided $p \in C\,[a, b]$ and $v \in BV\,[a, b]$, we have, by (4.40), that

$$|T\,(f, g; u)|$$

$$\leq \sup_{t \in [a,b]} \left| \left[f\,(t) - \frac{m + M}{2} \right] \left[g\,(t) - \frac{1}{u\,(b) - u\,(a)} \int_a^b g\,(s)\,du\,(s) \right] \right|$$

$$\cdot \frac{1}{|u\,(b) - u\,(a)|} \bigvee_a^b (u)$$

$$\left(\text{since } \left| f\,(t) - \frac{m + M}{2} \right| \leq \frac{M - m}{2} \text{ for any } t \in [a, b] \right)$$

$$\leq \frac{M - m}{2} \left\| g - \frac{1}{u\,(b) - u\,(a)} \int_a^b g\,(s)\,du\,(s) \right\|_\infty \cdot \frac{1}{|u\,(b) - u\,(a)|} \bigvee_a^b (u)$$

and the inequality (4.39) is proved.

To prove the sharpness of the constant $\frac{1}{2}$ in the inequality (4.39), we assume that it holds with a constant $C > 0$, namely,

$$|T\,(f, g; u)| \leq C\,(M - m)\,\frac{1}{|u\,(b) - u\,(a)|}$$

$$\times \left\| g - \frac{1}{u\,(b) - u\,(a)} \int_a^b g\,(s)\,du\,(s) \right\|_\infty \bigvee_a^b (u). \quad (4.42)$$

Let us consider the functions $f = g$, $f : [a, b] \to \mathbb{R}$, $f\,(t) = t$, $t \in [a, b]$, and $u : [a, b] \to \mathbb{R}$ given by

$$u\,(t) = \begin{cases} -1 & \text{if } t = a, \\ 0 & \text{if } t \in (a, b), \\ 1 & \text{if } t = b. \end{cases} \quad (4.43)$$

Then f, g are continuous on $[a, b]$, u is of bounded variation on $[a, b]$, and

$$\frac{1}{u\,(b) - u\,(a)} \int_a^b f\,(t)\,g\,(t)\,du\,(t) = \frac{b^2 + a^2}{2},$$

$$\frac{1}{u\,(b) - u\,(a)} \int_a^b f\,(t)\,du\,(t) = \frac{b + a}{2},$$

$$\left\| g - \frac{1}{u\,(b) - u\,(a)} \int_a^b g\,(s)\,du\,(s) \right\|_\infty = \sup_{t \in [a,b]} \left| t - \frac{a + b}{2} \right| = \frac{b - a}{2},$$

and

$$\bigvee_a^b (u) = 2, \quad M = b, \quad m = a.$$

Substituting these values in (4.42), we get

$$\left| \frac{a^2 + b^2}{2} - \frac{(a+b)^2}{4} \right| \leq C (b - a) \cdot \frac{1}{2} \cdot \frac{(b-a)}{2} \cdot 2,$$

giving $C \geq \frac{1}{2}$. This completes the proof. $\qquad\square$

The corresponding result for monotonic function u is incorporated in the following [78]:

Assume that f and g are as above. If $u : [a, b] \to \mathbb{R}$ is monotonic nondecreasing on $[a, b]$, then one has the inequality:

$$|T (f, g; u)| \leq \frac{1}{2} (M - m) \frac{1}{u(b) - u(a)}$$

$$\times \int_a^b \left| g(t) - \frac{1}{u(b) - u(a)} \int_a^b g(s) \, du(s) \right| du(t). \quad (4.44)$$

The constant $\frac{1}{2}$ is sharp in the sense that it cannot be replaced by a smaller constant.

PROOF Using the known inequality (4.41), we have (from the identity in Equation 4.40) that

$$|T (f, g; u)|$$

$$\leq \frac{1}{u(b) - u(a)} \int_a^b \left| f(t) - \frac{m + M}{2} \right|$$

$$\times \left| g(t) - \frac{1}{u(b) - u(a)} \int_a^b g(s) \, du(s) \right| du(t)$$

$$\leq \frac{1}{2} (M - m) \frac{1}{u(b) - u(a)} \int_a^b \left| g(t) - \frac{1}{u(b) - u(a)} \int_a^b g(s) \, du(s) \right| du(t).$$

Now, assume that the inequality (4.44) holds with a constant $D > 0$, instead of $\frac{1}{2}$, so that,

$$|T (f, g; u)| \leq D \cdot (M - m) \frac{1}{u(b) - u(a)}$$

$$\times \int_a^b \left| g(t) - \frac{1}{u(b) - u(a)} \int_a^b g(s) \, du(s) \right| du(t). \quad (4.45)$$

If we choose the same function as above, we observe that f, g are continuous and u is monotonic nondecreasing on $[a, b]$. Then, for these functions, we have

$$T (f, g; u) = \frac{a^2 + b^2}{2} - \frac{(a+b)^2}{4} = \frac{(b-a)^2}{4},$$

$$\int_a^b \left| g(t) - \frac{1}{u(b) - u(a)} \int_a^b g(s) \, du(s) \right| du(t) = \int_a^b \left| t - \frac{a+b}{2} \right| du(t)$$

$$= b - a,$$

so that by (4.45) we get

$$\frac{(b-a)^2}{4} \le D(b-a) \frac{1}{2}(b-a),$$

giving $D \ge \frac{1}{2}$. This completes the proof. □

The case when u is a Lipschitzian function is embodied in the following [78]:

Assume that $f, g : [a, b] \to \mathbb{R}$ are Riemann integrable functions on $[a, b]$ and f satisfies the condition (4.38). If $u : (a, b) \to \mathbb{R}$ $(u(b) \ne u(a))$ is Lipschitzian with the constant L, then we have the inequality

$$|T(f, g; u)| \le \frac{1}{2} L(M - m) \frac{1}{|u(b) - u(a)|}$$

$$\times \int_a^b \left| g(t) - \frac{1}{u(b) - u(a)} \int_a^b g(s) \, du(s) \right| dt. \quad (4.46)$$

The constant $\frac{1}{2}$ cannot be replaced by a smaller constant.

PROOF   It is well known that if $p : [a, b] \to \mathbb{R}$ is Riemann integrable on $[a, b]$ and $v : [a, b] \to \mathbb{R}$ is Lipschitzian with the constant L, then the Riemann-Stieltjes integral $\int_a^b p(t) \, dv(t)$ exists and

$$\left| \int_a^b p(t) \, dv(t) \right| \le L \int_a^b |p(t)| \, dt. \quad (4.47)$$

Using this fact and the identity (4.40), we deduce the following:

$$|T(f, g; u)|$$

$$\le \frac{L}{|u(b) - u(a)|} \int_a^b \left| f(t) - \frac{m + M}{2} \right|$$

$$\times \left| g(t) - \frac{1}{u(b) - u(a)} \int_a^b g(s) \, du(s) \right| dt$$

$$\le \frac{1}{2}(M - m) \frac{L}{|u(b) - u(a)|} \int_a^b \left| g(t) - \frac{1}{u(b) - u(a)} \int_a^b g(s) \, du(s) \right| dt$$

and the inequality (4.46) is proved.

Now, assume that (4.46) holds with a constant $E > 0$ instead of $\frac{1}{2}$, i.e.,

$$|T(f, g; u)| \le EL(M - m) \frac{1}{|u(b) - u(a)|}$$

$$\times \int_a^b \left| g(t) - \frac{1}{u(b) - u(a)} \int_a^b g(s) \, du(s) \right| dt. \quad (4.48)$$

Consider the function $f = g$, $f : [a, b] \to \mathbb{R}$ with

$$f(t) = \begin{cases} -1 \text{ if } t \in \left[a, \frac{a+b}{2}\right] \\ 1 \quad \text{if } t \in \left(\frac{a+b}{2}, b\right] \end{cases}$$

and $u : [a, b] \to \mathbb{R}$, $u(t) = t$. Then, obviously, f and g are Riemann integrable on $[a, b]$ and u is Lipschitzian with the constant $L = 1$.

Since

$$\frac{1}{u(b) - u(a)} \int_a^b f(t) g(t) \, du(t) = \frac{1}{b - a} \int_a^b dt = 1,$$

$$\frac{1}{u(b) - u(a)} \int_a^b f(t) \, du(t) = \frac{1}{u(b) - u(a)} \int_a^b g(t) \, du(t) = 0,$$

$$\int_a^b \left| g(t) - \frac{1}{u(b) - u(a)} \int_a^b g(s) \, du(s) \right| dt = \int_a^b dt = b - a$$

and

$$M = 1, \quad m = 1,$$

then, by (4.48), we deduce that $E \ge \frac{1}{2}$, and the result is completely proved. $\blacksquare$

Comments

For $f, g, w : [a, b] \to \mathbb{R}$, integrable and with the property that $\int_a^b w(t) \, dt \ne 0$, consider the weighted Čebyšev functional

$$T_w(f, g) := \frac{1}{\int_a^b w(t) \, dt} \int_a^b w(t) f(t) g(t) \, dt$$

$$- \frac{1}{\int_a^b w(t) \, dt} \int_a^b w(t) f(t) \, dt \cdot \frac{1}{\int_a^b w(t) \, dt} \int_a^b w(t) g(t) \, dt. \quad (4.49)$$

(a) If $f, g, w : [a, b] \to \mathbb{R}$ are continuous and there exist the real constants m, M such that

$$m \le f(t) \le M \text{ for each } t \in [a, b], \quad (4.50)$$

then one has the inequality

$$|T_w(f,g)| \leq \frac{1}{2}(M-m)\frac{1}{\left|\int_a^b w(s)\,ds\right|}$$

$$\times \left\|g - \frac{1}{\int_a^b w(s)\,ds}\int_a^b g(s)\,w(s)\,ds\right\|_{[a,b],\infty}\int_a^b |w(s)|\,ds. \quad (4.51)$$

The proof follows by (4.39) on choosing $u(t) = \int_a^t w(s)\,ds$.
(b) If f,g,w are as in (a) and $w(s) \geq 0$ for $s \in [a,b]$, then one has the inequality

$$|T_w(f,g)| \leq \frac{1}{2}(M-m)\frac{1}{\int_a^b w(s)\,ds}$$

$$\times \int_a^b \left|g(t) - \frac{1}{\int_a^b w(s)\,ds}\int_a^b g(s)\,w(s)\,ds\right|w(s)\,ds. \quad (4.52)$$

(c) If f,g are Riemann integrable on $[a,b]$ and f satisfies (4.50), and w is continuous on $[a,b]$, then one has the inequality

$$|T_w(f,g)| \leq \frac{1}{2}\|w\|_{[a,b],\infty}(M-m)\frac{1}{\left|\int_a^b w(s)\,ds\right|}$$

$$\times \int_a^b \left|g(t) - \frac{1}{\int_a^b w(s)\,ds}\int_a^b g(s)\,w(s)\,ds\right|ds. \quad (4.53)$$

4.7 Grüss Type Inequalities for the Stieltjes Integral of Lipschitzian Integrands

In the following, some Grüss type inequalities for the Stieltjes integral of Hölder and Lipshitz continuous integrands are presented [60]:

Let $f,g:[a,b] \to \mathbb{R}$ be such that f is of r-H-*Hölder type on* $[a,b]$, *i.e.*,

$$|f(t) - f(s)| \leq H|t-s|^r \quad \text{for any } t,s \in [a,b], \quad (4.54)$$

and g is continuous on $[a,b]$. If $u:[a,b] \to \mathbb{R}$ is of bounded variation on $[a,b]$ with $u(a) \neq u(b)$, then we have the inequality

$$|T(f,g;u)| \leq \frac{H(b-a)^r}{2^r}\cdot\frac{1}{|u(b)-u(a)|}$$

$$\times \left\|g - \frac{1}{u(b)-u(a)}\int_a^b g(s)\,du(s)\right\|_\infty \bigvee_a^b(u), \quad (4.55)$$

where $\bigvee_a^b (u)$ denotes the total variation of u on $[a, b]$.

PROOF   It is easy to see, by simple computation with the Stieltjes integral, that the following identity

$$T (f, g; u) = \frac{1}{u (b) - u (a)} \int_a^b \left[f (t) - f \left(\frac{a+b}{2} \right) \right]$$
$$\times \left[g (t) - \frac{1}{u (b) - u (a)} \int_a^b g (s) \, du (s) \right] du (t) \quad (4.56)$$

holds.

Using the known inequality (4.41), we have from (4.56) that

$$|T (f, g; u)| \leq \sup_{t \in [a,b]} \left| \left[f (t) - f \left(\frac{a+b}{2} \right) \right] \left[g (t) - \frac{1}{u (b) - u (a)} \int_a^b g (s) \, du (s) \right] \right|$$
$$\times \frac{1}{|u (b) - u (a)|} \bigvee_a^b (u)$$

$$\leq \sup_{t \in [a,b]} \left| f (t) - f \left(\frac{a+b}{2} \right) \right| \left\| g - \frac{1}{u (b) - u (a)} \int_a^b g (s) \, du (s) \right\|_\infty$$
$$\times \frac{1}{|u (b) - u (a)|} \bigvee_a^b (u)$$

$$\leq L \left(\frac{b-a}{2} \right)^r \left\| g - \frac{1}{u (b) - u (a)} \int_a^b g (s) \, du (s) \right\|_\infty$$
$$\times \frac{1}{|u (b) - u (a)|} \bigvee_a^b (u).$$

This completes the proof.   ☐

The following particular case is a Grüss type inequality for Lipschitz continuous integrands, and may be useful in applications [60]:

Let f be Lipschitzian with the constant $L > 0$, that is,

$$|f (t) - f (s)| \leq L |t - s| \quad \text{for any } t, s \in [a, b], \quad (4.57)$$

and u, g are as above. Then we have the inequality

$$|T (f, g; u)| \leq \frac{1}{2} \frac{L (b - a)}{|u (b) - u (a)|}$$
$$\times \left\| g - \frac{1}{u (b) - u (a)} \int_a^b g (s) \, du (s) \right\|_\infty \bigvee_a^b (u). \quad (4.58)$$

The constant $\frac{1}{2}$ cannot be replaced by a smaller constant.

PROOF   Inequality (4.58) follows by (4.55) for $r = 1$. It remains to prove only the sharpness of the constant $\frac{1}{2}$.

Consider the functions $f = g$, where $f : [a, b] \to \mathbb{R}$, $f(t) = t$, and $u : [a, b] \to \mathbb{R}$, given by

$$u(t) = \begin{cases} -1 & \text{if } t = a, \\ 0 & \text{if } t \in (a, b), \\ 1 & \text{if } t = b. \end{cases} \tag{4.59}$$

Then, f is Lipschitzian with the constant $L = 1$, g is continuous, and u is of bounded variation.

If we assume that the inequality (4.58) holds with a constant $C > 0$, namely,

$$|T(f, g; u)| \leq CL(b - a) \left\| g - \frac{1}{u(b) - u(a)} \int_a^b g(s) \, du(s) \right\|_\infty \bigvee_a^b (u), \tag{4.60}$$

and since

$$\frac{1}{u(b) - u(a)} \int_a^b f(t) g(t) \, du(t) = \frac{b^2 + a^2}{2},$$

$$\frac{1}{u(b) - u(a)} \int_a^b f(t) \, du(t) = \frac{1}{u(b) - u(a)} \int_a^b g(t) \, du(t) = \frac{b + a}{2},$$

$$\left\| g - \frac{1}{u(b) - u(a)} \int_a^b g(s) \, du(s) \right\|_\infty = \sup_{t \in [a,b]} \left| t - \frac{a + b}{2} \right| = \frac{b - a}{2}$$

and $\bigvee_a^b (u) = 2$, then, by (4.60), we have

$$\left| \frac{b^2 + a^2}{2} - \left(\frac{a + b}{2} \right)^2 \right| \leq C \frac{(b - a)}{2} \frac{b - a}{2} \cdot 2,$$

giving $C \geq \frac{1}{2}$.   ☐

The following result concerning monotonic function $u : [a, b] \to \mathbb{R}$ also holds [60]:

Assume that f and g are as above. If $u : [a, b] \to \mathbb{R}$ is monotonic nonde-

creasing on $[a, b]$ *with* $u(b) > u(a)$, *then we have the inequalities:*

$$|T(f, g; u)| \leq \frac{H}{u(b) - u(a)} \int_a^b \left| t - \frac{a+b}{2} \right|^r \tag{4.61}$$

$$\times \left| g(t) - \frac{1}{u(b) - u(a)} \int_a^b g(s) \, du(s) \right| du(t)$$

$$\leq \frac{H(b-a)^r}{2^r [u(b) - u(a)]}$$

$$\times \int_a^b \left| g(t) - \frac{1}{u(b) - u(a)} \int_a^b g(s) \, du(s) \right| du(t).$$

PROOF   Using the known inequality

$$\left| \int_a^b p(t) \, dv(t) \right| \leq \int_a^b |p(t)| \, dv(t), \tag{4.62}$$

provided $p \in C[a, b]$ *and* v *is monotonic nondecreasing on* $[a, b]$, we have, by (4.56), the following estimate:

$$|T(f, g; u)| \leq \frac{1}{u(b) - u(a)} \int_a^b \left| \left(f(t) - f\left(\frac{a+b}{2} \right) \right) \right.$$

$$\times \left(g(t) - \frac{1}{u(b) - u(a)} \int_a^b g(s) \, du(s) \right) \bigg| du(t)$$

$$\leq \frac{H}{u(b) - u(a)} \int_a^b \left| t - \frac{a+b}{2} \right|^r$$

$$\times \left| g(t) - \frac{1}{u(b) - u(a)} \int_a^b g(s) \, du(s) \right| du(t)$$

$$\leq \frac{H}{u(b) - u(a)} \sup_{t \in [a,b]} \left| t - \frac{a+b}{2} \right|^r$$

$$\times \int_a^b \left| g(t) - \frac{1}{u(b) - u(a)} \int_a^b g(s) \, du(s) \right| du(t)$$

which simply provides (4.61). ⬚

The particular case of Lipschitzian functions that is relevant for applications is embodied in the following result [60]:

Assume that f *is L-Lipschitzian,* g *is continuous, and* u *is monotonic non-*

decreasing on $[a, b]$ *with* $u(b) > u(a)$. *Then we have the inequalities*

$$|T(f, g; u)| \leq \frac{L}{u(b) - u(a)} \int_a^b \left| t - \frac{a+b}{2} \right| \qquad (4.63)$$

$$\times \left| g(t) - \frac{1}{u(b) - u(a)} \int_a^b g(s) \, du(s) \right| du(t)$$

$$\leq \frac{1}{2} \cdot \frac{L(b-a)}{u(b) - u(a)}$$

$$\times \int_a^b \left| g(t) - \frac{1}{u(b) - u(a)} \int_a^b g(s) \, du(s) \right| du(t).$$

The first inequality is sharp. The constant $\frac{1}{2}$ *in the second inequality cannot be replaced by a smaller constant.*

PROOF The inequality (4.63) follows by (4.61) on choosing $r = 1$ and $H \equiv L$. Assume that (4.63) holds with the constants $D, E > 0$, so that

$$|T(f, g; u)| \qquad (4.64)$$

$$\leq \frac{L \cdot D}{u(b) - u(a)} \int_a^b \left| t - \frac{a+b}{2} \right|$$

$$\times \left| g(t) - \frac{1}{u(b) - u(a)} \int_a^b g(s) \, du(s) \right| du(t)$$

$$\leq \frac{L \cdot E(b-a)}{u(b) - u(a)} \int_a^b \left| g(t) - \frac{1}{u(b) - u(a)} \int_a^b g(s) \, du(s) \right| du(t).$$

Consider the functions $f = g$, where $f : [a, b] \to \mathbb{R}$, $f(t) = t$, and u is as given by (4.59). Then, f is Lipschitzian with the constant $L = 1$, g is continuous, and u is monotonic nondecreasing on $[a, b]$.

Since

$$T(f, g; u) = \frac{(b-a)^2}{4}$$

and

$$\int_a^b \left| t - \frac{a+b}{2} \right| \left| g(t) - \frac{1}{u(b) - u(a)} \int_a^b g(s) \, du(s) \right| du(t) = \frac{(b-a)^2}{2},$$

$$\int_a^b \left| g(t) - \frac{1}{u(b) - u(a)} \int_a^b g(s) \, du(s) \right| du(t) = b - a,$$

then by (4.64) we deduce

$$\frac{(b-a)^2}{4} \leq \frac{D}{2} \cdot \frac{(b-a)^2}{2} \leq \frac{E(b-a)^2}{2},$$

giving $D \geq 1$ and $E \geq \frac{1}{2}$. ☐

Another natural possibility to obtain bounds for the functional $T(f, g; u)$, where u is Lipschitzian with the constant $K > 0$, is embodied in the following result [60]:

Assume that $f : [a, b] \to \mathbb{R}$ is of r-H-Hölder type on $[a, b]$. If $g : [a, b] \to \mathbb{R}$ is Riemann integrable on $[a, b]$ and $u : [a, b] \to \mathbb{R}$ is Lipschitzian with the constant $K > 0$ and $u(a) \neq u(b)$, then one has the inequalities:

$$|T(f, g; u)| \tag{4.65}$$

$$\leq \frac{HK}{|u(b) - u(a)|} \int_a^b \left| t - \frac{a+b}{2} \right|^r$$

$$\times \left| g(t) - \frac{1}{u(b) - u(a)} \int_a^b g(s)\, du(s) \right| dt$$

$$\leq \begin{cases} \frac{HK(b-a)^{r+1}}{2^r (r+1)|u(b)-u(a)|} \left\| g - \frac{1}{u(b)-u(a)} \int_a^b g(s)\, du(s) \right\|_\infty; \\[2ex] \frac{HK(b-a)^{r+\frac{1}{q}}}{2^r (qr+1)^{\frac{1}{q}} |u(b)-u(a)|} \left\| g - \frac{1}{u(b)-u(a)} \int_a^b g(s)\, du(s) \right\|_p \\[1ex] \qquad\qquad\qquad \text{if } p > 1, \ \frac{1}{p} + \frac{1}{q} = 1; \\[2ex] \frac{HK(b-a)^r}{2^r |u(b)-u(a)|} \left\| g - \frac{1}{u(b)-u(a)} \int_a^b g(s)\, du(s) \right\|_1. \end{cases}$$

PROOF Using the identity (4.56), we have successively

$$|T(f, g; u)| \leq \frac{K}{|u(b) - u(a)|} \int_a^b \left| f(t) - f\left(\frac{a+b}{2}\right) \right| \tag{4.66}$$

$$\times \left| g(t) - \frac{1}{u(b) - u(a)} \int_a^b g(s)\, du(s) \right| dt$$

$$\leq \frac{KH}{|u(b) - u(a)|} \int_a^b \left| t - \frac{a+b}{2} \right|^r$$

$$\times \left| g(t) - \frac{1}{u(b) - u(a)} \int_a^b g(s)\, du(s) \right| dt,$$

which proves the first inequality in (4.65).

Since

$$\int_a^b \left| t - \frac{a+b}{2} \right|^r \left| g(t) - \frac{1}{u(b)-u(a)} \int_a^b g(s)\, du(s) \right| dt$$

$$\leq \left\| g - \frac{1}{u(b)-u(a)} \int_a^b g(s)\, du(s) \right\|_\infty \int_a^b \left| t - \frac{a+b}{2} \right|^r dt$$

$$= \frac{(b-a)^{r+1}}{2^r(r+1)} \left\| g - \frac{1}{u(b)-u(a)} \int_a^b g(s)\, du(s) \right\|_\infty ,$$

then by (4.66) we deduce the first part in the second inequality in (4.65).

By Hölder's integral inequality we have

$$\int_a^b \left| t - \frac{a+b}{2} \right|^r \left| g(t) - \frac{1}{u(b)-u(a)} \int_a^b g(s)\, du(s) \right| dt$$

$$\leq \left(\int_a^b \left| t - \frac{a+b}{2} \right|^{qr} dt \right)^{\frac{1}{q}} \left(\int_a^b \left| g(t) - \frac{1}{u(b)-u(a)} \int_a^b g(s)\, du(s) \right|^p dt \right)^{\frac{1}{p}}$$

$$= \left[\frac{(b-a)^{qr+1}}{2^{qr}(qr+1)} \right]^{\frac{1}{q}} \left\| g - \frac{1}{u(b)-u(a)} \int_a^b g(s)\, du(s) \right\|_p$$

$$= \frac{(b-a)^{r+\frac{1}{q}}}{2^r(qr+1)^{\frac{1}{q}}} \left\| g - \frac{1}{u(b)-u(a)} \int_a^b g(s)\, du(s) \right\|_p .$$

Using (4.66), we deduce the second part of the second inequality in (4.65).

Finally, since

$$\left| t - \frac{a+b}{2} \right|^r \leq \left(\frac{b-a}{2} \right)^r, \quad t \in [a,b],$$

we deduce

$$\int_a^b \left| t - \frac{a+b}{2} \right|^r \left| g(t) - \frac{1}{u(b)-u(a)} \int_a^b g(s)\, du(s) \right| dt$$

$$\leq \frac{(b-a)^r}{2^r} \left\| g - \frac{1}{u(b)-u(a)} \int_a^b g(s)\, du(s) \right\|_1 .$$

This completes the proof. □

The following particular cases are useful in applications [60].

If f is Lipschitzian with the constant L and g and u are as above, then we have the inequalities:

$$|T\left(f,g;u\right)| \leq \frac{LK}{|u\left(b\right)-u\left(a\right)|}\int_a^b \left|t-\frac{a+b}{2}\right|$$

$$\times \left|g\left(t\right)-\frac{1}{u\left(b\right)-u\left(a\right)}\int_a^b g\left(s\right)du\left(s\right)\right| dt \quad (4.67)$$

$$\leq \begin{cases} \dfrac{LK\left(b-a\right)^2}{4\left|u\left(b\right)-u\left(a\right)\right|}\left\|g-\dfrac{1}{u\left(b\right)-u\left(a\right)}\int_a^b g\left(s\right)du\left(s\right)\right\|_\infty ; \\[3ex] \dfrac{LK\left(b-a\right)^{1+\frac{1}{q}}}{2\left(q+1\right)^{\frac{1}{q}}\left|u\left(b\right)-u\left(a\right)\right|}\left\|g-\dfrac{1}{u\left(b\right)-u\left(a\right)}\int_a^b g\left(s\right)du\left(s\right)\right\|_p \\[1ex] \qquad\qquad\qquad \text{if } p>1,\ \frac{1}{p}+\frac{1}{q}=1; \\[3ex] \dfrac{LK\left(b-a\right)}{2\left|u\left(b\right)-u\left(a\right)\right|}\left\|g-\dfrac{1}{u\left(b\right)-u\left(a\right)}\int_a^b g\left(s\right)du\left(s\right)\right\|_1 . \end{cases}$$

The first inequality in (4.67) is sharp.

The constants $\frac{1}{4}$ and $\frac{1}{2}$ in the first and second branches of the second inequality cannot be replaced by smaller constants.

PROOF   The inequality (4.67) follows readily from (4.65) on choosing $r = 1$.

Now, assume that the following inequalities hold:

$$|T\left(f,g;u\right)| \qquad\qquad\qquad\qquad\qquad\qquad\qquad\qquad (4.68)$$

$$\leq \frac{CLK}{|u\left(b\right)-u\left(a\right)|}\int_a^b \left|t-\frac{a+b}{2}\right|$$

$$\times \left|g\left(t\right)-\frac{1}{u\left(b\right)-u\left(a\right)}\int_a^b g\left(s\right)du\left(s\right)\right| dt$$

$$\leq \begin{cases} \dfrac{DLK\left(b-a\right)^2}{\left|u\left(b\right)-u\left(a\right)\right|}\left\|g-\dfrac{1}{u\left(b\right)-u\left(a\right)}\int_a^b g\left(s\right)du\left(s\right)\right\|_\infty ; \\[3ex] \dfrac{ELK\left(b-a\right)^{1+\frac{1}{q}}}{\left(q+1\right)^{\frac{1}{q}}\left|u\left(b\right)-u\left(a\right)\right|}\left\|g-\dfrac{1}{u\left(b\right)-u\left(a\right)}\int_a^b g\left(s\right)du\left(s\right)\right\|_p \\[1ex] \qquad\qquad\qquad \text{if } p>1,\ \frac{1}{p}+\frac{1}{q}=1; \end{cases}$$

with $C, D, E > 0$.

Consider the functions $f, g, u : [a, b] \rightarrow \mathbb{R}$, defined by $f(t) = t - \frac{a+b}{2}$, $u(t) = t$ and

$$g(t) = \begin{cases} -1 \text{ if } t \in \left[a, \frac{a+b}{2}\right], \\ 1 \quad \text{if } t \in \left(\frac{a+b}{2}, b\right]. \end{cases}$$

Then both f and u are Lipschitzian with the constant $L = K = 1$ and g is Riemann integrable on $[a, b]$.

We obviously have

$$|T(f, g; u)| = \frac{1}{b-a} \int_a^b f(t) g(t) \, dt - \frac{1}{b-a} \int_a^b f(t) \, dt \cdot \frac{1}{b-a} \int_a^b g(t) \, dt$$

$$= \frac{b-a}{4},$$

$$\int_a^b \left| t - \frac{a+b}{2} \right| \left| g(t) - \frac{1}{u(b) - u(a)} \int_a^b g(s) \, du(s) \right| dt = \frac{(b-a)^2}{4}$$

$$\left\| g - \frac{1}{u(b) - u(a)} \int_a^b g(s) \, du(s) \right\|_\infty = \|g\|_\infty = 1$$

and

$$\left\| g - \frac{1}{u(b) - u(a)} \int_a^b g(s) \, du(s) \right\|_p = \|g\|_p = (b-a)^{\frac{1}{p}}.$$

Consequently, by (4.68), one has

$$\frac{b-a}{4} \leq \frac{C}{b-a} \frac{(b-a)^2}{4} \leq \begin{cases} \frac{D(b-a)^2}{b-a} \cdot 1 \\ \frac{E(b-a)^2}{(q+1)^{\frac{1}{q}}(b-a)} \end{cases}$$

giving

$$\frac{1}{4} \leq \frac{C}{4} \leq \begin{cases} D \\ \frac{E}{(q+1)^{\frac{1}{q}}}, \quad q > 1, \end{cases}$$

from which we conclude that $C \geq 1$, $D \geq \frac{1}{4}$ and $E \geq \frac{(q+1)^{\frac{1}{q}}}{4}$. Letting $q \to 1+$, we deduce $E \geq \frac{1}{2}$ and the result is proved. $\square$

Comments

(a) If $f, g, w : [a, b] \to \mathbb{R}$ are continuous and f is of r-H-Hölder type, then one has the identity

$$|T_w (f, g)| \leq \frac{H \, |b - a|^r}{2^r} \cdot \frac{1}{\left| \int_a^b w(s) \, ds \right|}$$

$$\times \left\| g - \frac{1}{\int_a^b w(s) \, ds} \int_a^b g(s) w(s) \, ds \right\|_{[a,b],\infty} \int_a^b |w(s)| \, ds.$$

The proof follows from (4.55) on choosing $u(t) = \int_a^t w(s) \, ds$.

(b) If f, g, w are as in **(a)** and $w(s) \geq 0$ for $s \in [a, b]$, then one has the inequality

$$|T_w (f, g)| \tag{4.69}$$

$$\leq \frac{H}{\int_a^b w(s) \, ds} \int_a^b \left| t - \frac{a + b}{2} \right|^r$$

$$\times \left| g(t) - \frac{1}{\int_a^b w(s) \, ds} \int_a^b g(s) w(s) \, ds \right| w(s) \, ds$$

$$\leq \frac{H (b - a)^r}{2^r \int_a^b w(s) \, ds} \int_a^b \left| g(t) - \frac{1}{\int_a^b w(s) \, ds} \int_a^b g(s) w(s) \, ds \right| w(s) \, ds.$$

The proof follows from (4.61) on choosing $u(t) = \int_a^t w(s) \, ds$.

(c) If f is of r-H-Hölder type, g is Riemann integrable on $[a, b]$, and w is continuous on $[a, b]$, then one has the inequality

$$|T_w (f, g)| \tag{4.70}$$

$$\leq \frac{H \, \|w\|_{[a,b],\infty}}{\left| \int_a^b w(s) \, ds \right|} \int_a^b \left| t - \frac{a + b}{2} \right|^r \left| g(t) - \frac{1}{\int_a^b w(s) \, ds} \int_a^b g(s) w(s) \, ds \right| dt$$

$$\leq \begin{cases} \dfrac{H \, \|w\|_{[a,b],\infty} \, (b - a)^{r+1}}{2^r (r + 1) \left| \int_a^b w(s) \, ds \right|} \left\| g - \dfrac{1}{\int_a^b w(s) \, ds} \int_a^b g(s) w(s) \, ds \right\|_{[a,b],\infty} \; ; \\[4mm] \dfrac{H \, \|w\|_{[a,b],\infty} \, (b - a)^{r+\frac{1}{q}}}{2^r (qr + 1)^{\frac{1}{q}} \left| \int_a^b w(s) \, ds \right|} \left\| g - \dfrac{1}{\int_a^b w(s) \, ds} \int_a^b g(s) w(s) \, ds \right\|_{[a,b],p} \; , \\ \qquad\qquad\qquad\qquad\qquad\qquad\quad p > 1, \; \frac{1}{p} + \frac{1}{q} = 1; \\[4mm] \dfrac{H \, \|w\|_{[a,b],\infty} \, (b - a)^r}{2^r \left| \int_a^b w(s) \, ds \right|} \left\| g - \dfrac{1}{\int_a^b w(s) \, ds} \int_a^b g(s) w(s) \, ds \right\|_{[a,b],1} \; . \end{cases}$$

The proof follows from (4.65) on choosing $u(t) = \int_a^t w(s)\,ds$.

4.8 Other Grüss Type Inequalities for the Riemann-Stieltjes Integral

Dragomir and Fedotov [90] considered the following functional,

$$D(f;u) := \int_a^b f(x)\,du(x) - [u(b) - u(a)] \cdot \frac{1}{b-a} \int_a^b f(t)\,dt, \qquad (4.71)$$

with the provision that the involved integrals exist.

In the same article [90], the following result, in estimating the above functional, was obtained:

Let $f, u : [a,b] \to \mathbb{R}$ be such that u is L-Lipschitzian on $[a,b]$ so that

$$|u(x) - u(y)| \le L|x - y| \quad \text{for any } x, y \in [a,b] \quad (L > 0) \qquad (4.72)$$

and f is Riemann integrable on $[a,b]$. If $m, M \in \mathbb{R}$ are such that

$$m \le f(x) \le M \quad \text{for any } x, y \in [a,b], \qquad (4.73)$$

then we have the inequality

$$|D(f;u)| \le \frac{1}{2}L(M - m)(b - a). \qquad (4.74)$$

The constant $\frac{1}{2}$ is sharp in the sense that it cannot be replaced by a smaller constant.

In Dragomir and Fedotov [91], the following result complementing the above was obtained.

Let $f, u : [a,b] \to \mathbb{R}$ be such that $u : [a,b] \to \mathbb{R}$ is of bounded variation in $[a,b]$ and $f : [a,b] \to \mathbb{R}$ is K-Lipschitzian $(K > 0)$. Then we have the inequality

$$|D(f;u)| \le \frac{1}{2}K(b - a)\bigvee_a^b(u). \qquad (4.75)$$

The constant $\frac{1}{2}$ is sharp in the above sense.

The proofs are left to the reader.

In this section, we consider some Grüss type inequalities for the Riemann-Stieltjes integral. Before stating the results, we introduce the following identity (cf. Dragomir [59]).

Let $f, u : [a, b] \to \mathbb{R}$ be such that the Stieltjes integral $\int_a^b f(t)\, du(t)$ and the Riemann integral $\int_a^b f(t)\, dt$ exist. Then we have the identity

$$D(f; u) = \int_a^b \Phi(t)\, df(t) = \frac{1}{b-a} \int_a^b \Gamma(t)\, df(t) \qquad (4.76)$$

$$= \frac{1}{b-a} \int_a^b (t-a)(b-t)\, \Delta(t)\, df(t),$$

where

$$\Phi(t) := \frac{(t-a)u(b) + (b-t)u(a)}{b-a} - u(t), \quad t \in [a, b],$$

$$\Gamma(t) := (t-a)[u(b) - u(t)] - (b-t)[u(t) - u(a)], \quad t \in [a, b],$$

and

$$\Delta(t) := [u; b, t] - [u; t, a], \quad t \in (a, b),$$

where $[u; \alpha, \beta]$ is the divided difference, that is, we recall,

$$[u; \alpha, \beta] := \frac{u(\alpha) - u(\beta)}{\alpha - \beta}.$$

PROOF   We observe that

$$\int_a^b \Phi(t)\, df(t) = \int_a^b \left[\frac{(t-a)u(b) + (b-t)u(a)}{b-a} - u(t) \right] df(t)$$

$$= \left[\frac{(t-a)u(b) + (b-t)u(a)}{b-a} - u(t) \right] f(t) \Bigg|_a^b$$

$$- \int_a^b f(t)\, d\left[\frac{(t-a)u(b) + (b-t)u(a)}{b-a} - u(t) \right]$$

$$= -\int_a^b f(t) \left[\frac{u(b) - u(a)}{b-a} dt - du(t) \right]$$

$$= \int_a^b f(t)\, du(t) - \frac{u(b) - u(a)}{b-a} \int_a^b f(t)\, dt$$

and the first identity in (4.76) is proved.

The second and third identities are obvious.   □

If u is an integral, so that $u(t) = \int_a^t g(s)\, ds$, then from (4.76) we deduce Cerone's result [23] in

$$T(f, g) = \frac{1}{(b-a)} \int_a^b \Psi(t)\, df(t), \qquad (4.77)$$

where

$$\Psi(t) = \frac{t-a}{b-a} \int_a^b g(s)\,ds - \int_a^t g(s)\,ds \quad (t \in [a,b])$$

$$= \frac{1}{b-a} \left[(t-a) \int_t^b g(s)\,ds - (b-t) \int_a^t g(s)\,ds \right] \quad (t \in [a,b])$$

$$= \frac{(t-a)(b-t)}{b-a} \left[\frac{\int_t^b g(s)\,ds}{b-t} - \frac{\int_a^t g(s)\,ds}{t-a} \right] \quad (t \in (a,b)).$$

If $w : [a,b] \to \mathbb{R}$ is integrable and $\int_a^b w(t)\,dt \neq 0$, then the choice of

$$u(t) := \frac{\int_a^t w(s)\,g(s)\,ds}{\int_a^t w(s)\,ds}, \quad t \in [a,b] \tag{4.78}$$

produces

$$D_w(f;u) = \frac{\int_a^b w(s)\,f(s)\,g(s)\,ds}{\int_a^b w(s)\,ds} - \frac{\int_a^b w(s)\,g(s)\,ds}{\int_a^b w(s)\,ds} \cdot \frac{1}{b-a} \int_a^b f(t)\,dt$$

$$=: E(f,g;w).$$

The following weighted integral inequality is thus a natural application of the above result (4.76).

If w, f, g are Riemann integrable on $[a,b]$ and $\int_a^b w(t)\,dt \neq 0$, then

$$E(f,g;w) = \int_a^b \Phi_w(t)\,df(t) = \frac{1}{b-a} \int_a^b \Gamma_w(t)\,df(t) \tag{4.79}$$

$$= \frac{1}{b-a} \int_a^b (t-a)(b-t)\,\Delta_w(t)\,df(t),$$

where

$$\Phi_w(t) = \left(\frac{t-a}{b-a} \right) \cdot \frac{\int_a^b w(s)\,g(s)\,ds}{\int_a^b w(s)\,ds} - \frac{\int_a^t w(s)\,g(s)\,ds}{\int_a^b w(s)\,ds},$$

$$\Gamma_w(t) = (t-a) \frac{\int_t^b w(s)\,g(s)\,ds}{\int_a^b w(s)\,ds} - (b-t) \frac{\int_a^t w(s)\,g(s)\,ds}{\int_a^b w(s)\,ds},$$

$$\Delta_w(t) = \frac{\int_t^b w(s)\,g(s)\,ds}{(b-t) \int_a^b w(s)\,ds} - \frac{\int_a^t w(s)\,g(s)\,ds}{(t-a) \int_a^b w(s)\,ds}.$$

The following general result in bounding the functional $D(f;u)$ may be stated [59].

Let $f, u : [a, b] \to \mathbb{R}$.

(i) *If f is of bounded variation and u is continuous on $[a, b]$, then*

$$|D(f; u)| \leq \begin{cases} \sup_{t \in [a,b)} |\Phi(t)| V_a^b(f), \\[2ex] \frac{1}{b-a} \sup_{t \in [a,b]} |\Gamma(t)| V_a^b(f), \\[2ex] \frac{1}{b-a} \sup_{t \in (a,b)} [(t-a)(b-t)|\Delta(t)|] V_a^b(f). \end{cases} \tag{4.80}$$

(ii) *If f is L-Lipschitzian and u is Riemann integrable on $[a, b]$, then*

$$|D(f; u)| \leq \begin{cases} L \int_a^b |\Phi(t)| \, dt, \\[2ex] \frac{L}{b-a} \int_a^b |\Gamma(t)| \, dt, \\[2ex] \frac{L}{b-a} \int_a^b (t-a)(b-t)|\Delta(t)| \, dt. \end{cases} \tag{4.81}$$

(iii) *If f is monotonic nondecreasing on $[a, b]$ and u is continuous on $[a, b]$, then*

$$|D(f; u)| \leq \begin{cases} \int_a^b |\Phi(t)| \, df(t), \\[2ex] \frac{1}{b-a} \int_a^b |\Gamma(t)| \, df(t), \\[2ex] \frac{1}{b-a} \int_a^b (t-a)(b-t)|\Delta(t)| \, df(t). \end{cases} \tag{4.82}$$

The proof follows by the identity (4.76), and further details are omitted.

Comments

It is natural to consider the following particular cases, since they provide simpler bounds for the functional $D(f; u)$ in terms of Δ defined above [59].

If f is of bounded variation and u is continuous on $[a, b]$, then

$$|D(f; u)| \leq \frac{1}{b-a} \sup_{t \in [a,b]} [(t-a)(b-t)\Delta(t)] \bigvee_a^b (f) \tag{4.83}$$

$$\leq \frac{b-a}{4} \|\Delta\|_\infty \bigvee_a^b (f),$$

where $V_a^b(f)$ denotes the total variation of f on $[a, b]$.

If f is L-Lipschitzian and u is Riemann integrable on $[a, b]$, then

$$|D(f; u)| \qquad (4.84)$$

$$\leq \frac{L}{b-a} \int_a^b (t-a)(b-t) |\Delta(t)| \, dt$$

$$\leq \begin{cases} \frac{1}{6} L (b-a)^2 \|\Delta\|_\infty, \\ L (b-a)^{1+\frac{1}{q}} [B(q+1, q+1)]^{\frac{1}{q}} \|\Delta\|_p, \ p > 1, \ \frac{1}{p} + \frac{1}{q} = 1; \\ \frac{1}{4} L (b-a) \|\Delta\|_1, \end{cases}$$

where $B(\cdot, \cdot)$ is Euler's beta function (see Equation 3.33).

If f is monotonic nondecreasing and g is continuous, then

$$|D(f; u)| \qquad (4.85)$$

$$\leq \frac{1}{b-a} \int_a^b (t-a)(b-t) |\Delta(t)| \, dt$$

$$\leq \begin{cases} \frac{1}{4} (b-a) \int_a^b |\Delta(t)| \, df(t), \\ \frac{1}{b-a} \left(\int_a^b [(b-t)(t-a)]^q \, df(t) \right)^{\frac{1}{q}} \left(\int_a^b |\Delta(t)|^p \, df(t) \right)^{\frac{1}{p}}, \\ \qquad\qquad p > 1, \ \frac{1}{p} + \frac{1}{q} = 1; \\ \frac{1}{b-a} \|\Delta\|_\infty \int_a^b (t-a)(b-t) \, df(t). \end{cases}$$

If one chooses $u(t) = \int_a^t g(s) \, ds$ above, then the result incorporated in Theorems 4–6 of Cerone [23] are recaptured.

Finally, the following result on the positivity of the functional $D(f; u)$ holds [59]:

Let f be a monotonic nondecreasing function on $[a, b]$. If u is such that

$$\Delta(t) = \Delta(u; a, t, b) := [u; b, t] - [u; t, a] \geq 0 \qquad (4.86)$$

for any $t \in (a, b)$, then we have the inequality

$$D(f; u) \geq \frac{1}{b-a} \left| \int_a^b (t-a)(b-t) \left[|[u; b, t]| - |[u; t, a]| \right] df(t) \right| \geq 0. \quad (4.87)$$

The proof is similar to the case in Theorem 3 by Cerone and Dragomir [36]; and the details are left to the interested reader.

It is easy to see that a sufficient condition for (4.86) to hold is that $u : [a, b] \to \mathbb{R}$ is a convex function on $[a, b]$.

4.9 Inequalities for Monotonic Integrators

The following result holds [59]:

Let $f : [a, b] \to \mathbb{R}$ be L-Lipschitzian on $[a, b]$ and u monotonic nondecreasing on $[a, b]$. Then we have the inequality

$$|D (f; u)| \leq \frac{1}{2} L (b - a) [u (b) - u (a) - P (u)] \tag{4.88}$$

$$\leq \frac{1}{2} L (b - a) [u (b) - u (a)],$$

where $D (f; u)$ is as defined in (4.71) and

$$P (u) := \frac{4}{(b - a)^2} \int_a^b u (x) \left(x - \frac{a + b}{2} \right) dx \tag{4.89}$$

$$= \frac{4}{(b - a)^2} \int_a^b \left[u (x) - u \left(\frac{a + b}{2} \right) \right] \left(x - \frac{a + b}{2} \right) dx \geq 0.$$

The constant $\frac{1}{2}$ in both inequalities is sharp in the sense that it cannot be replaced by a smaller constant.

PROOF Since u is monotonic nondecreasing on $[a, b]$, then from (4.71)

$$|D(f, u)| = \left| \int_a^b f (x) \, du (x) - \frac{u (b) - u (a)}{b - a} \int_a^b f (t) \, dt \right| \tag{4.90}$$

$$= \left| \int_a^b \left(f (x) - \frac{1}{b - a} \int_a^b f (t) \, dt \right) du (x) \right|$$

$$\leq \int_a^b \left| f (x) - \frac{1}{b - a} \int_a^b f (t) \, dt \right| du (x).$$

Taking into account that f is L-Lipschitzian, we have the following Ostrowski type inequality (see for example, Dragomir [87]),

$$\left| f (x) - \frac{1}{b - a} \int_a^b f (t) \, dt \right| \leq L \left[\frac{1}{4} + \left(\frac{x - \frac{a+b}{2}}{b - a} \right)^2 \right] (b - a), \tag{4.91}$$

for all $x \in [a, b]$, from which we deduce

$$\int_a^b \left| f(x) - \frac{1}{b-a} \int_a^b f(t)\, dt \right| du(x)$$

$$\leq L(b-a) \int_a^b \left[\frac{1}{4} + \left(\frac{x - \frac{a+b}{2}}{b-a} \right)^2 \right] du(x). \quad (4.92)$$

Now, observe that, by the integration by parts formula for the Riemann-Stieltjes integral, we have

$$\int_a^b \left(x - \frac{a+b}{2} \right)^2 du(x) = u(x) \left(x - \frac{a+b}{2} \right)^2 \Big|_a^b - 2 \int_a^b u(x) \left(x - \frac{a+b}{2} \right) dx$$

$$= \frac{(b-a)^2}{4} [u(b) - u(a)] - 2 \int_a^b u(x) \left(x - \frac{a+b}{2} \right) dx$$

and then

$$\int_a^b \left[\frac{1}{4} + \left(\frac{x - \frac{a+b}{2}}{b-a} \right)^2 \right] du(x)$$

$$= \frac{1}{2} [u(b) - u(a)] - \frac{2}{(b-a)^2} \int_a^b u(x) \left(x - \frac{a+b}{2} \right) dx. \quad (4.93)$$

Using (4.90)–(4.93) we deduce the first part of (4.88).

The second part is obvious by (4.89) which follows by the monotonicity of u on $[a, b]$.

To prove the sharpness of the constant $\frac{1}{2}$, assume that (4.94) holds with the constants $C, D > 0$, so that

$$|D(f; u)| \leq CL(b-a) [u(b) - u(a) - P(u)] \quad (4.94)$$
$$\leq DL(b-a) [u(b) - u(a)].$$

Consider the functions $f, u : [a, b] \to \mathbb{R}$ given by $f(x) = x - \frac{a+b}{2}$ and

$$u(x) = \begin{cases} 0 \text{ if } x \in [a, b) \\ \\ 1 \text{ if } x = b. \end{cases}$$

Thus f is L-Lipschitzian with the constant $L = 1$ and u is monotonic nondecreasing.

We observe that for these particular choices of f and u,

$$D(f; u) = \int_a^b f(x)\, du(x) = f(x) u(x) \Big|_a^b - \int_a^b u(x)\, dx = \frac{b-a}{2},$$

$K(u) = 0$, and $u(b) - u(a) = 1$, giving in (4.94)

$$\frac{b-a}{2} \leq C(b-a) \leq D(b-a).$$

Thus, $C, D \geq \frac{1}{2}$, proving the sharpness of the constant $\frac{1}{2}$ in (4.88).   ☐

Another result of this type is the following one [59].

Let $u : [a, b] \to \mathbb{R}$ be monotonic nondecreasing on $[a, b]$ and $f : [a, b] \to \mathbb{R}$ be of bounded variation such that the Stieltjes integral $\int_a^b f(x) \, du(x)$ exists. Then we have the inequality

$$|D(f; u)| \leq [u(b) - u(a) - Q(u)] \bigvee_a^b (f) \tag{4.95}$$

$$\leq [u(b) - u(a)] \bigvee_a^b (f),$$

where

$$Q(u) := \frac{1}{b-a} \int_a^b \operatorname{sgn}\left(x - \frac{a+b}{2}\right) u(x) \, dx \tag{4.96}$$

$$= \frac{1}{b-a} \int_a^b \operatorname{sgn}\left(x - \frac{a+b}{2}\right) \left[u(x) - u\left(\frac{a+b}{2}\right)\right] dx \geq 0.$$

The first inequality in (4.95) is sharp in the sense that the constant 1 cannot be replaced by a smaller constant.

PROOF   Since u is monotonic nondecreasing, we have (see Equation 4.90) that

$$|D(f; u)| \leq \int_a^b \left| f(x) - \frac{1}{b-a} \int_a^b f(t) \, dt \right| du(x). \tag{4.97}$$

Using the following Ostrowski type inequality obtained by Dragomir [70],

$$\left| f(x) - \frac{1}{b-a} \int_a^b f(t) \, dt \right| \leq \left[\frac{1}{2} + \left| \frac{x - \frac{a+b}{2}}{b-a} \right| \right] \bigvee_a^b (f), \tag{4.98}$$

for any $x \in [a, b]$, we have

$$\int_a^b \left| f(x) - \frac{1}{b-a} \int_a^b f(t) \, dt \right| du(x)$$

$$\leq \bigvee_a^b (f) \int_a^b \left[\frac{1}{2} + \left| \frac{x - \frac{a+b}{2}}{b-a} \right| \right] du(x). \tag{4.99}$$

A simple calculation with the Riemann-Stieltjes integral gives that

$$\int_a^b \left| x - \frac{a+b}{2} \right| du\,(x) \tag{4.100}$$

$$= \int_a^{\frac{a+b}{2}} \left(\frac{a+b}{2} - x \right) du\,(x) + \int_{\frac{a+b}{2}}^b \left(x - \frac{a+b}{2} \right) du\,(x)$$

$$= u\,(x) \left(\frac{a+b}{2} - x \right) \Big|_a^{\frac{a+b}{2}} + \int_a^{\frac{a+b}{2}} u\,(x)\,dx$$

$$+ \left(x - \frac{a+b}{2} \right) u\,(x) \Big|_{\frac{a+b}{2}}^b - \int_{\frac{a+b}{2}}^b u\,(x)\,dx$$

$$= \frac{1}{2}\,(b-a)\,[u\,(b) - u\,(a)] - \int_a^b \mathrm{sgn}\left(x - \frac{a+b}{2} \right) u\,(x)\,dx$$

and then by (4.97)–(4.100) we deduce the first inequality in (4.95).

The second part of (4.95) follows by (4.96) which holds by the monotonicity property of u.

Now, assume that the first inequality in (4.95) holds with a constant $E > 0$, so that

$$|D\,(f;u)| \le E \bigvee_a^b (f)\,[u\,(b) - u\,(a) - Q\,(u)]. \tag{4.101}$$

Consider the mappings $f, u : [a, b] \to \mathbb{R}$, $f\,(x) = x - \frac{a+b}{2}$, and

$$u\,(x) = \begin{cases} 0 \ \text{if} \ x \in \left[a, \frac{a+b}{2} \right], \\[2mm] 1 \ \text{if} \ x \in \left(\frac{a+b}{2}, b \right]. \end{cases}$$

Then we have

$$D\,(f;u) = \int_a^b f\,(x)\,du\,(x) - \frac{u\,(b) - u\,(a)}{b-a} \int_a^b f\,(t)\,dt$$

$$= \int_a^b \left(x - \frac{a+b}{2} \right) du\,(x) = \left(x - \frac{a+b}{2} \right) u\,(x) \Big|_a^b - \int_a^b u\,(x)\,dx$$

$$= \frac{b-a}{2}\,[u\,(b) + u\,(a)] = \frac{b-a}{2}$$

and

$$\bigvee_a^b (f) \left[u(b) - u(a) - Q(u) \right]$$

$$= (b-a) \left[u(b) - u(a) - \left(\frac{1}{b-a} \int_a^{\frac{a+b}{2}} \mathrm{sgn}\left(x - \frac{a+b}{2} \right) u(x)\, dx \right.\right.$$

$$\left.\left. + \frac{1}{b-a} \int_{\frac{a+b}{2}}^b \mathrm{sgn}\left(x - \frac{a+b}{2} \right) u(x)\, dx \right) \right]$$

$$= \frac{b-a}{2}.$$

Thus, by (4.101) we obtain

$$\frac{b-a}{2} \le E \cdot \frac{b-a}{2},$$

showing that $E \ge 1$, and the result is proved. $\qquad\qquad\qquad \Box$

Comments
Similar results for composite rules in approximating the Riemann-Stieltjes integral may be stated, but we omit the details (see Dragomir [59]).

4.10   Generalisations of Steffensen's Inequality over Subintervals

The following inequality is due to Steffensen [162] (see also Mitrinović, Pečarić, and Fink [141, p. 181]).

Let $f, g : [a,b] \to \mathbb{R}$ be integrable mappings on $[a,b]$ such that f is nonincreasing and $0 \le g(t) \le 1$ for $t \in [a,b]$. Then

$$\int_{b-\lambda}^b f(t)\, dt \le \int_a^b f(t) g(t)\, dt \le \int_a^{a+\lambda} f(t)\, dt, \qquad (4.102)$$

where

$$\lambda = \int_a^b g(t)\, dt. \qquad (4.103)$$

Hayashi obtains a similar result (see Mitrinović, Pečarić, and Fink [141, p. 182]), which may ostensibly be achieved from (4.102) by replacing $g(t)$ with $\frac{g(t)}{A}$, where A is some positive constant.

For Steffensen type inequalities with integrals over a measure space, see the work of Gauchman [118].

It may be noted that Steffensen's inequality (4.102) and (4.103) involve integrals of functions and of products of functions.

The following result [26] will be useful for the subsequent work:

Let $f, g : [a, b] \to \mathbb{R}$ be integrable mappings on $[a, b]$. Further, let $[c, d] \subseteq [a, b]$ with $\lambda = d - c = \int_a^b g(t)\, dt$. Then the following identities hold. Namely,

$$\int_c^d f(t)\, dt - \int_a^b f(t) g(t)\, dt = \int_a^c (f(d) - f(t)) g(t)\, dt$$
$$+ \int_c^d (f(t) - f(d)) (1 - g(t))\, dt + \int_d^b (f(d) - f(t)) g(t)\, dt \quad (4.104)$$

and

$$\int_a^b f(t) g(t)\, dt - \int_c^d f(t)\, dt = \int_a^c (f(t) - f(c)) g(t)\, dt$$
$$+ \int_c^d (f(c) - f(t)) (1 - g(t))\, dt + \int_d^b (f(t) - f(c)) g(t)\, dt. \quad (4.105)$$

PROOF   We follow the proof by Cerone [26]. Let

$$S(c, d; a, b) = \int_c^d f(t)\, dt - \int_a^b f(t) g(t)\, dt, \quad a \leq c < d \leq b, \quad (4.106)$$

then

$$S(c, d; a, b) = \int_c^d (1 - g(t)) f(t)\, dt - \left[\int_a^c f(t) g(t)\, dt + \int_d^b f(t) g(t)\, dt \right]$$
$$= \int_c^d (1 - g(t)) (f(t) - f(d))\, dt + f(d) \int_c^d (1 - g(t))\, dt$$
$$+ \int_a^c (f(d) - f(t)) g(t)\, dt - f(d) \int_a^c g(t)\, dt$$
$$+ \int_d^b (f(d) - f(t)) g(t)\, dt - f(d) \int_d^b g(t)\, dt.$$

The identity (4.104) is readily obtained on noting that

$$f(d) \left[\int_c^d dt - \int_a^b g(t)\, dt \right] = 0.$$

Identity (4.105) follows immediately from (4.104) and (4.106) on realising that (4.105) is $S\left(d, c; b, a\right)$ or, equivalently, $-S\left(c, d; a, b\right)$. □

If $c = a$ in (4.104) and $d = b$ in (4.105), then the identities obtained by Mitrinović [139] are recaptured.

The following result was developed by Cerone [26].

Let $f, g : [a, b] \to \mathbb{R}$ be integrable mappings on $[a, b]$ and let f be nonincreasing. Further, let $0 \leq g\left(t\right) \leq 1$ and $\lambda = \int_a^b g\left(t\right) dt = d_i - c_i$, where $[c_i, d_i] \subset [a, b]$ for $i = 1, 2$ and $d_1 \leq d_2$. Then the result

$$\int_{c_2}^{d_2} f\left(t\right) dt - r\left(c_2, d_2\right) \leq \int_a^b f\left(t\right) g\left(t\right) dt \leq \int_{c_1}^{d_1} f\left(t\right) dt + R\left(c_1, d_1\right) \quad (4.107)$$

holds, where

$$r\left(c_2, d_2\right) = \int_{d_2}^b \left(f\left(c_2\right) - f\left(t\right)\right) g\left(t\right) dt \geq 0$$

and

$$R\left(c_1, d_1\right) = \int_a^{c_1} \left(f\left(t\right) - f\left(d_1\right)\right) g\left(t\right) dt \geq 0.$$

PROOF   From (4.104) and (4.106)

$$S\left(c_1, d_1; a, b\right) + \int_a^{c_1} \left(f\left(t\right) - f\left(d_1\right)\right) g\left(t\right) dt$$

$$= \int_{c_1}^{d_1} \left(f\left(t\right) - f\left(d_1\right)\right) \left(1 - g\left(t\right)\right) dt + \int_{d_1}^b \left(f\left(d_1\right) - f\left(t\right)\right) g\left(t\right) dt \geq 0$$

by the stated assumptions.

Hence, from (4.106)

$$\int_{c_1}^{d_1} f\left(t\right) dt + \int_a^{c_1} \left(f\left(t\right) - f\left(d_1\right)\right) g\left(t\right) dt - \int_a^b f\left(t\right) g\left(t\right) dt \geq 0.$$

Thus, the right inequality is valid.

Now, from (4.105) and (4.106), we have

$$-S\left(c_2, d_2; a, b\right) + \int_{d_2}^b \left(f\left(c_2\right) - f\left(t\right)\right) g\left(t\right) dt$$

$$= \int_a^{c_2} \left(f\left(t\right) - f\left(c_2\right)\right) g\left(t\right) dt + \int_{c_2}^{d_2} \left(f\left(c_2\right) - f\left(t\right)\right) \left(1 - g\left(t\right)\right) dt \geq 0$$

from the assumptions.

Thus, from (4.106) we obtain

$$\int_a^b f(t) g(t) \, dt - \left[\int_{c_2}^{d_2} f(t) \, dt - \int_{d_2}^b (f(c_2) - f(t)) g(t) \, dt \right] \geq 0,$$

giving the left inequality.

Both $r(c_2, d_2)$ and $R(c_1, d_1)$ are nonnegative since f is nonincreasing and g is nonnegative. The result is now completely proved. □

If in (4.107) we take $c_1 = a$ and so $d_1 = a + \lambda$, then $R(a, a + \lambda) = 0$. Further, taking $d_2 = b$ so that $c_2 = b - \lambda$ gives $r(b - \lambda, b) = 0$. The Steffensen inequality (4.102) is thus recaptured. Since (4.103) holds, then $c_2 \geq a$ and $d_1 \leq b$, giving $[c_i, d_i] \subset [a, b]$. The result (4.107) may thus be viewed as a generalisation of the Steffensen inequality as given in (4.102), which allows for two equal length subintervals that are not necessarily at the ends of $[a, b]$.

It may be advantageous at times to gain coarser bounds that may be more easily evaluated. The following bounds were obtained by Cerone [26].

Let the conditions leading to the result (4.107) *hold. Then*

$$\int_{c_2}^b f(t) \, dt - (b - d_2) f(c_2) \leq \int_a^b f(t) g(t) \, dt \qquad (4.108)$$

$$\leq \int_a^{d_1} f(t) \, dt - (c_1 - a) f(d_1).$$

PROOF From result (4.107) on using the fact that $0 \leq g(t) \leq 1$, we get

$$0 \leq r(c_2, d_2) = \int_{d_2}^b (f(c_2) - f(t)) g(t) \, dt$$

$$\leq \int_{d_2}^b (f(c_2) - f(t)) \, dt = (b - d_2) f(c_2) - \int_{d_2}^b f(t) \, dt.$$

This implies that

$$\int_{c_2}^{d_2} f(t) \, dt - r(c_2, d_2) \geq \int_{c_2}^{d_2} f(t) \, dt - (b - d_2) f(c_2) + \int_{d_2}^b f(t) \, dt.$$

Combining the two integrals produces the left inequality of (4.108). The proof for the right inequality is similar. □

If we take $c_1 = a$ and so $d_1 = a + \lambda$ and $d_2 = b$ such that $c_2 = b - \lambda$, then (4.108) again recaptures Steffensen's inequality.

The following results produce alternative identities to those given by (4.104) and (4.105). The current identities involve the integral mean of $f(\cdot)$ over the subinterval $[c, d]$ (see Cerone [26]).

Let $f, g : [a, b] \to \mathbb{R}$ be integrable mappings on $[a, b]$. Define

$$G(x) = \int_a^x g(t)\, dt$$

and

$$\lambda = G(b) = d - c$$

where $[c, d] \subset [a, b]$.
The following identities hold:

$$\int_a^b f(x) g(x)\, dx - \int_c^d f(y)\, dy$$

$$= \lambda \left[f(b) - \mathcal{M}(f; c, d) \right] - \int_a^b G(x)\, df(x) \qquad (4.109)$$

and

$$\int_c^d f(y)\, dy - \int_a^b f(x) g(x)\, dx$$

$$= \lambda \left[\mathcal{M}(f; c, d) - f(a) \right] - \int_a^b \left[\lambda - G(x) \right] df(x), \qquad (4.110)$$

where $\mathcal{M}(f; c, d)$ is the integral mean of $f(\cdot)$ over $[c, d]$.

PROOF Consider

$$L := \int_a^b f(x) g(x)\, dx - \int_c^d f(y)\, dy.$$

Then from the postulates $\frac{G(b)}{d - c} = 1$, giving

$$L = \int_a^b f(x) g(x)\, dx - \frac{1}{d - c} \int_a^b g(x)\, dx \int_c^d f(y)\, dy.$$

Combining the integrals gives

$$L = \int_a^b g(x) \left[f(x) - \mathcal{M}(f; c, d) \right] dx, \qquad (4.111)$$

where $\mathcal{M}(f; c, d)$ is the integral mean of f over $[c, d]$.
Integration by parts from (4.111) gives

$$L = G(x)\left[f(x) - \mathcal{M}(f;c,d)\right]\Big]_a^b - \int_a^b G(x)\,df(x)$$

and so

$$L = \lambda\left[f(b) - \mathcal{M}(f;c,d)\right] - \int_a^b G(x)\,df(x)$$

since $G(b) = \lambda$ and $G(a) = 0$.

The second identity follows along similar lines. □

The identities (4.109) and (4.110) were used to procure the following results (see Cerone [26]):

Let $f, g : [a, b] \to \mathbb{R}$ be integrable mappings on $[a, b]$ and f be nonincreasing. Further, let $g(t) \geq 0$ and $G(x) = \int_a^x g(t)\,dt$ with $\lambda = G(b) = d_i - c_i$ where $[c_i, d_i] \subset [a, b]$ for $i = 1, 2$ and $d_1 < d_2$. Then

$$\int_{c_2}^{d_2} f(y)\,dy - \lambda\left[\mathcal{M}(f;c_2,d_2) - f(b)\right]$$

$$\leq \int_a^b f(x)\,g(x)\,dx \leq \int_{c_1}^{d_1} f(y)\,dy + \lambda\left[f(a) - \mathcal{M}(f;c_1,d_1)\right] \quad (4.112)$$

where $d_2 > d_1$.

PROOF From (4.109), and using the facts that f is nonincreasing and $g(t) \geq 0$, we get [26]

$$-\int_a^b G(x)\,df(x) \geq 0.$$

This implies that

$$\int_a^b f(x)\,g(x)\,dx - \left[\int_{c_2}^{d_2} f(y)\,dy + \lambda\left[f(b) - \mathcal{M}(f;c_2,d_2)\right]\right] \geq 0;$$

and so the left inequality is obtained.

Similarly, from (4.110) and the postulates we have

$$-\int_a^b \left[\lambda - G(x)\right]df(x) \geq 0,$$

which gives

$$\int_{c_1}^{d_1} f(y)\,dy + \lambda\left[\mathcal{M}(f;c_1,d_1) - f(a)\right] - \int_a^b f(x)\,g(x)\,dx \geq 0.$$

□

Comments

(a) The lower and upper inequalities in (4.112) may be simplified to $\lambda f(b)$ and $\lambda f(a)$, respectively, since

$$\int_c^d f(y)\,dy = \lambda \mathcal{M}(f;c,d).$$

That is,

$$\lambda f(b) \leq \int_a^b f(x)g(x)\,dx \leq \lambda f(a). \tag{4.113}$$

The result should not be overly surprising, since it may be obtained directly from the postulates since

$$\inf_{x\in[a,b]} f(x) \int_a^b g(x)\,dx \leq \int_a^b f(x)g(x)\,dx \leq \sup_{x\in[a,b]} f(x)\int_a^b g(x)\,dx.$$

The result (4.113) readily follows on noting that

$$\int_a^b g(x)[f(x) - f(b)]\,dx \geq 0$$

and

$$\int_a^b f(x)[f(a) - f(x)]\,dx \geq 0.$$

(b) The following result expresses $\mathcal{S}$ as a double integral over a rectangular region to obtain bounds for the Steffensen functional.

Let $f, g : [a, b] \to \mathbb{R}$ be integrable mappings on $[a, b]$ such that f is non-increasing and $0 \leq g(t) \leq 1$ for $t \in [a, b]$. Further, let $[c, d] \subseteq [a, b]$ with $\lambda = d - c = \int_a^b g(t)\,dt$, then the following inequality holds,

$$|\mathcal{S}| := \left| \int_a^b f(x)g(x)\,dx - \int_c^d f(y)\,dy \right| \tag{4.114}$$

$$\leq (a + b + c + d)\mathcal{M}(f;c,d) - \frac{4}{d-c}\mu(f;c,d)$$

$$+ \int_a^c f(x)\,dx - \int_d^b f(x)\,dx$$

$$\leq (c - a)f(a) - (b - d)f(b) + (a + b + c + d)f(c)$$
$$- 2(d + c)f(d),$$

where $\mathcal{M}(f;c,d)$ is the integral mean and

$$\mu(f;c,d) = \int_c^d xf(x)\,dx. \tag{4.115}$$

PROOF   The following identity may easily be shown to hold:

$$S = \frac{1}{d-c} \int_a^b \int_c^d g(x) (f(x) - f(y)) \, dy dx. \tag{4.116}$$

Then

$$|S| \le \frac{\|g\|_\infty}{d-c} \int_a^b \int_c^d |f(x) - f(y)| \, dy dx,$$

where $\|g\|_\infty := \operatorname*{ess\,sup}_{x \in [a,b]} |g(x)| = 1$, from the postulates.

Thus,

$$|S| \le \frac{1}{d-c} \int_a^b \int_c^d |f(x) - f(y)| \, dy dx := I. \tag{4.117}$$

Now, using the fact that f is nonincreasing and that $[c, d] \subseteq [a, b]$, we have

$$(d-c) I = \int_a^c \int_c^d (f(x) - f(y)) \, dy dx + \int_c^d \int_c^x (f(y) - f(x)) \, dy dx$$

$$+ \int_c^d \int_x^d (f(x) - f(y)) \, dy dx + \int_d^b \int_c^d (f(y) - f(x)) \, dy dx.$$

Using the facts that

$$\int_c^d \int_c^x f(y) \, dy dx = \int_c^d (d-x) f(x) \, dx \quad \text{and}$$

$$\int_c^d \int_x^d f(y) \, dy dx = \int_c^d (x-c) f(x) \, dx,$$

then

$$(d-c) I = (d-c) \left[\int_a^c f(x) \, dx - \int_d^b f(x) \, dx \right]$$

$$+ \int_c^d [a+b+c+d-4x] f(x) \, dx, \tag{4.118}$$

from which the first inequality results.

The coarser inequality is obtained using the fact that f is nonincreasing, giving, from (4.118),

$$I \le (c-a) f(a) - (b-d) f(b) + (a+b+c+d) f(c) - \frac{4}{d-c} f(d) \int_c^d x \, dx,$$

which upon simplification gives the second inequality in (4.114).   ⬜

(**c**) The following result was obtained by Cerone [26]:

Let the conditions pertaining to the result (4.114) hold. Then

$$-2cM\left(f;c,d\right) - \phi\left(c,d\right) \leq \int_a^b f\left(x\right)g\left(x\right)dx \qquad (4.119)$$

$$\leq 2dM\left(f;c,d\right) + \phi\left(c,d\right),$$

where

$$\phi\left(c,d\right) = \left(a+b\right)M\left(f;c,d\right) + \int_a^c f\left(x\right)dx$$

$$- \int_d^b f\left(x\right)dx - \frac{4}{d-c}\mu\left(f;c,d\right) \qquad (4.120)$$

with $M\left(f;c,d\right)$ being the integral mean and $\mu\left(f;c,d\right)$ the mean of f over the subinterval $[c,d]$ given by (4.115).

When these results are specialised to refer to end intervals, then it is an open problem as to whether the bounds are better or worse than those provided by the Steffensen bounds (4.102).

Chapter 5

Inequalities in Inner Product Spaces

In Modern Functional Analysis the concept of a Hilbert space plays a fundamental role. It creates a natural background for solving numerous problems in mathematics, physics, engineering, and science.

The purpose of this chapter is to present some of the fundamental inequalities that involve inner products and norms. Various inequalities related to the Schwarz, triangle, and Bessel inequalities are surveyed. More recent results due to Boas-Bellman's and Bombieri's generalisations of the Bessel inequality are presented. The generalisations due to Kurepa, Buzano, and Precupanu of the Schwarz inequality as well as the Dunkl-Williams inequality are also given. Last but not least, an account on recent advancement of the Grüss inequality in inner product spaces is provided as well. These, as usual, have been complemented by numerous remarks and comments that engender further research and application of the results.

5.1 Schwarz's Inequality in Inner Product Spaces

Let $(H; \langle \cdot, \cdot \rangle)$ be an inner product space over $\mathbb{K}$, where $\mathbb{K} = \mathbb{R}$ or $\mathbb{C}$. Then

$$\|x\| \, \|y\| \geq |\langle x, y \rangle| \tag{5.1}$$

for all $x, y \in H$. Equality holds in (5.1) iff x and y are linearly dependent, that is, $x = \lambda y$ for some $\lambda \in \mathbb{K}$.

PROOF Observe that

$$0 \leq \left\| \|y\|^2 x - \overline{\langle x, y \rangle} y \right\|^2 \tag{5.2}$$

$$= \|y\|^4 \|x\|^2 - 2 \|y\|^2 \langle x, y \rangle \overline{\langle x, y \rangle} + |\langle x, y \rangle|^2 \|y\|^2$$

$$= \|y\|^2 \left(\|x\|^2 \|y\|^2 - |\langle x, y \rangle|^2 \right).$$

If $\|y\| = 0$, that is, $y = 0$, then the inequality (5.1) is obviously satisfied. Assume that $\|y\| \neq 0$. Then by (5.2) we get

$$\|x\|^2 \|y\|^2 \geq |\langle x, y \rangle|^2 \tag{5.3}$$

which is clearly equivalent to (5.1).

If $x = \lambda y$ $(\lambda \in \mathbb{K})$, then $\|x\| \|y\| = |\lambda| \|y\|^2$ and $|\langle x, y \rangle| = |\lambda| \|y\|^2$, and we have equality in (5.1).

Now, if in (5.1) we have equality, then by (5.2)

$$\left\| \|y\|^2 x - \overline{\langle x, y \rangle} y \right\| = 0$$

and thus, x and y are linearly dependent. This completes the proof. $\quad\square$

Comments

Assume that $\mathcal{H}$ is the class of all inner products which can be defined on H. Let the mapping $\mu : \mathcal{H} \times \mathcal{H}^2 \to [0, \infty]$ be defined by

$$\mu \left(\langle \cdot, \cdot \rangle, x, y \right) := \|x\| \|y\| - |\langle x, y \rangle|.$$

Then $\mu \left(\cdot, x, y \right)$ is *superadditive in* $\mathcal{H}$ [98], that is,

$$\mu \left(\langle \cdot, \cdot \rangle_1 + \langle \cdot, \cdot \rangle_2, x, y \right) \geq \mu \left(\langle \cdot, \cdot \rangle_1, x, y \right) + \mu \left(\langle \cdot, \cdot \rangle_2, x, y \right) \geq 0 \qquad (5.4)$$

for all $\langle \cdot, \cdot \rangle_i \in \mathcal{H}$ $(i = 1, 2)$.

The proof may be demonstrated as follows:

$$\mu \left(\langle \cdot, \cdot \rangle_1 + \langle \cdot, \cdot \rangle_2, x, y \right) \qquad (5.5)$$
$$= \left(\|x\|_1^2 + \|x\|_2^2 \right)^{\frac{1}{2}} \left(\|y\|_1^2 + \|y\|_2^2 \right)^{\frac{1}{2}} - |\langle x, y \rangle_1 + \langle x, y \rangle_2|$$
$$\geq \|x\|_1 \|y\|_1 + \|x\|_2 \|y\|_2 - |\langle x, y \rangle_1| - |\langle x, y \rangle_2|$$
$$= \mu \left(\langle \cdot, \cdot \rangle_1, x, y \right) + \mu \left(\langle \cdot, \cdot \rangle_2, x, y \right).$$

Note the use of the following elementary inequality for real numbers:

$$\left(a^2 + b^2 \right) \left(c^2 + d^2 \right) \geq (ac + bd)^2$$

for $a, b, c, d \geq 0$.

We say that the inner product $\langle \cdot, \cdot \rangle_2$ is greater than $\langle \cdot, \cdot \rangle_1$ and denote it by $\langle \cdot, \cdot \rangle_2 > \langle \cdot, \cdot \rangle_1$ if $\langle x, x \rangle_2 > \langle x, x \rangle_1$ for all $x \in H \backslash \{0\}$. If $\langle \cdot, \cdot \rangle_2 > \langle \cdot, \cdot \rangle_1$, then the mapping $\langle \cdot, \cdot \rangle_{2,1} := \langle \cdot, \cdot \rangle_2 - \langle \cdot, \cdot \rangle_1$ is an inner product on H.

Consequently, we have the following monotonicity property for $\mu \left(\cdot, x, y \right)$:

If $\langle \cdot, \cdot \rangle_2 > \langle \cdot, \cdot \rangle_1$, then $\mu \left(\langle \cdot, \cdot \rangle_2, x, y \right) \geq \mu \left(\langle \cdot, \cdot \rangle_1, x, y \right)$ *for all* $x, y \in H$ [98]. The proof is shown below:

$$\mu \left(\langle \cdot, \cdot \rangle_2, x, y \right) = \mu \left(\langle \cdot, \cdot \rangle_{2,1} + \langle \cdot, \cdot \rangle_1, x, y \right) \quad \text{(by 5.4)}$$
$$\geq \mu \left(\langle \cdot, \cdot \rangle_{2,1}, x, y \right) + \mu \left(\langle \cdot, \cdot \rangle_1, x, y \right) \geq \mu \left(\langle \cdot, \cdot \rangle_1, x, y \right).$$

Note the use of the fact that $\mu \left(\langle \cdot, \cdot \rangle_{2,1}, x, y \right) \geq 0$, which is Schwarz's inequality for the inner product $\langle \cdot, \cdot \rangle_{2,1}$.

5.2 A Conditional Refinement of the Schwarz Inequality

Let $(H; \langle \cdot, \cdot \rangle)$ be an inner product space over the real or complex number field $\mathbb{K}$ and $r_1, r_2 > 0$. If $x, y \in H$ such that

$$\|x - y\| \geq r_2 \geq r_1 \geq |\|x\| - \|y\||, \tag{5.6}$$

then we have the following refinement of Schwarz's inequality:

$$\|x\| \, \|y\| - \operatorname{Re} \langle x, y \rangle \geq \frac{1}{2} \left(r_2^2 - r_1^2 \right) (\geq 0). \tag{5.7}$$

The constant $\frac{1}{2}$ is best possible in the sense that it cannot be replaced by a larger quantity.

PROOF   We follow the proof by Dragomir [73].
From the first inequality in (5.6) we have

$$\|x\|^2 + \|y\|^2 \geq r_2^2 + 2 \operatorname{Re} \langle x, y \rangle. \tag{5.8}$$

Subtracting in (5.8) the quantity $2 \|x\| \, \|y\|$, we get

$$(\|x\| - \|y\|)^2 \geq r_2^2 - 2 (\|x\| \, \|y\| - \operatorname{Re} \langle x, y \rangle). \tag{5.9}$$

By the second inequality in (5.6) we have

$$r_1^2 \geq (\|x\| - \|y\|)^2. \tag{5.10}$$

Hence, from (5.9) and (5.10) we deduce the desired inequality (5.7).
To prove the sharpness of the constant $\frac{1}{2}$ in (5.7), let us assume that there is a constant $C > 0$ such that

$$\|x\| \, \|y\| - \operatorname{Re} \langle x, y \rangle \geq C \left(r_2^2 - r_1^2 \right), \tag{5.11}$$

provided that x and y satisfy (5.6).
Let $e \in H$ with $\|e\| = 1$ and for $r_2 > r_1 > 0$, define

$$x = \frac{r_2 + r_1}{2} \cdot e \quad \text{and} \quad y = \frac{r_1 - r_2}{2} \cdot e. \tag{5.12}$$

Then

$$x - y = r_2 e, \qquad \|x - y\| = r_2,$$
$$\|x\| - \|y\| = r_1, \quad \text{and} \quad |\|x\| - \|y\|| = r_1.$$

If we replace x and y as defined in (5.12) into inequality (5.11), we get

$$\frac{(r_2 + r_1)(r_2 - r_1)}{4} - \frac{(r_2 + r_1)(r_1 - r_2)}{4} \geq C \left(r_2^2 - r_1^2 \right).$$

Simplifying the above inequality gives

$$\frac{r_2^2 - r_1^2}{2} \geq C \left(r_2^2 - r_1^2 \right),$$

which implies that $C \leq \frac{1}{2}$. This completes the proof. ▯

The following triangle type inequality may be stated [73]:

With the assumptions of (5.6), we have the inequality:

$$\|x\| + \|y\| - \frac{\sqrt{2}}{2} \|x + y\| \geq \frac{\sqrt{2}}{2} \sqrt{r_2^2 - r_1^2}. \tag{5.13}$$

PROOF We have, by (5.7), that

$$\left(\|x\| + \|y\| \right)^2 - \|x + y\|^2 = 2 \left(\|x\| \, \|y\| - \operatorname{Re} \langle x, y \rangle \right) \geq r_2^2 - r_1^2 \geq 0,$$

which gives

$$\left(\|x\| + \|y\| \right)^2 \geq \|x + y\|^2 + \left(\sqrt{r_2^2 - r_1^2} \right)^2. \tag{5.14}$$

By employing the following elementary inequality

$$2 \left(\alpha^2 + \beta^2 \right) \geq \left(\alpha + \beta \right)^2, \qquad \alpha, \beta \geq 0,$$

we get

$$\|x + y\|^2 + \left(\sqrt{r_2^2 - r_1^2} \right)^2 \geq \frac{1}{2} \left(\|x + y\| + \sqrt{r_2^2 - r_1^2} \right)^2. \tag{5.15}$$

Utilising (5.14) and (5.15), we deduce the desired inequality (5.13). ▯

Comments
Assume that $(H; \langle \cdot, \cdot \rangle)$ is a Hilbert space over the real or complex number field. Suppose that $p_i \geq 0$, $i \in \mathbb{N}$ with $\sum_{i=1}^{\infty} p_i = 1$ and define

$$\ell_p^2 (H) := \left\{ \mathbf{x} := (x_i)_{i \in \mathbb{N}} \middle| \, x_i \in H, \ i \in \mathbb{N} \ \text{ and } \ \sum_{i=1}^{\infty} p_i \, \|x_i\|^2 < \infty \right\}.$$

It is well known that $\ell_p^2 (H)$ is endowed with the inner product $\langle \cdot, \cdot \rangle_p$ defined by

$$\langle \mathbf{x}, \mathbf{y} \rangle_p := \sum_{i=1}^{\infty} p_i \langle x_i, y_i \rangle$$

which induces the norm

$$\|\mathbf{x}\|_p := \left(\sum_{i=1}^{\infty} p_i \, \|x_i\|^2 \right)^{\frac{1}{2}}.$$

Furthermore, $\ell^2(H)$ is a Hilbert space over $\mathbb{K}$.

We may state the following discrete inequality improving the Cauchy-Bunyakovsky-Schwarz classical result [73]:

Let $(H; \langle \cdot, \cdot \rangle)$ be a Hilbert space and $p_i \geq 0$ $(i \in \mathbb{N})$ with $\sum_{i=1}^{\infty} p_i = 1$. Assume that $x, y \in \ell_p^2(H)$ and $r_1, r_2 > 0$ satisfy the condition

$$\|x_i - y_i\| \geq r_2 \geq r_1 \geq |\|x_i\| - \|y_i\|| \tag{5.16}$$

for each $i \in \mathbb{N}$. Then we have the following refinement of the Cauchy-Bunyakovsky-Schwarz inequality:

$$\left(\sum_{i=1}^{\infty} p_i \|x_i\|^2 \sum_{i=1}^{\infty} p_i \|y_i\|^2 \right)^{\frac{1}{2}} - \sum_{i=1}^{\infty} p_i \operatorname{Re} \langle x_i, y_i \rangle \geq \frac{1}{2} \left(r_2^2 - r_1^2 \right) \geq 0. \tag{5.17}$$

The constant $\frac{1}{2}$ is best possible.

PROOF   From the condition (5.16) we simply deduce

$$\sum_{i=1}^{\infty} p_i \|x_i - y_i\|^2 \geq r_2^2 \geq r_1^2 \geq \sum_{i=1}^{\infty} p_i (\|x_i\| - \|y_i\|)^2 \tag{5.18}$$

$$\geq \left[\left(\sum_{i=1}^{\infty} p_i \|x_i\|^2 \right)^{\frac{1}{2}} - \left(\sum_{i=1}^{\infty} p_i \|y_i\|^2 \right)^{\frac{1}{2}} \right]^2 .$$

Note the use of the Cauchy-Bunyakovsky-Schwarz inequality,

$$\left(\sum_{i=1}^{\infty} p_i \|x_i\|^2 \sum_{i=1}^{\infty} p_i \|y_i\|^2 \right)^{\frac{1}{2}} \geq \sum_{i=1}^{\infty} p_i \|x_i\| \|y_i\|,$$

in inequality (5.18).

In terms of the norm $\|\cdot\|_p$, the inequality (5.18) may be written as

$$\|\mathbf{x} - \mathbf{y}\|_p \geq r_2 \geq r_1 \geq \left| \|\mathbf{x}\|_p - \|\mathbf{y}\|_p \right|. \tag{5.19}$$

Utilising (5.7) for the Hilbert space $\left(\ell_p^2(H), \langle \cdot, \cdot \rangle_p \right)$, we deduce the desired inequality (5.17).

For $n = 1$ $(p_1 = 1)$, the inequality (5.17) reduces to (5.7), for which we have shown that $\frac{1}{2}$ is the best possible constant. ∎

By the use of (5.13), we may state the following result as well:

With the assumptions of (5.16), we have the inequality

$$\left(\sum_{i=1}^{\infty} p_i \|x_i\|^2\right)^{\frac{1}{2}} + \left(\sum_{i=1}^{\infty} p_i \|y_i\|^2\right)^{\frac{1}{2}} - \frac{\sqrt{2}}{2} \left(\sum_{i=1}^{\infty} p_i \|x_i + y_i\|^2\right)^{\frac{1}{2}}$$

$$\geq \frac{\sqrt{2}}{2} \sqrt{r_2^2 - r_1^2}. \quad (5.20)$$

5.3   The Duality Schwarz-Triangle Inequalities

Let $(H; \langle \cdot, \cdot \rangle)$ be an inner product space and $R \geq 1$. For $x, y \in H$, the subsequent statements are equivalent:

(i) *The following refinement of the triangle inequality holds:*

$$\|x\| + \|y\| \geq R \|x + y\| ; \quad (5.21)$$

(ii) *The following refinement of the Schwarz inequality holds:*

$$\|x\| \|y\| - \operatorname{Re} \langle x, y \rangle \geq \frac{1}{2} \left(R^2 - 1\right) \|x + y\|^2. \quad (5.22)$$

PROOF   We follow the proof by Dragomir [73].
Taking the square in (5.21), we have

$$2 \|x\| \|y\| \geq \left(R^2 - 1\right) \|x\|^2 + 2R^2 \operatorname{Re} \langle x, y \rangle + \left(R^2 - 1\right) \|y\|^2. \quad (5.23)$$

Subtracting from both sides of (5.23) the quantity $2 \operatorname{Re} \langle x, y \rangle$, we obtain

$$2 \left(\|x\| \|y\| - \operatorname{Re} \langle x, y \rangle\right) \geq \left(R^2 - 1\right) \left[\|x\|^2 + 2 \operatorname{Re} \langle x, y \rangle + \|y\|^2\right]$$

$$= \left(R^2 - 1\right) \|x + y\|^2,$$

which is clearly equivalent to (5.22). ⬜

By the use of the above equivalence, we may now state the following result concerning a refinement of the Schwarz inequality [73]:

Let $(H; \langle \cdot, \cdot \rangle)$ be an inner product space over the real or complex number field and $R \geq 1$, $r \geq 0$. If $x, y \in H$ are such that

$$\frac{1}{R} \left(\|x\| + \|y\|\right) \geq \|x + y\| \geq r, \quad (5.24)$$

then we have the following refinement of the Schwarz inequality:

$$\|x\| \|y\| - \operatorname{Re} \langle x, y \rangle \geq \frac{1}{2} \left(R^2 - 1\right) r^2. \quad (5.25)$$

The constant $\frac{1}{2}$ is best possible.

PROOF   The inequality (5.25) follows readily from (5.21). We only need to prove that $\frac{1}{2}$ is the best possible constant in (5.25).

Assume that there exists a $C > 0$ such that

$$\|x\| \, \|y\| - \operatorname{Re} \langle x, y \rangle \geq C \left(R^2 - 1 \right) r^2, \tag{5.26}$$

provided x, y, R, and r satisfy (5.24).

Consider $r = 1$, $R > 1$ and choose $x = \frac{1-R}{2} e$, $y = \frac{1+R}{2} e$ with $e \in H$, $\|e\| = 1$. Then

$$x + y = e, \qquad \frac{\|x\| + \|y\|}{R} = \frac{\frac{R-1}{2} + \frac{R+1}{2}}{R} = 1,$$

giving equality in (5.24).

From (5.26), for the above choices of x and y, we have

$$\frac{R^2 - 1}{4} - \frac{1 - R^2}{4} \geq C \left(R^2 - 1 \right),$$

giving $\frac{1}{2} \left(R^2 - 1 \right) \geq C \left(R^2 - 1 \right)$, which shows that $C \leq \frac{1}{2}$. ☐

The following result also holds [73]:

Let $(H; \langle \cdot, \cdot \rangle)$ be an inner product space over the real or complex number field K and $r \in (0, 1]$. For $x, y \in H$, the following statements are equivalent:

(i) *We have the inequality*

$$\left| \|x\| - \|y\| \right| \leq r \, \|x - y\| \, ; \tag{5.27}$$

(ii) *We have the following refinement of the Schwarz inequality:*

$$\|x\| \, \|y\| - \operatorname{Re} \langle x, y \rangle \geq \frac{1}{2} \left(1 - r^2 \right) \|x - y\|^2 . \tag{5.28}$$

The constant $\frac{1}{2}$ in (5.28) is best possible.

PROOF   By taking the square in (5.27), we have

$$\|x\|^2 - 2 \|x\| \, \|y\| + \|y\|^2 \leq r^2 \left(\|x\|^2 - 2 \operatorname{Re} \langle x, y \rangle + \|y\|^2 \right),$$

which is clearly equivalent to

$$\left(1 - r^2 \right) \left[\|x\|^2 - 2 \operatorname{Re} \langle x, y \rangle + \|y\|^2 \right] \leq 2 \left(\|x\| \, \|y\| - \operatorname{Re} \langle x, y \rangle \right)$$

or with (5.28).

Now, assume that (5.28) holds with a constant $E > 0$, i.e.,

$$\|x\| \, \|y\| - \operatorname{Re} \langle x, y \rangle \geq E \left(1 - r^2 \right) \|x - y\|^2 , \tag{5.29}$$

provided (5.27) holds.

Define $x = \frac{r+1}{2}e$, $y = \frac{r-1}{2}e$ with $e \in H$, $\|e\| = 1$. Then

$$\big|\|x\| - \|y\|\big| = r, \quad \|x - y\| = \|e\| = 1,$$

showing that (5.27) holds with equality.

If we replace x and y in (5.29), then we get $E\left(1 - r^2\right) \leq \frac{1}{2}\left(1 - r^2\right)$, which shows that $E \leq \frac{1}{2}$. □

Comments

The following result holds [73]:

Let $(H; \langle \cdot, \cdot \rangle)$ be a Hilbert space and $p_i \geq 0$ $(i \in \mathbb{N})$ with $\sum_{i=1}^{\infty} p_i = 1$. Assume that $\mathbf{x}, \mathbf{y} \in \ell_p^2(H)$ and $R \geq 1$, $r \geq 0$ satisfy the condition

$$\frac{1}{R}\left(\|x_i\| + \|y_i\|\right) \geq \|x_i + y_i\| \geq r \tag{5.30}$$

for each $i \in \mathbb{N}$. Then we have the following refinement of the Schwarz inequality:

$$\left(\sum_{i=1}^{\infty} p_i \|x_i\|^2 \sum_{i=1}^{\infty} p_i \|y_i\|^2\right)^{\frac{1}{2}} - \sum_{i=1}^{\infty} p_i \operatorname{Re} \langle x_i, y_i \rangle \geq \frac{1}{2}\left(R^2 - 1\right) r^2. \tag{5.31}$$

The constant $\frac{1}{2}$ is best possible in the sense that it cannot be replaced by a larger quantity.

PROOF By (5.30) we deduce

$$\frac{1}{R}\left[\sum_{i=1}^{\infty} p_i \left(\|x_i\| - \|y_i\|\right)^2\right]^{\frac{1}{2}} \geq \left(\sum_{i=1}^{\infty} p_i \|x_i + y_i\|^2\right)^{\frac{1}{2}} \geq r. \tag{5.32}$$

By the classical Minkowski inequality for nonnegative numbers, we have

$$\left(\sum_{i=1}^{\infty} p_i \|x_i\|^2\right)^{\frac{1}{2}} + \left(\sum_{i=1}^{\infty} p_i \|y_i\|^2\right)^{\frac{1}{2}} \geq \left[\sum_{i=1}^{\infty} p_i \left(\|x_i\| + \|y_i\|\right)^2\right]^{\frac{1}{2}}. \tag{5.33}$$

By utilising (5.32) and (5.33), we may state the following inequality in terms of $\|\cdot\|_p$:

$$\frac{1}{R}\left(\|\mathbf{x}\|_p + \|\mathbf{y}\|_p\right) \geq \|\mathbf{x} + \mathbf{y}\|_p \geq r. \tag{5.34}$$

By employing the inequality (5.25) for the Hilbert space $\ell_p^2(H)$ and the inequality (5.34), we deduce the desired result (5.31).

Since, for $p = 1$, $n = 1$, (5.31) reduces to (5.25), for which we have shown that $\frac{1}{2}$ is the best constant, we conclude that $\frac{1}{2}$ is the best constant in (5.31) as well. □

Finally, we may state and prove the following result incorporated by Dragomir [73]:

Let $(H; \langle \cdot, \cdot \rangle)$ be a Hilbert space and $p_i \geq 0$ $(i \in \mathbb{N})$ with $\sum_{i=1}^{\infty} p_i = 1$. Assume that $x, y \in \ell_p^2(H)$ and $r \in (0, 1]$ such that

$$|\|x_i\| - \|y_i\|| \leq r \|x_i - y_i\| \quad \text{for each } i \in \mathbb{N} \tag{5.35}$$

holds true. Then we have the following refinement of the Schwarz inequality:

$$\left(\sum_{i=1}^{\infty} p_i \|x_i\|^2 \sum_{i=1}^{\infty} p_i \|y_i\|^2 \right)^{\frac{1}{2}} - \sum_{i=1}^{\infty} p_i \operatorname{Re} \langle x_i, y_i \rangle$$

$$\geq \frac{1}{2} \left(1 - r^2 \right) \sum_{i=1}^{\infty} p_i \|x_i - y_i\|^2. \tag{5.36}$$

The constant $\frac{1}{2}$ is best possible in (5.36).

PROOF   From (5.35) we have

$$\left[\sum_{i=1}^{\infty} p_i \left(\|x_i\| + \|y_i\| \right)^2 \right]^{\frac{1}{2}} \leq r \left[\sum_{i=1}^{\infty} p_i \|x_i - y_i\|^2 \right]^{\frac{1}{2}}.$$

Utilising the following known result,

$$\left| \left(\sum_{i=1}^{\infty} p_i \|x_i\|^2 \right)^{\frac{1}{2}} - \left(\sum_{i=1}^{\infty} p_i \|y_i\|^2 \right)^{\frac{1}{2}} \right| \leq \left(\sum_{i=1}^{\infty} p_i \left(\|x_i\| + \|y_i\| \right)^2 \right)^{\frac{1}{2}},$$

derived from the Minkowski inequality, we may state that

$$\left| \|\mathbf{x}\|_p - \|\mathbf{y}\|_p \right| \leq r \|\mathbf{x} - \mathbf{y}\|_p.$$

Now, by making use of (5.28), we deduce the desired inequality (5.36) and the fact that $\frac{1}{2}$ is the best possible constant. We omit the details.   □

5.4   A Quadratic Reverse for the Schwarz Inequality

Let $(H; \langle \cdot, \cdot \rangle)$ be an inner product space over the real or complex number field $\mathbb{K}$ $(\mathbb{K} = \mathbb{R}, \mathbb{K} = \mathbb{C})$ and $x, a \in H$, $r > 0$ are such that

$$x \in \overline{B}(a, r) := \{ z \in H | \|z - a\| \leq r \}.$$

If $\|a\| > r$, then we have the inequalities

$$0 \leq \|x\|^2 \|a\|^2 - |\langle x, a \rangle|^2 \leq \|x\|^2 \|a\|^2 - [\operatorname{Re} \langle x, a \rangle]^2 \leq r^2 \|x\|^2. \quad (5.37)$$

The constant $C = 1$ in front of r^2 is best possible in the sense that it cannot be replaced by a smaller one.
    If $\|a\| = r$, then

$$\|x\|^2 \leq 2 \operatorname{Re} \langle x, a \rangle \leq 2 |\langle x, a \rangle|. \quad (5.38)$$

The constant 2 is best possible in both inequalities.
    If $\|a\| < r$, then

$$\|x\|^2 \leq r^2 - \|a\|^2 + 2 \operatorname{Re} \langle x, a \rangle \leq r^2 - \|a\|^2 + 2 |\langle x, a \rangle|. \quad (5.39)$$

Here the constant 2 is also best possible.

PROOF   We follow the proof by Dragomir [74].
    Since $x \in \overline{B}(a, r)$, it follows that $\|x - a\|^2 \leq r^2$, which is equivalent to

$$\|x\|^2 + \|a\|^2 - r^2 \leq 2 \operatorname{Re} \langle x, a \rangle. \quad (5.40)$$

If $\|a\| > r$, then we may divide (5.40) by $\sqrt{\|a\|^2 - r^2} > 0$ to get

$$\frac{\|x\|^2}{\sqrt{\|a\|^2 - r^2}} + \sqrt{\|a\|^2 - r^2} \leq \frac{2 \operatorname{Re} \langle x, a \rangle}{\sqrt{\|a\|^2 - r^2}}. \quad (5.41)$$

By using the elementary inequality

$$\alpha p + \frac{1}{\alpha} q \geq 2\sqrt{pq}, \quad \alpha > 0, \quad p, q \geq 0,$$

we may state that

$$2 \|x\| \leq \frac{\|x\|^2}{\sqrt{\|a\|^2 - r^2}} + \sqrt{\|a\|^2 - r^2}. \quad (5.42)$$

Making use of (5.41) and (5.42), we deduce that

$$\|x\| \sqrt{\|a\|^2 - r^2} \leq \operatorname{Re} \langle x, a \rangle. \quad (5.43)$$

Taking the square in (5.43) and re-arranging the terms, we deduce the third inequality in (5.37). The others are straightforward.
    To prove the sharpness of the constant, assume, under the stated hypotheses, that there exists a constant $c > 0$ such that

$$\|x\|^2 \|a\|^2 - [\operatorname{Re} \langle x, a \rangle]^2 \leq cr^2 \|x\|^2, \quad (5.44)$$

provided $x \in \overline{B}(a, r)$ and $\|a\| > r$.

Let $r = \sqrt{\varepsilon} > 0$, $\varepsilon \in (0, 1)$, $a, e \in H$ with $\|a\| = \|e\| = 1$ and $a \perp e$. Put $x = a + \sqrt{\varepsilon} e$. Then obviously $x \in \overline{B}(a, r)$, $\|a\| > r$ and $\|x\|^2 = \|a\|^2 + \varepsilon \|e\|^2 = 1 + \varepsilon$, $\mathrm{Re}\,\langle x, a \rangle = \|a\|^2 = 1$, and thus $\|x\|^2 \|a\|^2 - [\mathrm{Re}\,\langle x, a \rangle]^2 = \varepsilon$. By using (5.44), we may write that

$$\varepsilon \le c\varepsilon\,(1 + \varepsilon), \quad \varepsilon > 0,$$

giving

$$c + c\varepsilon \ge 1 \quad \text{for any } \varepsilon > 0. \tag{5.45}$$

Letting $\varepsilon \to 0+$, we get from (5.45) that $c \ge 1$, and the sharpness of the constant is proved.

The inequality (5.38) is obvious by (5.40) since $\|a\| = r$. The best constant follows in a similar way to the above.

The inequality (5.39) is obvious. The best constant may be proved in a similar way to the above. We omit the details.

$\square$

The following reverse of Schwarz's inequality holds [74]:

Let $(H; \langle \cdot, \cdot \rangle)$ be an inner product space over $\mathbb{K}$ and $x, y \in H$, $\gamma, \Gamma \in \mathbb{K}$ such that either

$$\mathrm{Re}\,\langle \Gamma y - x, x - \gamma y \rangle \ge 0, \tag{5.46}$$

or, equivalently,

$$\left\| x - \frac{\Gamma + \gamma}{2} y \right\| \le \frac{1}{2} |\Gamma - \gamma|\, \|y\|, \tag{5.47}$$

holds.

If $\mathrm{Re}\,(\Gamma \overline{\gamma}) > 0$, then we have the inequalities

$$\|x\|^2 \|y\|^2 \le \frac{1}{4} \cdot \frac{\{\mathrm{Re}\,[(\overline{\Gamma} + \overline{\gamma})\,\langle x, y \rangle]\}^2}{\mathrm{Re}\,(\Gamma \overline{\gamma})} \tag{5.48}$$

$$\le \frac{1}{4} \cdot \frac{|\Gamma + \gamma|^2}{\mathrm{Re}\,(\Gamma \overline{\gamma})}\, |\langle x, y \rangle|^2.$$

The constant $\frac{1}{4}$ is best possible in both inequalities.

If $\mathrm{Re}\,(\Gamma \overline{\gamma}) = 0$, then

$$\|x\|^2 \le \mathrm{Re}\,[(\overline{\Gamma} + \overline{\gamma})\,\langle x, y \rangle] \le |\Gamma + \gamma|\,|\langle x, y \rangle|.$$

If $\mathrm{Re}\,(\Gamma \overline{\gamma}) < 0$, then

$$\|x\|^2 \le -\mathrm{Re}\,(\Gamma \overline{\gamma})\,\|y\|^2 + \mathrm{Re}\,[(\overline{\Gamma} + \overline{\gamma})\,\langle x, y \rangle]$$

$$\le -\mathrm{Re}\,(\Gamma \overline{\gamma})\,\|y\|^2 + |\Gamma + \gamma|\,|\langle x, y \rangle|.$$

PROOF The proof of the equivalence between the inequalities (5.46) and (5.47) follows by the fact that in an inner product space, $\operatorname{Re} \langle Z - x, x - z \rangle \geq 0$ (for $x, z, Z \in H$) is equivalent to $\left\| x - \frac{z+Z}{2} \right\| \leq \frac{1}{2} \| Z - z \|$.

Consider for $y \neq 0$, $a = \frac{\gamma + \Gamma}{2} y$, and $r = \frac{1}{2} |\Gamma - \gamma| \, \|y\|$. Then

$$\|a\|^2 - r^2 = \frac{|\Gamma + \gamma|^2 - |\Gamma - \gamma|^2}{4} \|y\|^2 = \operatorname{Re}(\Gamma \overline{\gamma}) \|y\|^2 .$$

If $\operatorname{Re}(\Gamma \overline{\gamma}) > 0$, then by the second inequality in (5.37) we have

$$\|x\|^2 \frac{|\Gamma + \gamma|^2}{4} \|y\|^2 - \frac{1}{4} \left\{ \operatorname{Re} \left[(\overline{\Gamma} + \overline{\gamma}) \langle x, y \rangle \right] \right\}^2 \leq \frac{1}{4} |\Gamma - \gamma|^2 \|x\|^2 \|y\|^2$$

from which we derive

$$\frac{|\Gamma + \gamma|^2 - |\Gamma - \gamma|^2}{4} \|x\|^2 \|y\|^2 \leq \frac{1}{4} \left\{ \operatorname{Re} \left[(\overline{\Gamma} + \overline{\gamma}) \langle x, y \rangle \right] \right\}^2 ,$$

giving the first inequality in (5.48). The second inequality is obvious.

To prove the sharpness of the constant $\frac{1}{4}$, assume that the first inequality in (5.48) holds with a constant $c > 0$, i.e.,

$$\|x\|^2 \|y\|^2 \leq c \cdot \frac{\left\{ \operatorname{Re} \left[(\overline{\Gamma} + \overline{\gamma}) \langle x, y \rangle \right] \right\}^2}{\operatorname{Re}(\Gamma \overline{\gamma})}, \tag{5.49}$$

provided $\operatorname{Re}(\Gamma \overline{\gamma}) > 0$ and either (5.46) or (5.47) holds. Assume that $\Gamma, \gamma > 0$, and let $x = \gamma y$. Then (5.46) holds and by (5.49) we deduce

$$\gamma^2 \|y\|^4 \leq c \cdot \frac{(\Gamma + \gamma)^2 \gamma^2 \|y\|^4}{\Gamma \gamma},$$

giving

$$\Gamma \gamma \leq c \, (\Gamma + \gamma)^2 \quad \text{for any } \Gamma, \gamma > 0. \tag{5.50}$$

Let $\varepsilon \in (0, 1)$ and choose in (5.50), $\Gamma = 1 + \varepsilon$, $\gamma = 1 - \varepsilon > 0$ to get $1 - \varepsilon^2 \leq 4c$ for any $\varepsilon \in (0, 1)$. Letting $\varepsilon \to 0+$, we deduce $c \geq \frac{1}{4}$, and the sharpness of the constant is proved.

The other inequalities are obvious and we omit the details. ∎

The following result provides a reverse inequality for the additive version of Schwarz's inequality [74]:

With the above assumptions for x, y, γ, Γ and if $\operatorname{Re}(\Gamma \overline{\gamma}) > 0$, then we have the inequality:

$$0 \leq \|x\|^2 \|y\|^2 - |\langle x, y \rangle|^2 \leq \frac{1}{4} \cdot \frac{|\Gamma - \gamma|^2}{\operatorname{Re}(\Gamma \overline{\gamma})} |\langle x, y \rangle|^2 . \tag{5.51}$$

The constant $\frac{1}{4}$ is best possible in (5.51).

The proof is obvious from (5.48) on subtracting in both sides the same quantity $|\langle x, y \rangle|^2$. The sharpness of the constant may be proven in a similar manner to the one incorporated in the proof of (5.48). We omit the details.

Comments

Let (Ω, Σ, μ) be a measurable space consisting of a set Ω, a σ-algebra Σ of parts, and a countably additive and positive measure μ on Σ with values in $\mathbb{R} \cup \{\infty\}$. Let $\rho \geq 0$ be a g-measurable function on Ω with $\int_\Omega \rho(s) \, d\mu(s) = 1$. Denote by $L^2_\rho(\Omega, \mathbb{K})$ the Hilbert space of all real or complex valued functions defined on Ω and $2 - \rho$-integrable on Ω, that is,

$$\int_\Omega \rho(s) |f(s)|^2 \, d\mu(s) < \infty.$$

It is obvious that the following inner product,

$$\langle f, g \rangle_\rho := \int_\Omega \rho(s) f(s) \overline{g(s)} d\mu(s),$$

generates the norm $\|f\|_\rho := \left(\int_\Omega \rho(s) |f(s)|^2 \, d\mu(s) \right)^{\frac{1}{2}}$ of $L^2_\rho(\Omega, \mathbb{K})$, and all the above results may be stated for integrals.

It is important to observe that, if

$$\mathrm{Re} \left[f(s) \overline{g(s)} \right] \geq 0, \quad \text{for } \mu - \text{a.e. } s \in \Omega,$$

then, obviously,

$$\mathrm{Re} \langle f, g \rangle_\rho = \mathrm{Re} \left[\int_\Omega \rho(s) f(s) \overline{g(s)} d\mu(s) \right] \tag{5.52}$$

$$= \int_\Omega \rho(s) \mathrm{Re} \left[f(s) \overline{g(s)} \right] d\mu(s) \geq 0.$$

In general, the reverse is evidently not true.

Moreover, if the space is real, i.e., $\mathbb{K} = \mathbb{R}$, then a sufficient condition for (5.52) to hold is

$$f(s) \geq 0, \quad g(s) \geq 0, \quad \text{for } \mu - \text{a.e. } s \in \Omega.$$

We now provide, by the use of certain results obtained above, some integral inequalities that may be used in practical applications.

Let $f, g \in L^2_\rho(\Omega, \mathbb{K})$ and $r > 0$ with the properties that

$$|f(s) - g(s)| \leq r \leq |g(s)|, \quad \text{for } \mu - \text{a.e. } s \in \Omega. \tag{5.53}$$

Then we have the inequalities

$$0 \leq \int_\Omega \rho(s) |f(s)|^2 \, d\mu(s) \int_\Omega \rho(s) |g(s)|^2 \, d\mu(s) \tag{5.54}$$

$$- \left| \int_\Omega \rho(s) f(s) \overline{g(s)} d\mu(s) \right|^2$$

$$\leq \int_{\Omega} \rho(s) |f(s)|^2 d\mu(s) \int_{\Omega} \rho(s) |g(s)|^2 d\mu(s)$$

$$- \left[\int_{\Omega} \rho(s) \operatorname{Re}\left(f(s) \overline{g(s)} \right) d\mu(s) \right]^2$$

$$\leq r^2 \int_{\Omega} \rho(s) |g(s)|^2 d\mu(s).$$

The constant $c = 1$ in front of r^2 is best possible.

The proof follows by (5.37); and we omit the details [74].

Let $f, g \in L^2_\rho(\Omega, \mathbb{K})$ and $\gamma, \Gamma \in \mathbb{K}$ such that $\operatorname{Re}(\Gamma\bar{\gamma}) > 0$ and

$$\operatorname{Re}\left[(\Gamma g(s) - f(s)) \left(\overline{f(s)} - \overline{\gamma g(s)} \right) \right] \geq 0, \quad \text{for } \mu - \text{a.e. } s \in \Omega.$$

Then we have the inequalities

$$\int_{\Omega} \rho(s) |f(s)|^2 d\mu(s) \int_{\Omega} \rho(s) |g(s)|^2 d\mu(s) \tag{5.55}$$

$$\leq \frac{1}{4} \cdot \frac{\left\{ \operatorname{Re}\left[(\bar{\Gamma} + \bar{\gamma}) \int_{\Omega} \rho(s) f(s) \overline{g(s)} d\mu(s) \right] \right\}^2}{\operatorname{Re}(\Gamma\bar{\gamma})}$$

$$\leq \frac{1}{4} \cdot \frac{|\Gamma + \gamma|^2}{\operatorname{Re}(\Gamma\bar{\gamma})} \left| \int_{\Omega} \rho(s) f(s) \overline{g(s)} d\mu(s) \right|^2.$$

The constant $\frac{1}{4}$ is best possible in both inequalities.

The following result may be stated as well:

With the previous assumptions, we have the inequality

$$0 \leq \int_{\Omega} \rho(s) |f(s)|^2 d\mu(s) \int_{\Omega} \rho(s) |g(s)|^2 d\mu(s) \tag{5.56}$$

$$- \left| \int_{\Omega} \rho(s) f(s) \overline{g(s)} d\mu(s) \right|^2$$

$$\leq \frac{1}{4} \cdot \frac{|\Gamma - \gamma|^2}{\operatorname{Re}(\Gamma\bar{\gamma})} \left| \int_{\Omega} \rho(s) f(s) \overline{g(s)} d\mu(s) \right|^2.$$

The constant $\frac{1}{4}$ is best possible.

If the space is real ($\mathbb{K} = \mathbb{R}$) and we assume, for $M > m > 0$, that

$$mg(s) \leq f(s) \leq Mg(s), \quad \text{for } \mu - \text{a.e. } s \in \Omega,$$

then, by (5.55) and (5.56), we deduce the inequalities

$$\int_{\Omega} \rho(s) [f(s)]^2 d\mu(s) \int_{\Omega} \rho(s) [g(s)]^2 d\mu(s)$$

$$\leq \frac{1}{4} \cdot \frac{(M + m)^2}{mM} \left[\int_{\Omega} \rho(s) f(s) g(s) d\mu(s) \right]^2 \tag{5.57}$$

and

$$0 \le \int_{\Omega} \rho\left(s\right) \left[f\left(s\right)\right]^{2} d\mu\left(s\right) \int_{\Omega} \rho\left(s\right) \left[g\left(s\right)\right]^{2} d\mu\left(s\right) \tag{5.58}$$

$$- \left[\int_{\Omega} \rho\left(s\right) f\left(s\right) g\left(s\right) d\mu\left(s\right)\right]^{2}$$

$$\le \frac{1}{4} \cdot \frac{\left(M-m\right)^{2}}{mM} \left[\int_{\Omega} \rho\left(s\right) f\left(s\right) g\left(s\right) d\mu\left(s\right)\right]^{2}.$$

The inequality (5.57) is known in the literature as Cassels' inequality.

5.5 A Reverse of the Simple Schwarz Inequality

Let $(H; \langle \cdot, \cdot \rangle)$ be an inner product space over the real or complex number field $\mathbb{K}$, $x, a \in H$ *and* $r > 0$. *If*

$$x \in \bar{B}\left(a, r\right) := \left\{z \in H \mid \left\|z - a\right\| \le r\right\}, \tag{5.59}$$

then we have the inequalities:

$$0 \le \left\|x\right\| \left\|a\right\| - \left|\langle x, a \rangle\right| \le \left\|x\right\| \left\|a\right\| - \left|\operatorname{Re} \langle x, a \rangle\right| \tag{5.60}$$

$$\le \left\|x\right\| \left\|a\right\| - \operatorname{Re} \langle x, a \rangle \le \frac{1}{2} r^{2}.$$

The constant $\frac{1}{2}$ is best possible in (5.60) in the sense that it cannot be replaced by a smaller constant.

PROOF We follow the proof by Dragomir [61].
The condition (5.59) is equivalent to

$$\left\|x\right\|^{2} + \left\|a\right\|^{2} \le 2 \operatorname{Re} \langle x, a \rangle + r^{2}. \tag{5.61}$$

Using the elementary inequality

$$2 \left\|x\right\| \left\|a\right\| \le \left\|x\right\|^{2} + \left\|a\right\|^{2}, \quad a, x \in H$$

and (5.61), we deduce

$$2 \left\|x\right\| \left\|a\right\| \le 2 \operatorname{Re} \langle x, a \rangle + r^{2},$$

giving the last inequality in (5.60). The other inequalities are obvious.
To prove the sharpness of the constant $\frac{1}{2}$, assume that

$$0 \le \left\|x\right\| \left\|a\right\| - \operatorname{Re} \langle x, a \rangle \le c r^{2} \tag{5.62}$$

for any $x, a \in H$ and $r > 0$ satisfying (5.59).

Assume that $a, e \in H$, $\|a\| = \|e\| = 1$, and $e \perp a$. If $r = \sqrt{\varepsilon}$, $\varepsilon > 0$, and $x = a + \sqrt{\varepsilon}e$, then $\|x - a\| = \sqrt{\varepsilon} = r$, showing that the condition (5.59) is fulfilled.

On the other hand,

$$\|x\|\,\|a\| - \mathrm{Re}\,\langle x, a\rangle = \sqrt{\left\|a + \sqrt{\varepsilon}e\right\|^2} - \mathrm{Re}\,\langle a + \sqrt{\varepsilon}e, a\rangle$$

$$= \sqrt{\|a\|^2 + \varepsilon\,\|e\|^2} - \|a\|^2$$

$$= \sqrt{1 + \varepsilon} - 1.$$

Utilising (5.62), we conclude that

$$\sqrt{1 + \varepsilon} - 1 \le c\varepsilon \quad \text{for any } \varepsilon > 0. \tag{5.63}$$

Multiplying (5.63) by $\sqrt{1 + \varepsilon} + 1 > 0$ and then dividing by $\varepsilon > 0$, we get

$$\left(\sqrt{1 + \varepsilon} + 1\right)c \ge 1 \quad \text{for any } \varepsilon > 0. \tag{5.64}$$

Letting $\varepsilon \to 0+$ in (5.64), we deduce $c \ge \frac{1}{2}$, and the result is proved. $\quad\square$

The following result also holds [61]:

Let $(H; \langle \cdot, \cdot \rangle)$ be an inner product space over H and $x, y \in H$, $\gamma, \Gamma \in \mathbb{K}$ $(\Gamma \ne -\gamma)$ so that either

$$\mathrm{Re}\,\langle \Gamma y - x, x - \gamma y \rangle \ge 0 \tag{5.65}$$

or, equivalently,

$$\left\| x - \frac{\gamma + \Gamma}{2} y \right\| \le \frac{1}{2} |\Gamma - \gamma|\, \|y\| \tag{5.66}$$

holds. Then we have the inequalities

$$0 \le \|x\|\,\|y\| - |\langle x, y\rangle| \tag{5.67}$$

$$\le \|x\|\,\|y\| - \left| \mathrm{Re}\left[\frac{\bar{\Gamma} + \bar{\gamma}}{|\Gamma + \gamma|} \langle x, y\rangle \right] \right|$$

$$\le \|x\|\,\|y\| - \mathrm{Re}\left[\frac{\bar{\Gamma} + \bar{\gamma}}{|\Gamma + \gamma|} \langle x, y\rangle \right]$$

$$\le \frac{1}{4} \cdot \frac{|\Gamma - \gamma|^2}{|\Gamma + \gamma|} \|y\|^2.$$

The constant $\frac{1}{4}$ in the last inequality is the best possible.

PROOF Consider for $a, y \neq 0$, $a = \frac{\Gamma + \gamma}{2} \cdot y$, and $r = \frac{1}{2} |\Gamma - \gamma| \, \|y\|$. Thus from (5.60), we get

$$
\begin{aligned}
0 &\leq \|x\| \left| \frac{\Gamma + \gamma}{2} \right| \|y\| - \left| \frac{\Gamma + \gamma}{2} \right| |\langle x, y \rangle| \\
&\leq \|x\| \left| \frac{\Gamma + \gamma}{2} \right| \|y\| - \left| \mathrm{Re} \left[\frac{\bar{\Gamma} + \bar{\gamma}}{2} \langle x, y \rangle \right] \right| \\
&\leq \|x\| \left| \frac{\Gamma + \gamma}{2} \right| \|y\| - \mathrm{Re} \left[\frac{\bar{\Gamma} + \bar{\gamma}}{2} \langle x, y \rangle \right] \\
&\leq \frac{1}{8} \cdot |\Gamma - \gamma|^2 \, \|y\|^2 .
\end{aligned}
$$

Dividing by $\frac{1}{2} |\Gamma + \gamma| > 0$, we deduce the desired inequality (5.67).

To prove the sharpness of the constant $\frac{1}{4}$, assume that there exists a $c > 0$ such that:

$$
\|x\| \, \|y\| - \mathrm{Re} \left[\frac{\bar{\Gamma} + \bar{\gamma}}{|\Gamma + \gamma|} \langle x, y \rangle \right] \leq c \cdot \frac{|\Gamma - \gamma|^2}{|\Gamma + \gamma|} \|y\|^2 , \tag{5.68}
$$

provided either (5.65) or (5.66) holds.

Consider the real inner product space $\left(\mathbb{R}^2, \langle \cdot, \cdot \rangle \right)$ with $\langle \bar{\mathbf{x}}, \bar{\mathbf{y}} \rangle = x_1 y_1 + x_2 y_2$, $\bar{\mathbf{x}} = (x_1, x_2)$, $\bar{\mathbf{y}} = (y_1, y_2) \in \mathbb{R}^2$. Let $\bar{\mathbf{y}} = (1, 1)$ and $\Gamma, \gamma > 0$ with $\Gamma > \gamma$. Then, by (5.68), we deduce

$$
\sqrt{2} \sqrt{x_1^2 + x_2^2} - (x_1 + x_2) \leq 2c \cdot \frac{(\Gamma - \gamma)^2}{\Gamma + \gamma} . \tag{5.69}
$$

If $x_1 = \Gamma$, $x_2 = \gamma$, then

$$
\langle \Gamma \bar{\mathbf{y}} - \bar{\mathbf{x}}, \bar{\mathbf{x}} - \gamma \bar{\mathbf{y}} \rangle = (\Gamma - x_1)(x_1 - \gamma) + (\Gamma - x_2)(x_2 - \gamma) = 0,
$$

showing that the condition (5.65) is valid. By replacing x_1 and x_2 in (5.69), we deduce

$$
\sqrt{2} \sqrt{\Gamma^2 + \gamma^2} - (\Gamma + \gamma) \leq 2c \frac{(\Gamma - \gamma)^2}{\Gamma + \gamma} . \tag{5.70}
$$

If in (5.70) we choose $\Gamma = 1 + \varepsilon$, $\gamma = 1 - \varepsilon$ with $\varepsilon \in (0, 1)$, then we have

$$
2 \sqrt{1 + \varepsilon^2} - 2 \leq 2c \frac{4 \varepsilon^2}{2} ,
$$

giving

$$
\sqrt{1 + \varepsilon^2} - 1 \leq 2c \varepsilon^2 . \tag{5.71}
$$

Finally, by multiplying (5.71) with $\sqrt{1 + \varepsilon^2} + 1 > 0$ and then dividing by ε^2, we deduce

$$
1 \leq 2c \left(\sqrt{1 + \varepsilon^2} + 1 \right) \quad \text{for any } \varepsilon \in (0, 1). \tag{5.72}
$$

Letting $\varepsilon \to 0+$ in (5.72) we get $c \geq \frac{1}{4}$, and the sharpness of the constant is proved. □

Comments

We now provide, by using certain results obtained above, some integral inequalities that may be used in practical applications.

Let $f, g \in L_\rho^2(\Omega, \mathbb{K})$ *and* $r > 0$ *with the property that*

$$|f(s) - g(s)| \leq r \quad \text{for} \quad \mu - \text{a.e.} \quad s \in \Omega.$$

Then we have the inequalities

$$0 \leq \left[\int_\Omega \rho(s) |f(s)|^2 d\mu(s) \int_\Omega \rho(s) |g(s)|^2 d\mu(s) \right]^{\frac{1}{2}} \tag{5.73}$$
$$- \left| \int_\Omega \rho(s) f(s) \overline{g(s)} d\mu(s) \right|$$
$$\leq \left[\int_\Omega \rho(s) |f(s)|^2 d\mu(s) \int_\Omega \rho(s) |g(s)|^2 d\mu(s) \right]^{\frac{1}{2}}$$
$$- \left| \int_\Omega \rho(s) \operatorname{Re} \left[f(s) \overline{g(s)} \right] d\mu(s) \right|$$
$$\leq \left[\int_\Omega \rho(s) |f(s)|^2 d\mu(s) \int_\Omega \rho(s) |g(s)|^2 d\mu(s) \right]^{\frac{1}{2}}$$
$$- \int_\Omega \rho(s) \operatorname{Re} \left[f(s) \overline{g(s)} \right] d\mu(s)$$
$$\leq \frac{1}{2} r^2.$$

The constant $\frac{1}{2}$ *is best possible in* (5.73).

The proof follows by (5.60), and we omit the details.

Let $f, g \in L_\rho^2(\Omega, \mathbb{K})$ *and* $\gamma, \Gamma \in \mathbb{K}$ *so that* $\Gamma \neq -\gamma$, *and*

$$\operatorname{Re} \left[(\Gamma g(s) - f(s)) \left(\overline{f(s)} - \overline{\gamma} \overline{g(s)} \right) \right] \geq 0, \quad \text{for} \quad \mu - \text{a.e.} \quad s \in \Omega.$$

Then we have the inequalities

$$0 \leq \left[\int_\Omega \rho(s) |f(s)|^2 d\mu(s) \int_\Omega \rho(s) |g(s)|^2 d\mu(s) \right]^{\frac{1}{2}} \tag{5.74}$$
$$- \left| \int_\Omega \rho(s) f(s) \overline{g(s)} d\mu(s) \right|$$
$$\leq \left[\int_\Omega \rho(s) |f(s)|^2 d\mu(s) \int_\Omega \rho(s) |g(s)|^2 d\mu(s) \right]^{\frac{1}{2}}$$
$$- \left| \operatorname{Re} \left[\frac{\bar{\Gamma} + \bar{\gamma}}{|\Gamma + \gamma|} \int_\Omega \rho(s) f(s) \overline{g(s)} d\mu(s) \right] \right|$$

$$\leq \left[\int_\Omega \rho(s) |f(s)|^2 \, d\mu(s) \int_\Omega \rho(s) |g(s)|^2 \, d\mu(s) \right]^{\frac{1}{2}}$$

$$- \operatorname{Re} \left[\frac{\bar{\Gamma} + \bar{\gamma}}{|\Gamma + \gamma|} \int_\Omega \rho(s) f(s) \overline{g(s)} d\mu(s) \right]$$

$$\leq \frac{1}{4} \cdot \frac{|\Gamma - \gamma|^2}{|\Gamma + \gamma|} \int_\Omega \rho(s) |g(s)|^2 \, d\mu(s).$$

The constant $\frac{1}{4}$ is best possible.

If the space is real and we assume, for $M > m > 0$, that

$$mg(s) \leq f(s) \leq Mg(s), \quad \text{for} \quad \mu - \text{a.e.} \quad s \in \Omega, \tag{5.75}$$

then, by (5.74), we deduce the inequality

$$0 \leq \left[\int_\Omega \rho(s) |f(s)|^2 \, d\mu(s) \int_\Omega \rho(s) |g(s)|^2 \, d\mu(s) \right]^{\frac{1}{2}}$$

$$- \left| \int_\Omega \rho(s) f(s) \overline{g(s)} d\mu(s) \right|$$

$$\leq \frac{1}{4} \cdot \frac{(M-m)^2}{M+m} \int_\Omega \rho(s) |g(s)|^2 \, d\mu(s).$$

The constant $\frac{1}{4}$ is best possible.

5.6 A Reverse of Bessel's Inequality

Let $\{e_i\}_{i \in I}$ be a family of orthonormal vectors in H, F a finite part of I, and ϕ_i, Φ_i $(i \in F)$, real or complex numbers. The following statements are equivalent for $x \in H$:

(i) $\operatorname{Re} \left\langle \sum_{i \in F} \Phi_i e_i - x, x - \sum_{i \in F} \phi_i e_i \right\rangle \geq 0,$

(ii) $\left\| x - \sum_{i \in F} \frac{\phi_i + \Phi_i}{2} e_i \right\| \leq \frac{1}{2} \left(\sum_{i \in F} |\Phi_i - \phi_i|^2 \right)^{\frac{1}{2}}.$

PROOF It is easy to see that for $y, a, A \in H$, the following are equivalent:

(i) $\operatorname{Re} \langle A - y, y - a \rangle \geq 0$ and

(ii) $\left\| y - \frac{a+A}{2} \right\| \leq \frac{1}{2} \|A - a\|.$

Now, for $a = \sum_{i \in F} \phi_i e_i$ and $A = \sum_{i \in F} \Phi_i e_i$, we have

$$\|A - a\| = \left\| \sum_{i \in F} (\Phi_i - \phi_i)\, e_i \right\| = \left(\left\| \sum_{i \in F} (\Phi_i - \phi_i)\, e_i \right\|^2 \right)^{\frac{1}{2}}$$

$$= \left(\sum_{i \in F} |\Phi_i - \phi_i|^2 \, \|e_i\|^2 \right)^{\frac{1}{2}} = \left(\sum_{i \in F} |\Phi_i - \phi_i|^2 \right)^{\frac{1}{2}},$$

giving, for $y = x$, the desired equivalence. $\qquad\qquad\qquad\qquad$ ▯

The following reverse of Bessel's inequality holds [46]:

Let $\{e_i\}_{i \in I}$, F, ϕ_i, Φ_i, $i \in F$ and $x \in H$ such that either (i) or (ii) holds. Then we have the inequality

$$0 \le \|x\|^2 - \sum_{i \in F} |\langle x, e_i \rangle|^2 \quad \text{(Bessel's inequality)} \qquad (5.76)$$

$$\le \frac{1}{4} \sum_{i \in F} |\Phi_i - \phi_i|^2 - \mathrm{Re} \left\langle \sum_{i \in F} \Phi_i e_i - x, x - \sum_{i \in F} \phi_i e_i \right\rangle$$

$$\le \frac{1}{4} \sum_{i \in F} |\Phi_i - \phi_i|^2.$$

The constant $\frac{1}{4}$ is best in both inequalities.

PROOF   Define

$$I_1 := \sum_{i \in H} \mathrm{Re} \left[(\Phi_i - \langle x, e_i \rangle) \left(\overline{\langle x, e_i \rangle} - \overline{\phi_i} \right) \right]$$

and

$$I_2 := \mathrm{Re} \left[\left\langle \sum_{i \in H} \Phi_i e_i - x, x - \sum_{i \in H} \phi_i e_i \right\rangle \right].$$

Observe that

$$I_1 = \sum_{i \in H} \mathrm{Re} \left[\Phi_i \overline{\langle x, e_i \rangle} \right] + \sum_{i \in H} \mathrm{Re} \left[\overline{\phi_i} \langle x, e_i \rangle \right] - \sum_{i \in H} \mathrm{Re} \left[\Phi_i \overline{\phi_i} \right] - \sum_{i \in H} |\langle x, e_i \rangle|^2$$

and

$$I_2 = \mathrm{Re} \left[\sum_{i \in H} \Phi_i \overline{\langle x, e_i \rangle} + \sum_{i \in H} \overline{\phi_i} \langle x, e_i \rangle - \|x\|^2 - \sum_{i \in H} \sum_{j \in H} \Phi_i \overline{\phi_i} \langle e_i, e_j \rangle \right]$$

$$= \sum_{i \in H} \mathrm{Re} \left[\Phi_i \overline{\langle x, e_i \rangle} \right] + \sum_{i \in H} \mathrm{Re} \left[\overline{\phi_i} \langle x, e_i \rangle \right] - \|x\|^2 - \sum_{i \in H} \mathrm{Re} \left[\Phi_i \overline{\phi_i} \right].$$

Consequently, by subtracting I_2 from I_1, we deduce the following identity:

$$\|x\|^2 - \sum_{i \in F} |\langle x, e_i \rangle|^2 = \sum_{i \in H} \text{Re} \left[(\Phi_i - \langle x, e_i \rangle) \left(\overline{\langle x, e_i \rangle} - \overline{\phi_i} \right) \right]$$

$$- \text{Re} \left[\left\langle \sum_{i \in H} \Phi_i e_i - x, x - \sum_{i \in H} \phi_i e_i \right\rangle \right]. \quad (5.77)$$

Using the following elementary inequality for complex numbers,

$$\text{Re} \left[a\overline{b} \right] \le \frac{1}{4} |a + b|^2, \quad a, b \in \mathbb{K},$$

for the choices $a = \Phi_i - \langle x, e_i \rangle$, $b = \langle x, e_i \rangle - \phi_i$ $(i \in F)$, we deduce

$$\sum_{i \in H} \text{Re} \left[(\Phi_i - \langle x, e_i \rangle) \left(\overline{\langle x, e_i \rangle} - \overline{\phi_i} \right) \right] \le \frac{1}{4} \sum_{i \in H} |\Phi_i - \phi_i|^2. \quad (5.78)$$

Making use of (5.77), (5.78), and the assumption (i), we deduce (5.76).

The sharpness of the constant $\frac{1}{4}$ was proved for a single element e, $\|e\| = 1$ by Dragomir [47], and for the real case by Ujević [163].

We provide the following simple proof:

Assume that there is a $c > 0$ such that

$$0 \le \|x\|^2 - \sum_{i \in F} |\langle x, e_i \rangle|^2 \quad (5.79)$$

$$\le c \sum_{i \in F} |\Phi_i - \phi_i|^2 - \text{Re} \left\langle \sum_{i \in F} \Phi_i e_i - x, x - \sum_{i \in F} \phi_i e_i \right\rangle,$$

provided ϕ_i, Φ_i, x, and F satisfy (i) or (ii).

We choose $F = \{1\}$, $e_1 = e_2 = \left(\frac{1}{\sqrt{2}}, \frac{1}{\sqrt{2}} \right) \in \mathbb{R}^2$, $x = (x_1, x_2) \in \mathbb{R}^2$, $\Phi_1 = \Phi = m > 0$, $\phi_1 = \phi = -m$, $H = \mathbb{R}^2$ to get from (5.79) that

$$0 \le x_1^2 + x_2^2 - \frac{(x_1 + x_2)^2}{2} \quad (5.80)$$

$$\le 4cm^2 - \left(\frac{m}{\sqrt{2}} - x_1 \right) \left(x_1 + \frac{m}{\sqrt{2}} \right) - \left(\frac{m}{\sqrt{2}} - x_2 \right) \left(x_2 + \frac{m}{\sqrt{2}} \right),$$

provided

$$0 \le \langle me - x, x + me \rangle \quad (5.81)$$

$$= \left(\frac{m}{\sqrt{2}} - x_1 \right) \left(x_1 + \frac{m}{\sqrt{2}} \right) + \left(\frac{m}{\sqrt{2}} - x_2 \right) \left(x_2 + \frac{m}{\sqrt{2}} \right).$$

If we choose $x_1 = \frac{m}{\sqrt{2}}$, $x_2 = -\frac{m}{\sqrt{2}}$, then (5.81) is fulfilled and by (5.80) we get $m^2 \le 4cm^2$, giving $c \ge \frac{1}{4}$. $\quad \Box$

Comments

Consider the family $\{f_i\}_{i \in I}$ of functions in $L_\rho^2(\Omega, \mathbb{K})$ with the properties that

$$\int_\Omega \rho(s) f_i(s) \overline{f_j}(s) d\mu(s) = \delta_{ij}, \quad i, j \in I,$$

where δ_{ij} is zero if $i \neq j$ and $\delta_{ij} = 1$ if $i = j$. $\{f_i\}_{i \in I}$ is an orthonormal family in $L_\rho^2(\Omega, \mathbb{K})$.

The following proposition holds [46]:

Let $\{f_i\}_{i \in I}$ be an orthonormal family of functions in $L_\rho^2(\Omega, \mathbb{K})$, F a finite subset of I, $\phi_i, \Phi_i \in K$ $(i \in F)$, and $f \in L_\rho^2(\Omega, \mathbb{K})$, such that either

$$\int_\Omega \rho(s) \operatorname{Re}\left[\left(\sum_{i \in F} \Phi_i f_i(s) - f(s)\right)\left(\overline{f}(s) - \sum_{i \in F} \overline{\phi_i} \, \overline{f_i}(s)\right)\right] d\mu(s) \geq 0 \quad (5.82)$$

or, equivalently,

$$\int_\Omega \rho(s) \left| f(s) - \sum_{i \in F} \frac{\Phi_i + \phi_i}{2} f_i(s) \right|^2 d\mu(s) \leq \frac{1}{4} \sum_{i \in F} |\Phi_i - \phi_i|^2.$$

We then have the inequality

$$0 \leq \int_\Omega \rho(s) |f(s)|^2 d\mu(s) - \sum_{i \in F} \left| \int_\Omega \rho(s) f(s) \overline{f_i}(s) d\mu(s) \right|^2 \quad (5.83)$$

$$\leq \frac{1}{4} \sum_{i \in F} |\Phi_i - \phi_i|^2$$

$$\qquad - \int_\Omega \rho(s) \operatorname{Re}\left[\left(\sum_{i \in F} \Phi_i f_i(s) - f(s)\right)\left(\overline{f}(s) - \sum_{i \in F} \overline{\phi_i} \, \overline{f_i}(s)\right)\right] d\mu(s)$$

$$\leq \frac{1}{4} \sum_{i \in F} |\Phi_i - \phi_i|^2.$$

The constant $\frac{1}{4}$ is the best possible in both inequalities.

5.7 Reverses for the Triangle Inequality in Inner Product Spaces

The following inequality,

$$\left\| \sum_{k=1}^n x_k \right\| \leq \sum_{k=1}^n \|x_k\|,$$

is well known in the literature as the triangle inequality.

Let $(H; \langle \cdot, \cdot \rangle)$ *be a complex inner product space. Suppose that the vectors* $x_k \in H$, $k \in \{1, \ldots, n\}$ *satisfy the condition*

$$0 \le r_1 \|x_k\| \le \operatorname{Re} \langle x_k, e \rangle, \quad 0 \le r_2 \|x_k\| \le \operatorname{Im} \langle x_k, e \rangle \tag{5.84}$$

for each $k \in \{1, \ldots, n\}$, *where* $e \in H$ *is such that* $\|e\| = 1$ *and* $r_1, r_2 \ge 0$. *Then we have the inequality*

$$\sqrt{r_1^2 + r_2^2} \sum_{k=1}^{n} \|x_k\| \le \left\| \sum_{k=1}^{n} x_k \right\|, \tag{5.85}$$

where equality holds if and only if

$$\sum_{k=1}^{n} x_k = (r_1 + i r_2) \left(\sum_{k=1}^{n} \|x_k\| \right) e. \tag{5.86}$$

PROOF We follow the proof from Dragomir [86].

In view of the Schwarz inequality in the complex inner product space $(H; \langle \cdot, \cdot \rangle)$, we have

$$\left\| \sum_{k=1}^{n} x_k \right\|^2 = \left\| \sum_{k=1}^{n} x_k \right\|^2 \|e\|^2 \ge \left| \left\langle \sum_{k=1}^{n} x_k, e \right\rangle \right|^2 \tag{5.87}$$

$$= \left| \left\langle \sum_{k=1}^{n} x_k, e \right\rangle \right|^2$$

$$= \left| \sum_{k=1}^{n} \operatorname{Re} \langle x_k, e \rangle + i \left(\sum_{k=1}^{n} \operatorname{Im} \langle x_k, e \rangle \right) \right|^2$$

$$= \left(\sum_{k=1}^{n} \operatorname{Re} \langle x_k, e \rangle \right)^2 + \left(\sum_{k=1}^{n} \operatorname{Im} \langle x_k, e \rangle \right)^2.$$

Now, by hypothesis (5.84),

$$\left(\sum_{k=1}^{n} \operatorname{Re} \langle x_k, e \rangle \right)^2 \ge r_1^2 \left(\sum_{k=1}^{n} \|x_k\| \right)^2 \tag{5.88}$$

and

$$\left(\sum_{k=1}^{n} \operatorname{Im} \langle x_k, e \rangle \right)^2 \ge r_2^2 \left(\sum_{k=1}^{n} \|x_k\| \right)^2. \tag{5.89}$$

If we add (5.88) and (5.89) and use (5.87), then we deduce the desired inequality (5.85).

Now, if (5.86) holds, then

$$\left\|\sum_{k=1}^{n} x_k\right\| = |r_1 + ir_2| \left(\sum_{k=1}^{n} \|x_k\|\right) \|e\| = \sqrt{r_1^2 + r_2^2} \sum_{k=1}^{n} \|x_k\|.$$

The case of equality is valid in (5.85).

Before we prove the reverse implication, let us observe that for $x \in H$ and $e \in H$, $\|e\| = 1$, the following identity is true:

$$\|x - \langle x, e\rangle e\|^2 = \|x\|^2 - |\langle x, e\rangle|^2,$$

so that $\|x\| = |\langle x, e\rangle|$ if and only if $x = \langle x, e\rangle e$.

If we assume that equality holds in (5.85), then the case of equality must hold in all the inequalities required in the argument used to prove the inequality (5.85). Thus, we may state that

$$\left\|\sum_{k=1}^{n} x_k\right\| = \left|\left\langle \sum_{k=1}^{n} x_k, e \right\rangle\right|, \tag{5.90}$$

and

$$r_1 \|x_k\| = \operatorname{Re}\langle x_k, e\rangle, \quad r_2 \|x_k\| = \operatorname{Im}\langle x_k, e\rangle \tag{5.91}$$

for each $k \in \{1, \ldots, n\}$.

From (5.90) we deduce

$$\sum_{k=1}^{n} x_k = \left\langle \sum_{k=1}^{n} x_k, e \right\rangle e. \tag{5.92}$$

Further, from (5.91), by multiplying the second equation by i and summing both equations over k from 1 to n, we deduce

$$(r_1 + ir_2) \sum_{k=1}^{n} \|x_k\| = \left\langle \sum_{k=1}^{n} x_k, e \right\rangle. \tag{5.93}$$

Finally, by (5.93) and (5.92), we get the desired equality (5.86). □

The following result is of interest [86]:

Let e be a unit vector in the complex inner product space $(H; \langle \cdot, \cdot\rangle)$ and $\rho_1, \rho_2 \in (0, 1)$. If $x_k \in H$, $k \in \{1, \ldots, n\}$ are such that

$$\|x_k - e\| \le \rho_1, \quad \|x_k - ie\| \le \rho_2 \quad \text{for each } k \in \{1, \ldots, n\}, \tag{5.94}$$

then we have the inequality

$$\sqrt{2 - \rho_1^2 - \rho_2^2} \sum_{k=1}^{n} \|x_k\| \le \left\|\sum_{k=1}^{n} x_k\right\|, \tag{5.95}$$

with equality if and only if

$$\sum_{k=1}^{n} x_k = \left(\sqrt{1 - \rho_1^2} + i\sqrt{1 - \rho_2^2} \right) \left(\sum_{k=1}^{n} \|x_k\| \right) e. \qquad (5.96)$$

PROOF From the first inequality in (5.94) we deduce

$$0 \le \sqrt{1 - \rho_1^2} \, \|x_k\| \le \operatorname{Re} \langle x_k, e \rangle \qquad (5.97)$$

for each $k \in \{1, \ldots, n\}$.

From the second inequality in (5.94) we deduce

$$0 \le \sqrt{1 - \rho_2^2} \, \|x_k\| \le \operatorname{Re} \langle x_k, ie \rangle$$

for each $k \in \{1, \ldots, n\}$. Since

$$\operatorname{Re} \langle x_k, ie \rangle = \operatorname{Im} \langle x_k, e \rangle,$$

we get

$$0 \le \sqrt{1 - \rho_2^2} \, \|x_k\| \le \operatorname{Im} \langle x_k, e \rangle \qquad (5.98)$$

for each $k \in \{1, \ldots, n\}$.

Now, observe from (5.97) and (5.98) that the condition (5.84) is satisfied for $r_1 = \sqrt{1 - \rho_1^2}$, $r_2 = \sqrt{1 - \rho_2^2} \in (0, 1)$. Thus the result is proved. ∎

The following more practical result may be stated as well [86]:

Let e be a unit vector in the complex inner product space $(H; \langle \cdot, \cdot \rangle)$ and $M_1 \ge m_1 > 0$, $M_2 \ge m_2 > 0$. If $x_k \in H$, $k \in \{1, \ldots, n\}$ are such that either

$$\operatorname{Re} \langle M_1 e - x_k, x_k - m_1 e \rangle \ge 0, \qquad (5.99)$$
$$\operatorname{Re} \langle M_2 ie - x_k, x_k - m_2 ie \rangle \ge 0$$

or, equivalently,

$$\left\| x_k - \frac{M_1 + m_1}{2} e \right\| \le \frac{1}{2} (M_1 - m_1), \qquad (5.100)$$
$$\left\| x_k - \frac{M_2 + m_2}{2} ie \right\| \le \frac{1}{2} (M_2 - m_2),$$

for each $k \in \{1, \ldots, n\}$, then we have the inequality

$$2 \left[\frac{m_1 M_1}{(M_1 + m_1)^2} + \frac{m_2 M_2}{(M_2 + m_2)^2} \right]^{\frac{1}{2}} \sum_{k=1}^{n} \|x_k\| \le \left\| \sum_{k=1}^{n} x_k \right\|. \qquad (5.101)$$

The equality holds in (5.101) if and only if

$$\sum_{k=1}^{n} x_k = 2 \left(\frac{\sqrt{m_1 M_1}}{M_1 + m_1} + i \frac{\sqrt{m_2 M_2}}{M_2 + m_2} \right) \left(\sum_{k=1}^{n} \|x_k\| \right) e. \qquad (5.102)$$

PROOF   From the first inequality in (5.99),

$$0 \le \frac{2\sqrt{m_1 M_1}}{M_1 + m_1} \|x_k\| \le \mathrm{Re} \langle x_k, e \rangle \qquad (5.103)$$

for each $k \in \{1, \ldots, n\}$.

The proof follows the same path as the one above. We omit the details. ☐

Comments

The following reverse of the generalised triangle inequality with a clear geometric meaning may be stated [86]:

Let $z_1, \ldots, z_n$ be complex numbers with the property that

$$0 \le \varphi_1 \le \arg(z_k) \le \varphi_2 < \frac{\pi}{2} \qquad (5.104)$$

for each $k \in \{1, \ldots, n\}$. We then have the inequality

$$\sqrt{\sin^2 \varphi_1 + \cos^2 \varphi_2} \sum_{k=1}^{n} |z_k| \le \left| \sum_{k=1}^{n} z_k \right|. \qquad (5.105)$$

Equality holds in (5.105) if and only if

$$\sum_{k=1}^{n} z_k = (\cos \varphi_2 + i \sin \varphi_1) \sum_{k=1}^{n} |z_k|. \qquad (5.106)$$

PROOF   Let $z_k = a_k + ib_k$. We may assume that $b_k \ge 0$, $a_k > 0$, $k \in \{1, \ldots, n\}$, since, by (5.104), $\frac{b_k}{a_k} = \tan[\arg(z_k)] \in [0, \frac{\pi}{2})$, $k \in \{1, \ldots, n\}$. By (5.104) we have

$$0 \le \tan^2 \varphi_1 \le \frac{b_k^2}{a_k^2} \le \tan^2 \varphi_2, \qquad k \in \{1, \ldots, n\}$$

from which we get

$$\frac{b_k^2 + a_k^2}{a_k^2} \le \frac{1}{\cos^2 \varphi_2}, \qquad k \in \{1, \ldots, n\}, \ \varphi_2 \in \left(0, \frac{\pi}{2} \right)$$

and

$$\frac{a_k^2 + b_k^2}{a_k^2} \le \frac{1 + \tan^2 \varphi_1}{\tan^2 \varphi_1} = \frac{1}{\sin^2 \varphi_1}, \qquad k \in \{1, \ldots, n\}, \ \varphi_1 \in \left(0, \frac{\pi}{2} \right),$$

giving the inequalities

$$|z_k| \cos \varphi_2 \leq \mathrm{Re}\,(z_k) \quad \text{and} \quad |z_k| \sin \varphi_1 \leq \mathrm{Im}\,(z_k)$$

for each $k \in \{1, \ldots, n\}$.

Now, by applying (5.85) for the complex inner product C endowed with the inner product $\langle z, w \rangle = z \cdot \bar{w}$ for $x_k = z_k$, $r_1 = \cos \varphi_2$, $r_2 = \sin \varphi_1$ and $e = 1$, we deduce the desired inequality (5.105). The case of equality is also obvious. We omit the details. ▯

Another result that has an obvious geometrical interpretation is stated in the following:

Let $c \in C$ with $|z| = 1$ and $\rho_1, \rho_2 \in (0, 1)$. If $z_k \in C$, $k \in \{1, \ldots, n\}$ are such that

$$|z_k - c| \leq \rho_1, \quad |z_k - ic| \leq \rho_2 \quad \text{for each } k \in \{1, \ldots, n\}, \tag{5.107}$$

then we have the inequality

$$\sqrt{2 - \rho_1^2 - \rho_2^2} \sum_{k=1}^{n} |z_k| \leq \left| \sum_{k=1}^{n} z_k \right|, \tag{5.108}$$

with equality if and only if

$$\sum_{k=1}^{n} z_k = \left(\sqrt{1 - \rho_1^2} + i\sqrt{1 - \rho_2^2} \right) \left(\sum_{k=1}^{n} |z_k| \right) c. \tag{5.109}$$

If we choose $c = 1$, and for $\rho_1, \rho_2 \in (0, 1)$ we define

$$\bar{D}\,(1, \rho_1) := \{z \in \mathbb{C} | \, |z - 1| \leq \rho_1\},$$
$$\bar{D}\,(i, \rho_2) := \{z \in \mathbb{C} | \, |z - i| \leq \rho_2\},$$

then obviously the intersection

$$S_{\rho_1, \rho_2} := \bar{D}\,(1, \rho_1) \cap \bar{D}\,(i, \rho_2)$$

is nonempty if and only if $\rho_1 + \rho_2 > \sqrt{2}$.

If $z_k \in S_{\rho_1, \rho_2}$ for $k \in \{1, \ldots, n\}$, then (5.108) holds true. The equality holds in (5.108) if and only if

$$\sum_{k=1}^{n} z_k = \left(\sqrt{1 - \rho_1^2} + i\sqrt{1 - \rho_2^2} \right) \sum_{k=1}^{n} |z_k|.$$

5.8 The Boas-Bellman Inequality

Boas in 1941 [11] and Bellman in 1944 [7] independently proved the following generalisation of Bessel's inequality (see also Mitrinović, Pečarić, and Fink [141, p. 392]):

If $x, y_1, \ldots, y_n$ *are vectors in an inner product space* $(H; \langle \cdot, \cdot \rangle)$, *then the following inequality*,

$$\sum_{i=1}^{n} |\langle x, y_i \rangle|^2 \leq \|x\|^2 \left[\max_{1 \leq i \leq n} \|y_i\|^2 + \left(\sum_{1 \leq i \neq j \leq n} |\langle y_i, y_j \rangle|^2 \right)^{\frac{1}{2}} \right], \qquad (5.110)$$

holds.

A recent generalisation of the Boas-Bellman result was given in Mitrinović, Pečarić, and Fink [141, p. 392] where they proved the following:

If $x, y_1, \ldots, y_n$ *are as above and* $c_1, \ldots, c_n \in \mathbb{K}$, *then one has the inequality:*

$$\left| \sum_{i=1}^{n} c_i \langle x, y_i \rangle \right|^2$$

$$\leq \|x\|^2 \sum_{i=1}^{n} |c_i|^2 \left[\max_{1 \leq i \leq n} \|y_i\|^2 + \left(\sum_{1 \leq i \neq j \leq n} |\langle y_i, y_j \rangle|^2 \right)^{\frac{1}{2}} \right]. \qquad (5.111)$$

They also noted that if in (5.111) one chooses $c_i = \overline{\langle x, y_i \rangle}$, then this inequality becomes (5.110).

In the following [64], we point out some results that may be related to both the Mitrinović-Pečarić-Fink and Boas-Bellman inequalities:

Let $z_1, \ldots, z_n \in H$ *and* $\alpha_1, \ldots, \alpha_n \in \mathbb{K}$. *Then one has the inequality:*

$$\left\| \sum_{i=1}^{n} \alpha_i z_i \right\|^2 \leq \begin{cases} \max\limits_{1 \leq i \leq n} |\alpha_i|^2 \sum_{i=1}^{n} \|z_i\|^2; \\[2ex] \left(\sum_{i=1}^{n} |\alpha_i|^{2\alpha} \right)^{\frac{1}{\alpha}} \left(\sum_{i=1}^{n} \|z_i\|^{2\beta} \right)^{\frac{1}{\beta}}, \quad \text{where } \alpha > 1, \\ \qquad\qquad\qquad\qquad\qquad\qquad\qquad \frac{1}{\alpha} + \frac{1}{\beta} = 1; \\[2ex] \sum_{i=1}^{n} |\alpha_i|^2 \max\limits_{1 \leq i \leq n} \|z_i\|^2, \end{cases}$$

$$+ \begin{cases} \max\limits_{1\le i\ne j\le n} \{|\alpha_i\alpha_j|\} \sum\limits_{1\le i\ne j\le n} |\langle z_i, z_j\rangle| \,; \\[2ex] \left[\left(\sum_{i=1}^n |\alpha_i|^\gamma\right)^2 - \left(\sum_{i=1}^n |\alpha_i|^{2\gamma}\right)\right]^{\frac{1}{\gamma}} \left(\sum_{1\le i\ne j\le n} |\langle z_i, z_j\rangle|^\delta\right)^{\frac{1}{\delta}}, \\ \hspace{4cm} \text{where } \gamma > 1, \ \frac{1}{\gamma} + \frac{1}{\delta} = 1; \\[2ex] \left[\left(\sum_{i=1}^n |\alpha_i|\right)^2 - \sum_{i=1}^n |\alpha_i|^2\right] \max\limits_{1\le i\ne j\le n} |\langle z_i, z_j\rangle| \,. \end{cases} \quad (5.112)$$

PROOF We follow the proof from Dragomir [64].

Observe that

$$\left\|\sum_{i=1}^n \alpha_i z_i\right\|^2 = \left(\sum_{i=1}^n \alpha_i z_i, \sum_{j=1}^n \alpha_j z_j\right) \qquad (5.113)$$

$$= \sum_{i=1}^n \sum_{j=1}^n \alpha_i \overline{\alpha_j} \langle z_i, z_j\rangle = \left|\sum_{i=1}^n \sum_{j=1}^n \alpha_i \overline{\alpha_j} \langle z_i, z_j\rangle\right|$$

$$\le \sum_{i=1}^n \sum_{j=1}^n |\alpha_i| \, |\overline{\alpha_j}| \, |\langle z_i, z_j\rangle|$$

$$= \sum_{i=1}^n |\alpha_i|^2 \, \|z_i\|^2 + \sum_{1\le i\ne j\le n} |\alpha_i| \, |\alpha_j| \, |\langle z_i, z_j\rangle| \,.$$

Using Hölder's inequality, we may write that

$$\sum_{i=1}^n |\alpha_i|^2 \, \|z_i\|^2$$

$$\le \begin{cases} \max\limits_{1\le i\le n} |\alpha_i|^2 \sum\limits_{i=1}^n \|z_i\|^2 \,; \\[2ex] \left(\sum_{i=1}^n |\alpha_i|^{2\alpha}\right)^{\frac{1}{\alpha}} \left(\sum_{i=1}^n \|z_i\|^{2\beta}\right)^{\frac{1}{\beta}}, \quad \text{where } \alpha > 1, \frac{1}{\alpha} + \frac{1}{\beta} = 1; \\[2ex] \sum_{i=1}^n |\alpha_i|^2 \max\limits_{1\le i\le n} \|z_i\|^2 \,. \end{cases} \quad (5.114)$$

By Hölder's inequality for double sums we also have

$$\sum_{1\le i\ne j\le n} |\alpha_i| \, |\alpha_j| \, |\langle z_i, z_j\rangle| \qquad (5.115)$$

$$\leq \begin{cases} \max_{1\leq i\neq j\leq n} |\alpha_i\alpha_j| \sum_{1\leq i\neq j\leq n} |\langle z_i, z_j\rangle| \, ; \\[2ex] \left(\sum_{1\leq i\neq j\leq n} |\alpha_i|^\gamma |\alpha_j|^\gamma\right)^{\frac{1}{\gamma}} \left(\sum_{1\leq i\neq j\leq n} |\langle z_i, z_j\rangle|^\delta\right)^{\frac{1}{\delta}}, \\[1ex] \qquad \text{where } \gamma > 1, \ \frac{1}{\gamma} + \frac{1}{\delta} = 1; \\[2ex] \sum_{1\leq i\neq j\leq n} |\alpha_i| |\alpha_j| \max_{1\leq i\neq j\leq n} |\langle z_i, z_j\rangle| , \end{cases}$$

$$= \begin{cases} \max_{1\leq i\neq j\leq n} \{|\alpha_i\alpha_j|\} \sum_{1\leq i\neq j\leq n} |\langle z_i, z_j\rangle| \, ; \\[2ex] \left[\left(\sum_{i=1}^{n} |\alpha_i|^\gamma\right)^2 - \left(\sum_{i=1}^{n} |\alpha_i|^{2\gamma}\right)\right]^{\frac{1}{\gamma}} \left(\sum_{1\leq i\neq j\leq n} |\langle z_i, z_j\rangle|^\delta\right)^{\frac{1}{\delta}}, \\[1ex] \qquad \text{where } \gamma > 1, \ \frac{1}{\gamma} + \frac{1}{\delta} = 1; \\[2ex] \left[\left(\sum_{i=1}^{n} |\alpha_i|\right)^2 - \sum_{i=1}^{n} |\alpha_i|^2\right] \max_{1\leq i\neq j\leq n} |\langle z_i, z_j\rangle| . \end{cases}$$

Utilising (5.114) and (5.115) in (5.113), we may deduce the desired result (5.112). □

Inequality (5.112) in fact contains nine different inequalities which may be obtained by combining the first three inequalities with the last three.

A particular case that may be related to the Boas-Bellman result is embodied in the following inequality [64]:

With the above assumptions we have

$$\left\|\sum_{i=1}^{n} \alpha_i z_i\right\|^2 \tag{5.116}$$

$$\leq \sum_{i=1}^{n} |\alpha_i|^2$$

$$\cdot \left\{\max_{1\leq i\leq n} \|z_i\|^2 + \frac{\left[\left(\sum_{i=1}^{n} |\alpha_i|^2\right)^2 - \sum_{i=1}^{n} |\alpha_i|^4\right]^{\frac{1}{2}}}{\sum_{i=1}^{n} |\alpha_i|^2} \left(\sum_{1\leq i\neq j\leq n} |\langle z_i, z_j\rangle|^2\right)^{\frac{1}{2}}\right\}$$

$$\leq \sum_{i=1}^{n} |\alpha_i|^2 \left\{\max_{1\leq i\leq n} \|z_i\|^2 + \left(\sum_{1\leq i\neq j\leq n} |\langle z_i, z_j\rangle|^2\right)^{\frac{1}{2}}\right\}.$$

The first inequality follows by taking the third branch in the first curly bracket with the second branch in the second curly bracket for $\gamma = \delta = 2$.

The second inequality in (5.116) follows by the fact that

$$\left[\left(\sum_{i=1}^{n} |\alpha_i|^2\right)^2 - \sum_{i=1}^{n} |\alpha_i|^4\right]^{\frac{1}{2}} \le \sum_{i=1}^{n} |\alpha_i|^2.$$

Applying the following Cauchy-Bunyakovsky-Schwarz type inequality,

$$\left(\sum_{i=1}^{n} a_i\right)^2 \le n \sum_{i=1}^{n} a_i^2, \quad a_i \in \mathbb{R}_+, \ 1 \le i \le n,$$

we may write that

$$\left(\sum_{i=1}^{n} |\alpha_i|^\gamma\right)^2 - \sum_{i=1}^{n} |\alpha_i|^{2\gamma} \le (n-1) \sum_{i=1}^{n} |\alpha_i|^{2\gamma} \quad (n \ge 1) \qquad (5.117)$$

and

$$\left(\sum_{i=1}^{n} |\alpha_i|\right)^2 - \sum_{i=1}^{n} |\alpha_i|^2 \le (n-1) \sum_{i=1}^{n} |\alpha_i|^2 \quad (n \ge 1). \qquad (5.118)$$

Also, it is obvious that:

$$\max_{1 \le i \ne j \le n} \{|\alpha_i \alpha_j|\} \le \max_{1 \le i \le n} |\alpha_i|^2. \qquad (5.119)$$

Consequently, we may state the following coarser upper bounds for $\|\sum_{i=1}^{n} \alpha_i z_i\|^2$ that may be useful in applications [64]:

With the above assumptions we have the inequalities:

$$\left\|\sum_{i=1}^{n} \alpha_i z_i\right\|^2$$

$$\le \begin{cases} \max_{1 \le i \le n} |\alpha_i|^2 \sum_{i=1}^{n} \|z_i\|^2; \\ \left(\sum_{i=1}^{n} |\alpha_i|^{2\alpha}\right)^{\frac{1}{\alpha}} \left(\sum_{i=1}^{n} \|z_i\|^{2\beta}\right)^{\frac{1}{\beta}}, \quad \text{where } \alpha > 1, \frac{1}{\alpha} + \frac{1}{\beta} = 1; \\ \sum_{i=1}^{n} |\alpha_i|^2 \max_{1 \le i \le n} \|z_i\|^2, \end{cases}$$

$$+ \begin{cases} \max_{1 \le i \le n} |\alpha_i|^2 \sum_{1 \le i \ne j \le n} |\langle z_i, z_j \rangle|; \\ (n-1)^{\frac{1}{\gamma}} \left(\sum_{i=1}^{n} |\alpha_i|^{2\gamma}\right)^{\frac{1}{\gamma}} \left(\sum_{1 \le i \ne j \le n} |\langle z_i, z_j \rangle|^\delta\right)^{\frac{1}{\delta}}, \\ \qquad\qquad\qquad \text{where } \gamma > 1, \frac{1}{\gamma} + \frac{1}{\delta} = 1; \\ (n-1) \sum_{i=1}^{n} |\alpha_i|^2 \max_{1 \le i \ne j \le n} |\langle z_i, z_j \rangle|. \end{cases} \qquad (5.120)$$

The following inequalities which are incorporated in (5.120) are special cases of the more complicated results above:

$$\left\|\sum_{i=1}^{n} \alpha_i z_i\right\|^2 \leq \max_{1 \leq i \leq n} |\alpha_i|^2 \left[\sum_{i=1}^{n} \|z_i\|^2 + \sum_{1 \leq i \neq j \leq n} |\langle z_i, z_j\rangle|\right]; \qquad (5.121)$$

$$\left\|\sum_{i=1}^{n} \alpha_i z_i\right\|^2 \leq \left(\sum_{i=1}^{n} |\alpha_i|^{2p}\right)^{\frac{1}{p}} \left[\left(\sum_{i=1}^{n} \|z_i\|^{2q}\right)^{\frac{1}{q}}\right.$$

$$\left. + (n-1)^{\frac{1}{p}} \left(\sum_{1 \leq i \neq j \leq n} |\langle z_i, z_j\rangle|^q\right)^{\frac{1}{q}}\right], \qquad (5.122)$$

where $p > 1$, $\frac{1}{p} + \frac{1}{q} = 1$; and

$$\left\|\sum_{i=1}^{n} \alpha_i z_i\right\|^2 \leq \sum_{i=1}^{n} |\alpha_i|^2 \left[\max_{1 \leq i \leq n} \|z_i\|^2 + (n-1) \max_{1 \leq i \neq j \leq n} |\langle z_i, z_j\rangle|\right]. \qquad (5.123)$$

We are now able to present the following result obtained by Dragomir [64], which complements the inequality (5.111) due to Mitrinović, Pečarić, and Fink [141, p. 392]:

Let $x, y_1, \ldots, y_n$ be vectors of an inner product space $(H; \langle \cdot, \cdot\rangle)$ and $c_1, \ldots, c_n \in K$ $(\mathbb{K} = \mathbb{C}, \mathbb{R})$. Then one has the inequalities:

$$\left|\sum_{i=1}^{n} c_i \langle x, y_i\rangle\right|^2$$

$$\leq \|x\|^2 \times \begin{cases} \max\limits_{1 \leq i \leq n} |c_i|^2 \sum_{i=1}^{n} \|y_i\|^2; \\[2mm] \left(\sum_{i=1}^{n} |c_i|^{2\alpha}\right)^{\frac{1}{\alpha}} \left(\sum_{i=1}^{n} \|y_i\|^{2\beta}\right)^{\frac{1}{\beta}}, \quad \text{where } \alpha > 1, \\[2mm] \hspace{5cm} \frac{1}{\alpha} + \frac{1}{\beta} = 1; \\[2mm] \sum_{i=1}^{n} |c_i|^2 \max\limits_{1 \leq i \leq n} \|y_i\|^2, \end{cases}$$

$$+ \|x\|^2 \times \begin{cases} \max\limits_{1 \leq i \neq j \leq n} \{|c_i c_j|\} \sum_{1 \leq i \neq j \leq n} |\langle y_i, y_j\rangle|; \\[2mm] \left[\left(\sum_{i=1}^{n} |c_i|^{\gamma}\right)^2 - \left(\sum_{i=1}^{n} |c_i|^{2\gamma}\right)\right]^{\frac{1}{\gamma}} \\[2mm] \hspace{1cm} \times \left(\sum_{1 \leq i \neq j \leq n} |\langle y_i, y_j\rangle|^{\delta}\right)^{\frac{1}{\delta}}, \\[2mm] \hspace{2cm} \text{where } \gamma > 1, \ \frac{1}{\gamma} + \frac{1}{\delta} = 1; \\[2mm] \left[\left(\sum_{i=1}^{n} |c_i|\right)^2 - \sum_{i=1}^{n} |c_i|^2\right] \max\limits_{1 \leq i \neq j \leq n} |\langle y_i, y_j\rangle|. \end{cases} \qquad (5.124)$$

PROOF   We note that

$$\sum_{i=1}^{n} c_i \langle x, y_i \rangle = \left(x, \sum_{i=1}^{n} \overline{c_i} y_i \right).$$

Using Schwarz's inequality in inner product spaces, we have:

$$\left| \sum_{i=1}^{n} c_i \langle x, y_i \rangle \right|^2 \leq \|x\|^2 \left\| \sum_{i=1}^{n} \overline{c_i} y_i \right\|^2.$$

Now, by using (5.120) with $\alpha_i = \overline{c_i}$, $z_i = y_i$ $(i = 1, \ldots, n)$, we deduce the desired inequality (5.124). □

The following particular inequalities may be obtained from the above results [64].

With the above assumptions one has the inequalities:

$$\left| \sum_{i=1}^{n} c_i \langle x, y_i \rangle \right|^2$$

$$\leq \|x\|^2 \times \begin{cases} \sum_{i=1}^{n} |c_i|^2 \left\{ \max\limits_{1 \leq i \leq n} \|y_i\|^2 + \left(\sum_{1 \leq i \neq j \leq n} |\langle y_i, y_j \rangle|^2 \right)^{\frac{1}{2}} \right\}; \\[2mm] \max\limits_{1 \leq i \leq n} |c_i|^2 \left\{ \sum_{i=1}^{n} \|y_i\|^2 + \sum_{1 \leq i \neq j \leq n} |\langle y_i, y_j \rangle| \right\}; \\[2mm] \left(\sum_{i=1}^{n} |c_i|^{2p} \right)^{\frac{1}{p}} \left\{ \left(\sum_{i=1}^{n} \|y_i\|^{2q} \right)^{\frac{1}{q}} \right. \\[2mm] \qquad \left. + (n-1)^{\frac{1}{p}} \left(\sum_{1 \leq i \neq j \leq n} |\langle y_i, y_j \rangle|^q \right)^{\frac{1}{q}} \right\}, \\[2mm] \qquad\qquad\qquad \text{where } p > 1, \frac{1}{p} + \frac{1}{q} = 1; \\[2mm] \sum_{i=1}^{n} |c_i|^2 \left\{ \max\limits_{1 \leq i \leq n} \|y_i\|^2 + (n-1) \max\limits_{1 \leq i \neq j \leq n} |\langle y_i, y_j \rangle| \right\}. \end{cases} \tag{5.125}$$

Note that the first inequality in (5.125) is the result obtained by Mitrinović, Pečarić, and Fink [141]. The other three provide similar bounds in terms of the p-norms of the vector $\left(|c_1|^2, \ldots, |c_n|^2 \right)$.

If one chooses $c_i = \overline{\langle x, y_i \rangle}$ $(i = 1, \ldots, n)$ in (5.124), then it is possible to obtain nine different inequalities between the Fourier coefficients $\langle x, y_i \rangle$ and the norms and inner products of the vectors y_i $(i = 1, \ldots, n)$. We restrict ourselves only to those inequalities that may be obtained from (5.125).

Comments

As Mitrinović, Pečarić, and Fink noted [141, p. 392], the first inequality in (5.125) for the above selection of c_i will produce the Boas-Bellman inequality (5.110).

From the second inequality in (5.125) for $c_i = \overline{\langle x, y_i \rangle}$, we get

$$\left(\sum_{i=1}^{n} |\langle x, y_i \rangle|^2 \right)^2 \leq \|x\|^2 \max_{1 \leq i \leq n} |\langle x, y_i \rangle|^2 \left\{ \sum_{i=1}^{n} \|y_i\|^2 + \sum_{1 \leq i \neq j \leq n} |\langle y_i, y_j \rangle| \right\}.$$

By taking the square root in this inequality we obtain:

$$\sum_{i=1}^{n} |\langle x, y_i \rangle|^2 \leq \|x\| \max_{1 \leq i \leq n} |\langle x, y_i \rangle| \left\{ \sum_{i=1}^{n} \|y_i\|^2 + \sum_{1 \leq i \neq j \leq n} |\langle y_i, y_j \rangle| \right\}^{\frac{1}{2}} \quad (5.126)$$

for any $x, y_1, \ldots, y_n$ vectors in the inner product space $(H; \langle \cdot, \cdot \rangle)$.

If we assume that $(e_i)_{1 \leq i \leq n}$ is an orthonormal family in H, then by (5.126) we have

$$\sum_{i=1}^{n} |\langle x, e_i \rangle|^2 \leq \sqrt{n} \, \|x\| \max_{1 \leq i \leq n} |\langle x, e_i \rangle|, \quad x \in H.$$

From the third inequality in (5.125) for $c_i = \overline{\langle x, y_i \rangle}$, we deduce

$$\left(\sum_{i=1}^{n} |\langle x, y_i \rangle|^2 \right)^2 \leq \|x\|^2 \left(\sum_{i=1}^{n} |\langle x, y_i \rangle|^{2p} \right)^{\frac{1}{p}}$$

$$\times \left\{ \left(\sum_{i=1}^{n} \|y_i\|^{2q} \right)^{\frac{1}{q}} + (n-1)^{\frac{1}{p}} \left(\sum_{1 \leq i \neq j \leq n} |\langle y_i, y_j \rangle|^q \right)^{\frac{1}{q}} \right\},$$

for $p > 1$, $\frac{1}{p} + \frac{1}{q} = 1$.

By taking the square root in this inequality we get

$$\sum_{i=1}^{n} |\langle x, y_i \rangle|^2 \leq \|x\| \left(\sum_{i=1}^{n} |\langle x, y_i \rangle|^{2p} \right)^{\frac{1}{2p}}$$

$$\times \left\{ \left(\sum_{i=1}^{n} \|y_i\|^{2q} \right)^{\frac{1}{q}} + (n-1)^{\frac{1}{p}} \left(\sum_{1 \leq i \neq j \leq n} |\langle y_i, y_j \rangle|^q \right)^{\frac{1}{q}} \right\}^{\frac{1}{2}} \quad (5.127)$$

for any $x, y_1, \ldots, y_n \in H$, $p > 1$, $\frac{1}{p} + \frac{1}{q} = 1$.

The above inequality (5.127) becomes, for an orthonormal family $(e_i)_{1 \leq i \leq n}$,

$$\sum_{i=1}^{n} |\langle x, e_i \rangle|^2 \leq n^{\frac{1}{q}} \|x\| \left(\sum_{i=1}^{n} |\langle x, e_i \rangle|^{2p} \right)^{\frac{1}{2p}}, \quad x \in H.$$

Finally, the choice $c_i = \overline{\langle x, y_i \rangle}$ $(i = 1, \ldots, n)$ produces

$$\left(\sum_{i=1}^{n} |\langle x, y_i \rangle|^2 \right)^2 \leq \|x\|^2 \sum_{i=1}^{n} |\langle x, y_i \rangle|^2 \left\{ \max_{1 \leq i \leq n} \|y_i\|^2 + (n-1) \max_{1 \leq i \neq j \leq n} |\langle y_i, y_j \rangle| \right\}$$

in the last inequality in (5.125); giving the following Boas-Bellman type inequality,

$$\sum_{i=1}^{n} |\langle x, y_i \rangle|^2 \leq \|x\|^2 \left\{ \max_{1 \leq i \leq n} \|y_i\|^2 + (n-1) \max_{1 \leq i \neq j \leq n} |\langle y_i, y_j \rangle| \right\}, \qquad (5.128)$$

for any $x, y_1, \ldots, y_n \in H$.

It is obvious that for orthonormal families, (5.128) will provide the well-known Bessel inequality.

5.9 The Bombieri Inequality

In 1971, Bombieri [12] (see also Mitrinović and Pečarić [140, p. 394]) gave the following generalisation of Bessel's inequality.

If $x, y_1, \ldots, y_n$ are vectors in the inner product space $(H; \langle \cdot, \cdot \rangle)$, then the inequality

$$\sum_{i=1}^{n} |\langle x, y_i \rangle|^2 \leq \|x\|^2 \max_{1 \leq i \leq n} \left\{ \sum_{j=1}^{n} |\langle y_i, y_j \rangle| \right\} \qquad (5.129)$$

holds.

It is obvious that if $(y_i)_{1 \leq i \leq n}$ are orthonormal, then from (5.129) one can deduce Bessel's inequality.

In 1992, Pečarić [149] (see also Mitrinović and Pečarić [140, p. 394]) proved the following general inequality in inner product spaces:

Let $x, y_1, \ldots, y_n \in H$ and $c_1, \ldots, c_n \in \mathbb{K}$. Then

$$\left| \sum_{i=1}^{n} c_i \langle x, y_i \rangle \right|^2 \leq \|x\|^2 \sum_{i=1}^{n} |c_i|^2 \left(\sum_{j=1}^{n} |\langle y_i, y_j \rangle| \right) \qquad (5.130)$$

$$\leq \|x\|^2 \sum_{i=1}^{n} |c_i|^2 \max_{1 \leq i \leq n} \left\{ \sum_{j=1}^{n} |\langle y_i, y_j \rangle| \right\}.$$

He showed that the Bombieri inequality (5.129) may be obtained from (5.130) for the choice $c_i = \overline{\langle x, y_i \rangle}$ (using the second inequality).

In this section, we consider a generalisation of the Bombieri inequality which also compliments (5.130).

We start with the following general results for norm inequalities, which are also interesting in themselves [65].

Let $z_1, \ldots, z_n \in H$ and $\alpha_1, \ldots, \alpha_n \in \mathbb{K}$. Then one has the inequality:

$$\left\| \sum_{i=1}^{n} \alpha_i z_i \right\|^2 \leq \begin{cases} A \\ B \\ C \end{cases}, \tag{5.131}$$

where

$$A := \begin{cases} \max_{1 \leq k \leq n} |\alpha_k|^2 \sum_{i,j=1}^{n} |\langle z_i, z_j \rangle| \, ; \\[3mm] \max_{1 \leq k \leq n} |\alpha_k| \left(\sum_{i=1}^{n} |\alpha_i|^r \right)^{\frac{1}{r}} \left(\sum_{i=1}^{n} \left(\sum_{j=1}^{n} |\langle z_i, z_j \rangle| \right)^s \right)^{\frac{1}{s}}, \quad r > 1, \ \frac{1}{r} + \frac{1}{s} = 1; \\[3mm] \max_{1 \leq k \leq n} |\alpha_k| \sum_{k=1}^{n} |\alpha_k| \max_{1 \leq i \leq n} \left(\sum_{j=1}^{n} |\langle z_i, z_j \rangle| \right) ; \end{cases}$$

$$B := \begin{cases} \left(\sum_{k=1}^{n} |\alpha_k|^p \right)^{\frac{1}{p}} \max_{1 \leq i \leq n} |\alpha_i| \left(\sum_{i=1}^{n} \left(\sum_{j=1}^{n} |\langle z_i, z_j \rangle| \right)^q \right)^{\frac{1}{q}}, \quad p > 1, \ \frac{1}{p} + \frac{1}{q} = 1; \\[3mm] \left(\sum_{k=1}^{n} |\alpha_k|^p \right)^{\frac{1}{p}} \left(\sum_{i=1}^{n} |\alpha_i|^t \right)^{\frac{1}{t}} \left[\sum_{i=1}^{n} \left(\sum_{j=1}^{n} |\langle z_i, z_j \rangle|^q \right)^{\frac{u}{q}} \right]^{\frac{1}{u}}, \quad \begin{array}{l} p > 1, \ \frac{1}{p} + \frac{1}{q} = 1; \\ t > 1, \ \frac{1}{t} + \frac{1}{u} = 1; \end{array} \\[3mm] \left(\sum_{k=1}^{n} |\alpha_k|^p \right)^{\frac{1}{p}} \sum_{i=1}^{n} |\alpha_i| \max_{1 \leq i \leq n} \left\{ \left(\sum_{j=1}^{n} |\langle z_i, z_j \rangle|^q \right)^{\frac{1}{q}} \right\}, \quad p > 1, \ \frac{1}{p} + \frac{1}{q} = 1; \end{cases}$$

and

$$C := \begin{cases} \sum_{k=1}^{n} |\alpha_k| \max_{1 \leq i \leq n} |\alpha_i| \sum_{i=1}^{n} \left[\max_{1 \leq j \leq n} |\langle z_i, z_j \rangle| \right] ; \\[3mm] \sum_{k=1}^{n} |\alpha_k| \left(\sum_{i=1}^{n} |\alpha_i|^m \right)^{\frac{1}{m}} \left(\sum_{i=1}^{n} \left[\max_{1 \leq j \leq n} |\langle z_i, z_j \rangle| \right]^l \right)^{\frac{1}{t}}, \quad m > 1, \ \frac{1}{m} + \frac{1}{l} = 1; \\[3mm] \left(\sum_{k=1}^{n} |\alpha_k| \right)^2 \max_{i, 1 \leq j \leq n} |\langle z_i, z_j \rangle|. \end{cases}$$

PROOF   We follow the proof from Dragomir [65].

Observe that

$$\left\|\sum_{i=1}^{n}\alpha_i z_i\right\|^2 = \left(\sum_{i=1}^{n}\alpha_i z_i, \sum_{j=1}^{n}\alpha_j z_j\right)$$

$$= \sum_{i=1}^{n}\sum_{j=1}^{n}\alpha_i\overline{\alpha_j}\langle z_i, z_j\rangle = \left|\sum_{i=1}^{n}\sum_{j=1}^{n}\alpha_i\overline{\alpha_j}\langle z_i, z_j\rangle\right|$$

$$\le \sum_{i=1}^{n}\sum_{j=1}^{n}|\alpha_i|\,|\alpha_j|\,|\langle z_i, z_j\rangle| = \sum_{i=1}^{n}|\alpha_i|\left(\sum_{j=1}^{n}|\alpha_j|\,|\langle z_i, z_j\rangle|\right)$$

$$:= M.$$

Using Hölder's inequality, we may write that

$$\sum_{j=1}^{n}|\alpha_j|\,|\langle z_i, z_j\rangle| \le \begin{cases} \displaystyle\max_{1\le k\le n}|\alpha_k|\sum_{j=1}^{n}|\langle z_i, z_j\rangle| \\[2ex] \displaystyle\left(\sum_{k=1}^{n}|\alpha_k|^p\right)^{\frac{1}{p}}\left(\sum_{j=1}^{n}|\langle z_i, z_j\rangle|^q\right)^{\frac{1}{q}}, \quad p>1,\ \frac{1}{p}+\frac{1}{q}=1; \\[2ex] \displaystyle\sum_{k=1}^{n}|\alpha_k|\max_{1\le j\le n}|\langle z_i, z_j\rangle| \end{cases}$$

for any $i \in \{1,\ldots,n\}$, giving

$$M \le \begin{cases} \displaystyle\max_{1\le k\le n}|\alpha_k|\sum_{i=1}^{n}|\alpha_i|\sum_{j=1}^{n}|\langle z_i, z_j\rangle| =: M_1; \\[2ex] \displaystyle\left(\sum_{k=1}^{n}|\alpha_k|^p\right)^{\frac{1}{p}}\sum_{i=1}^{n}|\alpha_i|\left(\sum_{j=1}^{n}|\langle z_i, z_j\rangle|^q\right)^{\frac{1}{q}} := M_p, \\[1ex] \hspace{5cm} p>1,\ \frac{1}{p}+\frac{1}{q}=1; \\[2ex] \displaystyle\sum_{k=1}^{n}|\alpha_k|\sum_{i=1}^{n}|\alpha_i|\max_{1\le j\le n}|\langle z_i, z_j\rangle| =: M_\infty. \end{cases}$$

By Hölder's inequality, we also have:

$$\sum_{i=1}^{n}|\alpha_i|\left(\sum_{j=1}^{n}|\langle z_i, z_j\rangle|\right)$$

$$
\leq
\begin{cases}
\displaystyle\max_{1\leq i\leq n}|\alpha_i|\sum_{i,j=1}^{n}|\langle z_i, z_j\rangle|\,; \\[2em]
\displaystyle\left(\sum_{i=1}^{n}|\alpha_i|^r\right)^{\frac{1}{r}}\left(\sum_{i=1}^{n}\left(\sum_{j=1}^{n}|\langle z_i, z_j\rangle|\right)^s\right)^{\frac{1}{s}}, \\[1em]
\qquad\qquad\qquad r>1,\ \frac{1}{r}+\frac{1}{s}=1\,; \\[2em]
\displaystyle\sum_{i=1}^{n}|\alpha_i|\max_{1\leq i\leq n}\left(\sum_{j=1}^{n}|\langle z_i, z_j\rangle|\right).
\end{cases}
$$

Hence,

$$
M_1 \leq
\begin{cases}
\displaystyle\max_{1\leq k\leq n}|\alpha_k|^2\sum_{i,j=1}^{n}|\langle z_i, z_j\rangle|\,; \\[2em]
\displaystyle\max_{1\leq k\leq n}|\alpha_k|\left(\sum_{i=1}^{n}|\alpha_i|^r\right)^{\frac{1}{r}}\left(\sum_{i=1}^{n}\left(\sum_{j=1}^{n}|\langle z_i, z_j\rangle|\right)^s\right)^{\frac{1}{s}}, \\[1em]
\qquad\qquad\qquad r>1,\ \frac{1}{r}+\frac{1}{s}=1\,; \\[2em]
\displaystyle\max_{1\leq k\leq n}|\alpha_k|\sum_{i=1}^{n}|\alpha_i|\max_{1\leq i\leq n}\left(\sum_{j=1}^{n}|\langle z_i, z_j\rangle|\right)\,;
\end{cases}
$$

and the first three inequalities in (5.131) are obtained.

By Hölder's inequality we also have:

$$
M_p \leq \left(\sum_{k=1}^{n}|\alpha_k|^p\right)^{\frac{1}{p}}\times
\begin{cases}
\displaystyle\max_{1\leq i\leq n}|\alpha_i|\sum_{i=1}^{n}\left(\sum_{j=1}^{n}|\langle z_i, z_j\rangle|^q\right)^{\frac{1}{q}}\,; \\[2em]
\displaystyle\left(\sum_{i=1}^{n}|\alpha_i|^t\right)^{\frac{1}{t}}\left(\sum_{i=1}^{n}\left(\sum_{j=1}^{n}|\langle z_i, z_j\rangle|^q\right)^{\frac{u}{q}}\right)^{\frac{1}{u}}, \\[1em]
\qquad\qquad\qquad t>1,\ \frac{1}{t}+\frac{1}{u}=1\,; \\[2em]
\displaystyle\sum_{i=1}^{n}|\alpha_i|\max_{1\leq i\leq n}\left\{\left(\sum_{j=1}^{n}|\langle z_i, z_j\rangle|^q\right)^{\frac{1}{q}}\right\}\,;
\end{cases}
$$

and the next three inequalities in (5.131) are proved.

Finally, by the same Hölder inequality, we may state that:

$$M_\infty \leq \sum_{k=1}^{n} |\alpha_k| \times \begin{cases} \max\limits_{1 \leq i \leq n} |\alpha_i| \sum\limits_{i=1}^{n} \left(\max\limits_{1 \leq j \leq n} |\langle z_i, z_j \rangle| \right); \\[3mm] \left(\sum\limits_{i=1}^{n} |\alpha_i|^m \right)^{\frac{1}{m}} \left(\sum\limits_{i=1}^{n} \left(\max\limits_{1 \leq j \leq n} |\langle z_i, z_j \rangle| \right)^l \right)^{\frac{1}{t}}, \\[3mm] \qquad\qquad\qquad\qquad m > 1, \ \frac{1}{m} + \frac{1}{l} = 1; \\[3mm] \sum\limits_{i=1}^{n} |\alpha_i| \max\limits_{1 \leq i, j \leq n} |\langle z_i, z_j \rangle|; \end{cases}$$

and the last three inequalities in (5.131) are proven. ☐

If we would like to have some bounds for $\left\| \sum_{i=1}^{n} \alpha_i z_i \right\|^2$ in terms of $\sum_{i=1}^{n} |\alpha_i|^2$, then the following results may be used.

Let $z_1, \ldots, z_n$ and $\alpha_1, \ldots, \alpha_n$ be as above. If $1 < p \leq 2$, $1 < t \leq 2$, then one has the inequality

$$\left\| \sum_{i=1}^{n} \alpha_i z_i \right\|^2 \leq n^{\frac{1}{p} + \frac{1}{t} - 1} \sum_{k=1}^{n} |\alpha_k|^2 \left[\sum_{i=1}^{n} \left(\sum_{j=1}^{n} |\langle z_i, z_j \rangle|^q \right)^{\frac{u}{q}} \right]^{\frac{1}{u}}, \qquad (5.132)$$

where $\frac{1}{p} + \frac{1}{q} = 1$, $\frac{1}{t} + \frac{1}{u} = 1$.

PROOF By the monotonicity of power means, we may write that

$$\left(\frac{\sum_{k=1}^{n} |\alpha_k|^p}{n} \right)^{\frac{1}{p}} \leq \left(\frac{\sum_{k=1}^{n} |\alpha_k|^2}{n} \right)^{\frac{1}{2}}; \quad 1 < p \leq 2$$

and

$$\left(\frac{\sum_{k=1}^{n} |\alpha_k|^t}{n} \right)^{\frac{1}{t}} \leq \left(\frac{\sum_{k=1}^{n} |\alpha_k|^2}{n} \right)^{\frac{1}{2}}; \quad 1 < t \leq 2.$$

This implies that

$$\left(\sum_{k=1}^{n} |\alpha_k|^p \right)^{\frac{1}{p}} \leq n^{\frac{1}{p} - \frac{1}{2}} \left(\sum_{k=1}^{n} |\alpha_k|^2 \right)^{\frac{1}{2}}$$

and

$$\left(\sum_{k=1}^{n} |\alpha_k|^t \right)^{\frac{1}{t}} \leq n^{\frac{1}{t} - \frac{1}{2}} \left(\sum_{k=1}^{n} |\alpha_k|^2 \right)^{\frac{1}{2}}.$$

By using the fifth inequality in (5.131), we then deduce (5.132). ☐

An interesting particular case is the one for $p = q = t = u = 2$, giving

$$\left\|\sum_{i=1}^{n} \alpha_i z_i\right\|^2 \leq \sum_{k=1}^{n} |\alpha_k|^2 \left(\sum_{i,j=1}^{n} |\langle z_i, z_j\rangle|^2\right)^{\frac{1}{2}}.$$

With the above assumptions and if $1 < p \leq 2$, then

$$\left\|\sum_{i=1}^{n} \alpha_i z_i\right\|^2 \leq n^{\frac{1}{p}} \sum_{k=1}^{n} |\alpha_k|^2 \max_{1 \leq i \leq n} \left[\left(\sum_{j=1}^{n} |\langle z_i, z_j\rangle|^q\right)^{\frac{1}{q}}\right], \tag{5.133}$$

where $\frac{1}{p} + \frac{1}{q} = 1$.

PROOF Since

$$\left(\sum_{k=1}^{n} |\alpha_k|^p\right)^{\frac{1}{p}} \leq n^{\frac{1}{p} - \frac{1}{2}} \left(\sum_{k=1}^{n} |\alpha_k|^2\right)^{\frac{1}{2}}$$

and

$$\sum_{k=1}^{n} |\alpha_k| \leq n^{\frac{1}{2}} \left(\sum_{k=1}^{n} |\alpha_k|^2\right)^{\frac{1}{2}},$$

then inequality (5.133) follows from the sixth inequality in (5.131). ☐

In a similar fashion, one may prove the following two results:

With the above assumptions and if $1 < m \leq 2$, then

$$\left\|\sum_{i=1}^{n} \alpha_i z_i\right\|^2 \leq n^{\frac{1}{m}} \sum_{k=1}^{n} |\alpha_k|^2 \left(\sum_{i=1}^{n} \left[\max_{1 \leq j \leq n} |\langle z_i, z_j\rangle|\right]^l\right)^{\frac{1}{l}},$$

where $\frac{1}{m} + \frac{1}{l} = 1$.

With the assumptions, we have:

$$\left\|\sum_{i=1}^{n} \alpha_i z_i\right\|^2 \leq n \sum_{k=1}^{n} |\alpha_k|^2 \max_{1 \leq i,j \leq n} |\langle z_i, z_j\rangle|.$$

The following result, which also complements the inequality obtained by Pečarić [149], may be of interest as well [65]:

With the above assumptions, one has the inequalities

$$\left\| \sum_{i=1}^{n} \alpha_i z_i \right\|^2 \leq \sum_{i=1}^{n} |\alpha_i|^2 \sum_{j=1}^{n} |\langle z_i, z_j \rangle| \tag{5.134}$$

$$\leq \begin{cases} \sum_{i=1}^{n} |\alpha_i|^2 \max_{1 \leq i \leq n} \left[\sum_{j=1}^{n} |\langle z_i, z_j \rangle| \right]; \\[2mm] \left(\sum_{i=1}^{n} |\alpha_i|^{2p} \right)^{\frac{1}{p}} \left(\left(\sum_{j=1}^{n} |\langle z_i, z_j \rangle| \right)^q \right)^{\frac{1}{q}}, \\[1mm] \qquad\qquad\qquad p > 1, \ \frac{1}{p} + \frac{1}{q} = 1; \\[3mm] \max_{1 \leq i \leq n} |\alpha_i|^2 \sum_{i,j=1}^{n} |\langle z_i, z_j \rangle|. \end{cases}$$

PROOF   From the above, we know that

$$\left\| \sum_{i=1}^{n} \alpha_i z_i \right\|^2 \leq \sum_{i=1}^{n} \sum_{j=1}^{n} |\alpha_i| |\alpha_j| |\langle z_i, z_j \rangle|.$$

Using the simple observation that (see also Mitrinović and Pečarić [140, p. 394])

$$|\alpha_i| |\alpha_j| \leq \frac{1}{2} \left(|\alpha_i|^2 + |\alpha_j|^2 \right), \quad i, j \in \{1, \ldots, n\},$$

we have

$$\sum_{i=1}^{n} \sum_{j=1}^{n} |\alpha_i| |\alpha_j| |\langle z_i, z_j \rangle| \leq \frac{1}{2} \sum_{i,j=1}^{n} \left(|\alpha_i|^2 + |\alpha_j|^2 \right) |\langle z_i, z_j \rangle|$$

$$= \frac{1}{2} \left[\sum_{i,j=1}^{n} |\alpha_i|^2 |\langle z_i, z_j \rangle| + \sum_{i,j=1}^{n} |\alpha_j|^2 |\langle z_i, z_j \rangle| \right]$$

$$= \sum_{i,j=1}^{n} |\alpha_i|^2 |\langle z_i, z_j \rangle|,$$

which proves the first inequality in (5.134).

The second part follows from Hölder's inequality, and we omit the details.
⬜

The first part in (5.134) is the inequality obtained by Pečarić [149].

We are now able to present the following result obtained by Dragomir [65], which complements inequality (5.130) due to Pečarić [149] (see also Mitrinović and Pečarić [140, p. 394]).

Let $x, y_1, \ldots, y_n$ be vectors of an inner product space $(H; \langle \cdot, \cdot \rangle)$ and $c_1, \ldots, c_n \in \mathbb{K}$. Then one has the inequalities:

$$\left| \sum_{i=1}^{n} c_i \langle x, y_i \rangle \right|^2 \le \|x\|^2 \times \begin{cases} D \\ E, \\ F \end{cases} \qquad (5.135)$$

where

$$D := \begin{cases} \max\limits_{1 \le k \le n} |c_k|^2 \sum\limits_{i,j=1}^{n} |\langle y_i, y_j \rangle|; \\[3mm] \max\limits_{1 \le k \le n} |c_k| \left(\sum\limits_{i=1}^{n} |c_i|^r \right)^{\frac{1}{r}} \left[\sum\limits_{i=1}^{n} \left(\sum\limits_{j=1}^{n} |\langle y_i, y_j \rangle| \right)^s \right]^{\frac{1}{s}}, \quad r > 1, \ \frac{1}{r} + \frac{1}{s} = 1; \\[4mm] \max\limits_{1 \le k \le n} |c_k| \sum\limits_{k=1}^{n} |c_k| \max\limits_{1 \le i \le n} \left(\sum\limits_{j=1}^{n} |\langle y_i, y_j \rangle| \right); \end{cases}$$

$$E := \begin{cases} \left(\sum\limits_{k=1}^{n} |c_k|^p \right)^{\frac{1}{p}} \max\limits_{1 \le i \le n} |c_i| \left(\sum\limits_{i=1}^{n} \left(\sum\limits_{j=1}^{n} |\langle y_i, y_j \rangle| \right)^q \right)^{\frac{1}{q}}, \quad p > 1, \ \frac{1}{p} + \frac{1}{q} = 1; \\[4mm] \left(\sum\limits_{k=1}^{n} |c_k|^p \right)^{\frac{1}{p}} \left(\sum\limits_{i=1}^{n} |c_i|^t \right)^{\frac{1}{t}} \left[\sum\limits_{i=1}^{n} \left(\sum\limits_{j=1}^{n} |\langle y_i, y_j \rangle|^q \right)^{\frac{u}{q}} \right]^{\frac{1}{u}}, p > 1, \ \frac{1}{p} + \frac{1}{q} = 1; \\[2mm] \hspace{7cm} t > 1, \ \frac{1}{t} + \frac{1}{u} = 1; \\[3mm] \left(\sum\limits_{k=1}^{n} |c_k|^p \right)^{\frac{1}{p}} \sum\limits_{i=1}^{n} |c_i| \max\limits_{1 \le i \le n} \left\{ \left(\sum\limits_{j=1}^{n} |\langle y_i, y_j \rangle|^q \right)^{\frac{1}{q}} \right\}, \quad p > 1, \ \frac{1}{p} + \frac{1}{q} = 1; \end{cases}$$

and

$$F := \begin{cases} \sum\limits_{k=1}^{n} |c_k| \max\limits_{1 \le i \le n} |c_i| \sum\limits_{i=1}^{n} \left[\max\limits_{1 \le j \le n} |\langle y_i, y_j \rangle| \right]; \\[3mm] \sum\limits_{k=1}^{n} |c_k| \left(\sum\limits_{i=1}^{n} |c_i|^m \right)^{\frac{1}{m}} \left(\sum\limits_{i=1}^{n} \left[\max\limits_{1 \le j \le n} |\langle y_i, y_j \rangle| \right]^l \right)^{\frac{1}{l}}, \\[2mm] \hspace{5cm} m > 1, \ \frac{1}{m} + \frac{1}{l} = 1; \\[3mm] \left(\sum\limits_{k=1}^{n} |c_k| \right)^2 \max\limits_{i, 1 \le j \le n} |\langle y_i, y_j \rangle|. \end{cases}$$

PROOF We note that

$$\sum_{i=1}^{n} c_i \langle x, y_i \rangle = \left(x, \sum_{i=1}^{n} \overline{c_i} y_i \right).$$

Using Schwarz's inequality in inner product spaces, we have

$$\left| \sum_{i=1}^{n} c_i \langle x, y_i \rangle \right|^2 \le \|x\|^2 \left\| \sum_{i=1}^{n} \overline{c_i} y_i \right\|^2 .$$

Finally, by using the above result with $\alpha_i = \overline{c_i}$ and $z_i = y_i$ $(i = 1, \ldots, n)$, we deduce the desired inequality (5.135). We omit the details. □

The following results may be useful if one needs bounds in terms of $\sum_{i=1}^{n} |c_i|^2$:

With the previous assumptions and if $1 < p \le 2$, $1 < t \le 2$, $\frac{1}{p} + \frac{1}{q} = 1$, $\frac{1}{t} + \frac{1}{u} = 1$, *one has the inequality:*

$$\left| \sum_{i=1}^{n} c_i \langle x, y_i \rangle \right|^2 \le \|x\|^2 \, n^{\frac{1}{p} + \frac{1}{t} - 1} \sum_{i=1}^{n} |c_i|^2 \left[\sum_{i=1}^{n} \left(\sum_{j=1}^{n} |\langle y_i, y_j \rangle|^q \right)^{\frac{u}{q}} \right]^{\frac{1}{u}} , \quad (5.136)$$

and, in particular, for $p = q = t = u = 2$,

$$\left| \sum_{i=1}^{n} c_i \langle x, y_i \rangle \right|^2 \le \|x\|^2 \sum_{i=1}^{n} |c_i|^2 \left(\sum_{i,j=1}^{n} |\langle y_i, y_j \rangle|^2 \right)^{\frac{1}{2}} .$$

If $1 < p \le 2$, *then*

$$\left| \sum_{i=1}^{n} c_i \langle x, y_i \rangle \right|^2 \le \|x\|^2 \, n^{\frac{1}{p}} \sum_{k=1}^{n} |c_k|^2 \, \max_{1 \le i \le n} \left[\sum_{j=1}^{n} |\langle y_i, y_j \rangle|^q \right]^{\frac{1}{q}} ,$$

where $\frac{1}{p} + \frac{1}{q} = 1$.

The following two inequalities also hold:

With the above assumptions for x, y_i, c_i *and if* $1 < m \le 2$, *then*

$$\left| \sum_{i=1}^{n} c_i \langle x, y_i \rangle \right|^2 \le \|x\|^2 \, n^{\frac{1}{m}} \sum_{k=1}^{n} |c_k|^2 \left(\sum_{i=1}^{n} \left[\max_{1 \le j \le n} |\langle y_i, y_j \rangle| \right]^l \right)^{\frac{1}{t}} , \quad (5.137)$$

where $\frac{1}{m} + \frac{1}{l} = 1$.

With the above assumptions for x, y_i, c_i, *one has*

$$\left| \sum_{i=1}^{n} c_i \langle x, y_i \rangle \right|^2 \le \|x\|^2 \, n \sum_{k=1}^{n} |c_k|^2 \, \max_{1 \le j \le n} |\langle y_i, y_j \rangle| . \quad (5.138)$$

We may state the following result as well:

With the previous assumptions one has the inequalities:

$$\left|\sum_{i=1}^{n} c_i \langle x, y_i \rangle\right|^2 \le \|x\|^2 \sum_{i=1}^{n} |c_i|^2 \sum_{j=1}^{n} |\langle y_i, y_j \rangle|$$

$$\le \|x\|^2 \begin{cases} \sum_{i=1}^{n} |c_i|^2 \ \max_{1 \le i \le n} \left[\sum_{j=1}^{n} |\langle y_i, y_j \rangle|\right] ; \\[2ex] \left(\sum_{i=1}^{n} |c_i|^{2p}\right)^{\frac{1}{p}} \left(\sum_{i=1}^{n} \left(\sum_{j=1}^{n} |\langle y_i, y_j \rangle|\right)^q\right)^{\frac{1}{q}}, \\[1ex] \qquad\qquad\qquad\qquad p > 1, \ \frac{1}{p} + \frac{1}{q} = 1; \\[2ex] \max_{1 \le i \le n} |c_i|^2 \ \sum_{i,j=1}^{n} |\langle y_i, y_j \rangle| ; \end{cases}$$

that provide some alternatives to Pečarić's result (5.130).

Comments

We point out some inequalities of the Bombieri type that may be obtained from (5.135) on choosing $c_i = \overline{\langle x, y_i \rangle}$ $(i = 1, \dots, n)$.

If the above choice was made in the first inequality in (5.135), then one can obtain

$$\left(\sum_{i=1}^{n} |\langle x, y_i \rangle|^2\right)^2 \le \|x\|^2 \max_{1 \le i \le n} |\langle x, y_i \rangle|^2 \sum_{i,j=1}^{n} |\langle y_i, y_j \rangle|,$$

giving the following, by taking the square root:

$$\sum_{i=1}^{n} |\langle x, y_i \rangle|^2 \le \|x\| \max_{1 \le i \le n} |\langle x, y_i \rangle| \left(\sum_{i,j=1}^{n} |\langle y_i, y_j \rangle|\right)^{\frac{1}{2}}, \quad x \in H. \qquad (5.139)$$

If the same choice for c_i is made in the second inequality in (5.135), then one can get

$$\left(\sum_{i=1}^{n} |\langle x, y_i \rangle|^2\right)^2 \le \|x\|^2 \max_{1 \le i \le n} |\langle x, y_i \rangle| \left(\sum_{i=1}^{n} |\langle x, y_i \rangle|^r\right)^{\frac{1}{r}}$$

$$\times \left[\sum_{i=1}^{n} \left(\sum_{j=1}^{n} |\langle y_i, y_j \rangle|\right)^s\right]^{\frac{1}{s}}.$$

This implies that

$$\sum_{i=1}^{n} |\langle x, y_i \rangle|^2 \le \|x\| \max_{1 \le i \le n} |\langle x, y_i \rangle|^{\frac{1}{2}} \left(\sum_{i=1}^{n} |\langle x, y_i \rangle|^r \right)^{\frac{1}{2r}}$$

$$\times \left[\sum_{i=1}^{n} \left(\sum_{j=1}^{n} |\langle y_i, y_j \rangle| \right)^s \right]^{\frac{1}{2s}}, \quad (5.140)$$

where $\frac{1}{r} + \frac{1}{s} = 1$, $s > 1$.

The other inequalities in (5.135) will produce the following results, respectively:

$$\sum_{i=1}^{n} |\langle x, y_i \rangle|^2 \le \|x\| \max_{1 \le i \le n} |\langle x, y_i \rangle|^{\frac{1}{2}} \left(\sum_{i=1}^{n} |\langle x, y_i \rangle| \right)^{\frac{1}{2}}$$

$$\times \left[\max_{1 \le i \le n} \left(\sum_{j=1}^{n} |\langle y_i, y_j \rangle| \right) \right]; \quad (5.141)$$

$$\sum_{i=1}^{n} |\langle x, y_i \rangle|^2 \le \|x\| \max_{1 \le i \le n} |\langle x, y_i \rangle|^{\frac{1}{2}} \left(\sum_{i=1}^{n} |\langle x, y_i \rangle|^p \right)^{\frac{1}{2p}}$$

$$\times \left[\sum_{i=1}^{n} \left(\sum_{j=1}^{n} |\langle y_i, y_j \rangle|^q \right)^{\frac{1}{q}} \right]^{\frac{1}{2}}, \quad (5.142)$$

where $p > 1$, $\frac{1}{p} + \frac{1}{q} = 1$;

$$\sum_{i=1}^{n} |\langle x, y_i \rangle|^2 \le \|x\| \left(\sum_{i=1}^{n} |\langle x, y_i \rangle|^p \right)^{\frac{1}{2p}} \left(\sum_{i=1}^{n} |\langle x, y_i \rangle|^t \right)^{\frac{1}{2t}}$$

$$\times \left[\sum_{i=1}^{n} \left(\sum_{j=1}^{n} |\langle y_i, y_j \rangle|^q \right)^{\frac{u}{q}} \right]^{\frac{1}{2u}}, \quad (5.143)$$

where $p > 1$, $\frac{1}{p} + \frac{1}{q} = 1$, $t > 1$, $\frac{1}{t} + \frac{1}{u} = 1$;

$$\sum_{i=1}^{n} |\langle x, y_i \rangle|^2 \le \|x\| \left(\sum_{i=1}^{n} |\langle x, y_i \rangle|^p \right)^{\frac{1}{2p}} \left(\sum_{i=1}^{n} |\langle x, y_i \rangle| \right)^{\frac{1}{2}}$$

$$\times \max_{1 \le i \le n} \left\{ \left(\sum_{j=1}^{n} |\langle y_i, y_j \rangle|^q \right)^{\frac{1}{2q}} \right\}, \quad (5.144)$$

where $p > 1$, $\frac{1}{p} + \frac{1}{q} = 1$;

$$\sum_{i=1}^{n} |\langle x, y_i \rangle|^2 \leq \|x\| \left[\sum_{i=1}^{n} |\langle x, y_i \rangle| \right]^{\frac{1}{2}} \max_{1 \leq i \leq n} |\langle x, y_i \rangle|^{\frac{1}{2}}$$

$$\times \left(\sum_{i=1}^{n} \left[\max_{1 \leq j \leq n} |\langle y_i, y_j \rangle| \right] \right)^{\frac{1}{2}}; \quad (5.145)$$

$$\sum_{i=1}^{n} |\langle x, y_i \rangle|^2 \leq \|x\| \left[\sum_{i=1}^{n} |\langle x, y_i \rangle|^m \right]^{\frac{1}{2m}} \left[\sum_{i=1}^{n} \left[\max_{1 \leq j \leq n} |\langle y_i, y_j \rangle|^l \right] \right]^{\frac{1}{2l}}, \quad (5.146)$$

where $m > 1$, $\frac{1}{m} + \frac{1}{l} = 1$; and

$$\sum_{i=1}^{n} |\langle x, y_i \rangle|^2 \leq \|x\| \sum_{i=1}^{n} |\langle x, y_i \rangle| \max_{1 \leq j \leq n} |\langle y_i, y_j \rangle|^{\frac{1}{2}}. \quad (5.147)$$

If in the above inequalities we assume that $(y_i)_{1 \leq i \leq n} = (e_i)_{1 \leq i \leq n}$, where $(e_i)_{1 \leq i \leq n}$ are orthonormal vectors in the inner product space $(H, \langle \cdot, \cdot \rangle)$, then from (5.139)–(5.147) we may deduce the following inequalities, which are similar in a sense to Bessel's inequality:

$$\sum_{i=1}^{n} |\langle x, e_i \rangle|^2 \leq \sqrt{n} \, \|x\| \max_{1 \leq i \leq n} \{ |\langle x, e_i \rangle| \};$$

$$\sum_{i=1}^{n} |\langle x, e_i \rangle|^2 \leq n^{\frac{1}{2s}} \|x\| \max_{1 \leq i \leq n} \left\{ |\langle x, e_i \rangle|^{\frac{1}{2}} \right\} \left(\sum_{i=1}^{n} |\langle x, e_i \rangle|^r \right)^{\frac{1}{2r}},$$

where $r > 1$, $\frac{1}{r} + \frac{1}{s} = 1$;

$$\sum_{i=1}^{n} |\langle x, e_i \rangle|^2 \leq \|x\| \max_{1 \leq i \leq n} \left\{ |\langle x, e_i \rangle|^{\frac{1}{2}} \right\} \left(\sum_{i=1}^{n} |\langle x, e_i \rangle| \right)^{\frac{1}{2}},$$

$$\sum_{i=1}^{n} |\langle x, e_i \rangle|^2 \leq \sqrt{n} \, \|x\| \max_{1 \leq i \leq n} \left\{ |\langle x, e_i \rangle|^{\frac{1}{2}} \right\} \left(\sum_{i=1}^{n} |\langle x, e_i \rangle|^p \right)^{\frac{1}{2p}},$$

where $p > 1$;

$$\sum_{i=1}^{n} |\langle x, e_i \rangle|^2 \leq n^{\frac{1}{2u}} \|x\| \left(\sum_{i=1}^{n} |\langle x, e_i \rangle|^p \right)^{\frac{1}{2p}} \left(\sum_{i=1}^{n} |\langle x, e_i \rangle|^t \right)^{\frac{1}{2t}},$$

where $p > 1$, $t > 1$, $\frac{1}{t} + \frac{1}{u} = 1$;

$$\sum_{i=1}^{n} |\langle x, e_i \rangle|^2 \leq \|x\| \left(\sum_{i=1}^{n} |\langle x, e_i \rangle|^p \right)^{\frac{1}{2p}} \left(\sum_{i=1}^{n} |\langle x, e_i \rangle| \right)^{\frac{1}{2}}, \quad p > 1;$$

$$\sum_{i=1}^{n} |\langle x, e_i \rangle|^2 \leq \sqrt{n} \, \|x\| \left(\sum_{i=1}^{n} |\langle x, e_i \rangle| \right)^{\frac{1}{2}} \max_{1 \leq i \leq n} \left\{ |\langle x, e_i \rangle|^{\frac{1}{2}} \right\};$$

$$\sum_{i=1}^{n} |\langle x, e_i \rangle|^2 \leq n^{\frac{1}{2l}} \, \|x\| \left[\sum_{i=1}^{n} |\langle x, e_i \rangle|^m \right]^{\frac{1}{m}}, \quad m > 1, \ \frac{1}{m} + \frac{1}{l} = 1;$$

$$\sum_{i=1}^{n} |\langle x, e_i \rangle|^2 \leq \|x\| \sum_{i=1}^{n} |\langle x, e_i \rangle|.$$

Observe that some of the above results will produce the following inequalities which do not contain the Fourier coefficients in the right side of the inequality.

Indeed, if one chooses $c_i = \overline{\langle x, y_i \rangle}$ in (5.136), then

$$\left(\sum_{i=1}^{n} |\langle x, y_i \rangle|^2 \right)^2 \leq \|x\|^2 \, n^{\frac{1}{p} + \frac{1}{t} - 1} \sum_{i=1}^{n} |\langle x, y_i \rangle|^2 \left[\sum_{i=1}^{n} \left(\sum_{j=1}^{n} |\langle y_i, y_j \rangle|^q \right)^{\frac{u}{q}} \right]^{\frac{1}{u}},$$

giving the following Bombieri type inequality:

$$\sum_{i=1}^{n} |\langle x, y_i \rangle|^2 \leq n^{\frac{1}{p} + \frac{1}{t} - 1} \|x\|^2 \left[\sum_{i=1}^{n} \left(\sum_{j=1}^{n} |\langle y_i, y_j \rangle|^q \right)^{\frac{u}{q}} \right]^{\frac{1}{u}},$$

where $1 < p \leq 2$, $1 < t \leq 2$, $\frac{1}{p} + \frac{1}{q} = 1$, $\frac{1}{t} + \frac{1}{u} = 1$.

If in this inequality we take $p = q = t = u = 2$, then

$$\sum_{i=1}^{n} |\langle x, y_i \rangle|^2 \leq \|x\|^2 \left(\sum_{i,j=1}^{n} |\langle y_i, y_j \rangle|^2 \right)^{\frac{1}{2}}.$$

For a different proof of this result see also Dragomir, Mond, and Pečarić [101].

In a similar way, if $c_i = \overline{\langle x, y_i \rangle}$ in (5.137), then

$$\sum_{i=1}^{n} |\langle x, y_i \rangle|^2 \leq n^{\frac{1}{m}} \|x\|^2 \left(\sum_{i=1}^{n} \left[\max_{1 \leq j \leq n} |\langle y_i, y_j \rangle| \right]^l \right)^{\frac{1}{t}},$$

where $m > 1$, $\frac{1}{m} + \frac{1}{l} = 1$.

Finally, if $c_i = \overline{\langle x, y_i \rangle}$ $(i = 1, \ldots, n)$ is taken in (5.138), then

$$\sum_{i=1}^{n} |\langle x, y_i \rangle|^2 \leq n \, \|x\|^2 \max_{1 \leq i,j \leq n} |\langle y_i, y_j \rangle|.$$

5.10   Kurepa's Inequality

In 1960, de Bruijn proved the following refinement of the celebrated Cauchy-Bunyakovsky-Schwarz (CBS) inequality for a sequence of real numbers and the second of complex numbers (see de Bruijn [13] or Dragomir [53, p. 48]):

Let $(a_1, \ldots, a_n)$ be an n-tuple of real numbers and $(z_1, \ldots, z_n)$ be an n-tuple of complex numbers. Then

$$\left| \sum_{k=1}^{n} a_k z_k \right|^2 \le \frac{1}{2} \sum_{k=1}^{n} a_k^2 \left[\sum_{k=1}^{n} |z_k|^2 + \left| \sum_{k=1}^{n} z_k^2 \right| \right] \tag{5.148}$$

$$\left(\le \sum_{k=1}^{n} a_k^2 \cdot \sum_{k=1}^{n} |z_k|^2 \right).$$

Equality holds in (5.148) if and only if, for $k \in \{1, \ldots, n\}$, $a_k = \operatorname{Re}(\lambda z_k)$, where λ is a complex number such that $\lambda^2 \sum_{k=1}^{n} z_n^2$ is a nonnegative real number.

In 1966, in an effort to extend this result to inner products, Kurepa [130] obtained the following refinement for the complexification of a real inner product space $(H; \langle \cdot, \cdot \rangle)$.

Let $(H; \langle \cdot, \cdot \rangle)$ be a real inner product space and $(H_{\mathbb{C}}, \langle \cdot, \cdot \rangle_{\mathbb{C}})$ its complexification. For any $a \in H$ and $z \in H_{\mathbb{C}}$ we have the inequality:

$$|\langle z, a \rangle_{\mathbb{C}}|^2 \le \frac{1}{2} \|a\|^2 \left[\|z\|_{\mathbb{C}}^2 + |\langle z, \bar{z} \rangle_{\mathbb{C}}| \right] \tag{5.149}$$

$$\left(\le \|a\|^2 \|z\|_{\mathbb{C}}^2 \right).$$

To be comprehensive, we define in the following the concept of complexification for a real inner product space.

Let H be a real vector space with the inner product $\langle \cdot, \cdot \rangle$ and the norm $\|\cdot\|$. The *complexification* $H_{\mathbb{C}}$ of H is defined as a complex linear space $H \times H$ of all ordered pairs (x, y) $(x, y \in H)$ endowed with the operations

$$(x, y) + (x', y') := (x + x', y + y'), \qquad x, x', y, y' \in H;$$
$$(\sigma + i\tau) \cdot (x, y) := (\sigma x - \tau y, \tau x + \sigma y), \qquad x, y \in H \text{ and } \sigma, \tau \in \mathbb{R}.$$

On $H_{\mathbb{C}}$ one can canonically consider the *scalar product* $\langle \cdot, \cdot \rangle_{\mathbb{C}}$ defined by:

$$\langle z, z' \rangle_{\mathbb{C}} := \langle x, x' \rangle + \langle y, y' \rangle + i \left[\langle y, x' \rangle - \langle x, y' \rangle \right]$$

where $z = (x, y)$, $z' = (x', y') \in H_{\mathbb{C}}$. Obviously,

$$\|z\|_{\mathbb{C}}^2 = \|x\|^2 + \|y\|^2,$$

where $z = (x, y)$.

The conjugate of a vector $z = (x, y) \in H_{\mathbb{C}}$ is defined by $\bar{z} := (x, -y)$.

It is easy to see that the elements of $H_{\mathbb{C}}$ under defined operations behave as formal "complex" combinations $x + iy$ with $x, y \in H$. Because of this, we may write $z = x + iy$ instead of $z = (x, y)$. Thus, $\bar{z} = x - iy$.

The following elementary inequality is of interest [48].

Let $f : [0, 2\pi] \to \mathbb{R}$ given by

$$f(\alpha) = \lambda \sin^2 \alpha + 2\beta \sin \alpha \cos \alpha + \alpha \cos^2 \alpha, \tag{5.150}$$

where $\lambda, \beta, \gamma \in \mathbb{R}$. Then

$$\sup_{\alpha \in [0, 2\pi]} f(\alpha) = \frac{1}{2} (\lambda + \gamma) + \frac{1}{2} \left[(\gamma - \lambda)^2 + 4\beta^2 \right]^{\frac{1}{2}}. \tag{5.151}$$

PROOF Since

$$\sin^2 \alpha = \frac{1 - \cos 2\alpha}{2}, \quad \cos^2 \alpha = \frac{1 + \cos 2\alpha}{2}, \quad \text{and} \quad 2 \sin \alpha \cos \alpha = \sin 2\alpha,$$

we may write f as

$$f(\alpha) = \frac{1}{2} (\lambda + \gamma) + \frac{1}{2} (\gamma - \lambda) \cos 2\alpha + \beta \sin 2\alpha. \tag{5.152}$$

If $\beta = 0$, then (5.152) becomes

$$f(\alpha) = \frac{1}{2} (\lambda + \gamma) + \frac{1}{2} (\gamma - \lambda) \cos 2\alpha.$$

Obviously, in this case

$$\sup_{\alpha \in [0, 2\pi]} f(\alpha) = \frac{1}{2} (\lambda + \gamma) + \frac{1}{2} |\gamma - \lambda| = \max (\gamma, \lambda).$$

If $\beta \neq 0$, then (5.152) becomes

$$f(\alpha) = \frac{1}{2} (\lambda + \gamma) + \beta \left[\sin 2\alpha + \frac{(\gamma - \lambda)}{\beta} \cos 2\alpha \right].$$

Let $\varphi \in \left(-\frac{\pi}{2}, \frac{\pi}{2} \right)$ for which $\tan \varphi = \frac{\gamma - \lambda}{2\beta}$. Then f can be written as

$$f(\alpha) = \frac{1}{2} (\lambda + \gamma) + \frac{\beta}{\cos \varphi} \sin (2\alpha + \varphi).$$

For this function we have

$$\sup_{\alpha \in [0, 2\pi]} f(\alpha) = \frac{1}{2} (\lambda + \gamma) + \frac{|\beta|}{|\cos \varphi|}. \tag{5.153}$$

Since

$$\frac{\sin^2 \varphi}{\cos^2 \varphi} = \frac{(\gamma - \lambda)^2}{4\beta^2},$$

we get

$$\frac{1}{|\cos \varphi|} = \frac{\left[(\gamma - \lambda)^2 + 4\beta^2\right]^{\frac{1}{2}}}{2|\beta|}.$$

From (5.153) we deduce the desired result (5.151).   ⬜

The following result holds [48].

Let $(H; \langle \cdot, \cdot \rangle)$ be a complex inner product space. If $x, y, z \in H$ are such that

$$\mathrm{Im} \langle x, z \rangle = \mathrm{Im} \langle y, z \rangle = 0, \tag{5.154}$$

then we have the inequality:

$$\mathrm{Re}^2 \langle x, z \rangle + \mathrm{Re}^2 \langle y, z \rangle \tag{5.155}$$
$$= |\langle x + iy, z \rangle|^2$$
$$\leq \frac{1}{2} \left\{ \|x\|^2 + \|y\|^2 + \left[\left(\|x\|^2 - \|y\|^2 \right)^2 + 4\mathrm{Re}^2 \langle x, y \rangle \right]^{\frac{1}{2}} \right\} \|z\|^2$$
$$\leq \left(\|x\|^2 + \|y\|^2 \right) \|z\|^2.$$

PROOF   Obviously, by (5.154) we have

$$\langle x + iy, z \rangle = \mathrm{Re} \langle x, z \rangle + i \, \mathrm{Re} \langle y, z \rangle.$$

Therefore, the first part of (5.155) holds true.

Now, let $\varphi \in [0, 2\pi]$ be such that $\langle x + iy, z \rangle = e^{i\varphi} |\langle x + iy, z \rangle|$. Then

$$|\langle x + iy, z \rangle| = e^{-i\varphi} \langle x + iy, z \rangle = \langle e^{-i\varphi} (x + iy), z \rangle.$$

By utilising the above identity, we can write:

$$|\langle x + iy, z \rangle| = \mathrm{Re} \langle e^{-i\varphi} (x + iy), z \rangle$$
$$= \mathrm{Re} \langle (\cos \varphi - i \sin \varphi)(x + iy), z \rangle$$
$$= \mathrm{Re} \langle \cos \varphi \cdot x + \sin \varphi \cdot y - i \sin \varphi \cdot x + i \cos \varphi \cdot y, z \rangle$$
$$= \mathrm{Re} \langle \cos \varphi \cdot x + \sin \varphi \cdot y, z \rangle + \mathrm{Im} \langle \sin \varphi \cdot x - \cos \varphi \cdot y, z \rangle$$
$$= \mathrm{Re} \langle \cos \varphi \cdot x + \sin \varphi \cdot y, z \rangle + \sin \varphi \, \mathrm{Im} \langle x, z \rangle - \cos \varphi \, \mathrm{Im} \langle y, z \rangle$$
$$= \mathrm{Re} \langle \cos \varphi \cdot x + \sin \varphi \cdot y, z \rangle.$$

Note that for the last equality we have used the assumption (5.154).

Taking the square and using the Schwarz inequality for the inner product $\langle \cdot, \cdot \rangle$, we have

$$|\langle x + iy, z \rangle|^2 = [\mathrm{Re}\,\langle \cos\varphi \cdot x + \sin\varphi \cdot y, z \rangle]^2 \qquad (5.156)$$
$$\leq \|\cos\varphi \cdot x + \sin\varphi \cdot y\|^2 \|z\|^2.$$

On making use of (5.151), we have

$$\sup_{\alpha \in [0,2\pi]} \|\cos\varphi \cdot x + \sin\varphi \cdot y\|^2$$

$$= \sup_{\alpha \in [0,2\pi]} \left[\|x\|^2 \cos^2\varphi + 2\,\mathrm{Re}\,\langle x, y \rangle \sin\varphi \cos\varphi + \|y\|^2 \sin^2\varphi \right]$$

$$= \frac{1}{2} \left\{ \|x\|^2 + \|y\|^2 + \left[\left(\|x\|^2 - \|y\|^2 \right)^2 + 4\,\mathrm{Re}^2\,\langle x, y \rangle \right]^{\frac{1}{2}} \right\}$$

and the first inequality in (5.155) is proved.

Observe that

$$\left(\|x\|^2 - \|y\|^2 \right)^2 + 4\,\mathrm{Re}^2\,\langle x, y \rangle = \left(\|x\|^2 + \|y\|^2 \right)^2 - 4 \left[\|x\|^2 \|y\|^2 - \mathrm{Re}^2\,\langle x, y \rangle \right]$$

$$\leq \left(\|x\|^2 + \|y\|^2 \right)^2,$$

which proves the last part of (5.155). □

Comments

Observe that if $(H, \langle \cdot, \cdot \rangle)$ is a real inner product space, then for any $x, y, z \in H$ one has

$$\langle x, z \rangle^2 + \langle y, z \rangle^2 \qquad (5.157)$$

$$\leq \frac{1}{2} \left\{ \|x\|^2 + \|y\|^2 + \left[\left(\|x\|^2 - \|y\|^2 \right)^2 + 4\,\langle x, y \rangle^2 \right] \right\}^{\frac{1}{2}} \|z\|^2$$

$$\leq \left(\|x\|^2 + \|y\|^2 \right) \|z\|^2.$$

If H is a real space, with the inner product $\langle \cdot, \cdot \rangle$, $H_{\mathbb{C}}$ its complexification, and $\langle \cdot, \cdot \rangle_{\mathbb{C}}$ the corresponding complexification for $\langle \cdot, \cdot \rangle$, then for $x, y \in H$ and $w := x + iy \in H_{\mathbb{C}}$ and for $e \in H$ we have

$$\mathrm{Im}\,\langle x, e \rangle_{\mathbb{C}} = \mathrm{Im}\,\langle y, e \rangle_{\mathbb{C}} = 0,$$

$$\|w\|_{\mathbb{C}}^2 = \|x\|^2 + \|y\|^2, \qquad |\langle w, \bar{w} \rangle_{\mathbb{C}}| = \left(\|x\|^2 - \|y\|^2 \right)^2 + 4\,\langle x, y \rangle^2,$$

where $\bar{w} = x - iy \in H_{\mathbb{C}}$.

Applying (5.155) for the complex space $H_{\mathbb{C}}$ and complex inner product $\langle \cdot, \cdot \rangle_{\mathbb{C}}$, we deduce

$$|\langle w, e \rangle_{\mathbb{C}}|^2 \le \frac{1}{2} \|e\|^2 \left[\|w\|_{\mathbb{C}}^2 + |\langle w, \bar{w} \rangle_{\mathbb{C}}| \right] \le \|e\|^2 \|w\|_{\mathbb{C}}^2, \tag{5.158}$$

which is Kurepa's inequality (5.149).

Let x, y, z satisfy (5.154). In addition, if $\operatorname{Re} \langle x, y \rangle = 0$, then

$$\left[\operatorname{Re}^2 \langle x, z \rangle + \operatorname{Re}^2 \langle y, z \rangle \right]^{\frac{1}{2}} \le \|z\| \cdot \max \{ \|x\|, \|y\| \}. \tag{5.159}$$

If H is a real space equipped with an inner product $\langle \cdot, \cdot \rangle$, then for any $x, y, z \in H$ with $\langle x, y \rangle = 0$ we have

$$\left[\langle x, z \rangle^2 + \langle y, z \rangle^2 \right]^{\frac{1}{2}} \le \|z\| \cdot \max \{ \|x\|, \|y\| \}. \tag{5.160}$$

5.11   Buzano's Inequality

Buzano [14] obtained the following extension of the celebrated Schwarz inequality in a real or complex inner product space $(H; \langle \cdot, \cdot \rangle)$:

$$|\langle a, x \rangle \langle x, b \rangle| \le \frac{1}{2} \left[\|a\| \cdot \|b\| + |\langle a, b \rangle| \right] \|x\|^2, \tag{5.161}$$

for any $a, b, x \in H$.

It is clear that for $a = b$, the above inequality becomes the standard Schwarz inequality

$$|\langle a, x \rangle|^2 \le \|a\|^2 \|x\|^2, \qquad a, x \in H \tag{5.162}$$

with equality if and only if there exists a scalar $\lambda \in \mathbb{K}$ ($\mathbb{K} = \mathbb{R}$ or $\mathbb{C}$) such that $x = \lambda a$.

As noted by Fujii and Kubo [117], where they provided a simple proof of (5.161) by utilising orthogonal projection arguments, the case of equality holds in (5.161) if

$$x = \begin{cases} \alpha \left(\frac{a}{\|a\|} + \frac{\langle a, b \rangle}{|\langle a, b \rangle|} \cdot \frac{b}{\|b\|} \right), & \text{when } \langle a, b \rangle \ne 0 \\[2mm] \alpha \left(\frac{a}{\|a\|} + \beta \cdot \frac{b}{\|b\|} \right), & \text{when } \langle a, b \rangle = 0, \end{cases}$$

where $\alpha, \beta \in \mathbb{K}$.

It might be useful to observe that, from (5.161), one may get the following discrete inequality:

$$\left| \sum_{i=1}^{n} p_i a_i \overline{x_i} \sum_{i=1}^{n} p_i x_i \overline{b_i} \right|$$

$$\leq \frac{1}{2} \left[\left(\sum_{i=1}^{n} p_i \left| a_i \right|^2 \sum_{i=1}^{n} p_i \left| b_i \right|^2 \right)^{\frac{1}{2}} + \left| \sum_{i=1}^{n} p_i a_i \overline{b_i} \right| \right] \sum_{i=1}^{n} p_i \left| x_i \right|^2, \quad (5.163)$$

where $p_i \geq 0$, $a_i, x_i, b_i \in \mathbb{C}$, $i \in \{1, \ldots, n\}$.

If one takes in (5.163) $b_i = \overline{a_i}$ for $i \in \{1, \ldots, n\}$, then one obtains

$$\left| \sum_{i=1}^{n} p_i a_i \overline{x_i} \sum_{i=1}^{n} p_i a_i x_i \right| \leq \frac{1}{2} \left[\sum_{i=1}^{n} p_i \left| a_i \right|^2 + \left| \sum_{i=1}^{n} p_i a_i^2 \right| \right] \sum_{i=1}^{n} p_i \left| x_i \right|^2, \quad (5.164)$$

for any $p_i \geq 0$, $a_i, x_i, b_i \in \mathbb{C}$, $i \in \{1, \ldots, n\}$.

Note that, if x_i, $i \in \{1, \ldots, n\}$ are real numbers, then from (5.164) we may deduce the de Bruijn refinement of the celebrated Cauchy-Bunyakovsky-Schwarz inequality [13],

$$\left| \sum_{i=1}^{n} p_i x_i z_i \right|^2 \leq \frac{1}{2} \sum_{i=1}^{n} p_i x_i^2 \left[\sum_{i=1}^{n} p_i \left| z_i \right|^2 + \left| \sum_{i=1}^{n} p_i z_i^2 \right| \right], \quad (5.165)$$

where $z_i \in \mathbb{C}$, $i \in \{1, \ldots, n\}$. In this way, Buzano's result may be regarded as a generalisation of de Bruijn's inequality.

Similar comments obviously apply for integrals, but, for the sake of brevity we do not mention them here.

The following result may be stated [71]:

Let $(H; \langle \cdot, \cdot \rangle)$ be an inner product space over the real or complex number field K. For all $\alpha \in \mathbb{K} \setminus \{0\}$ and $x, a, b \in H$, $\alpha \neq 0$, one has the inequality

$$\left| \frac{\langle a, x \rangle \langle x, b \rangle}{\|x\|^2} - \frac{\langle a, b \rangle}{\alpha} \right|$$

$$\leq \frac{\|b\|}{|\alpha| \, \|x\|} \left[|\alpha - 1|^2 \, |\langle a, x \rangle|^2 + \|x\|^2 \, \|a\|^2 - |\langle a, x \rangle|^2 \right]. \quad (5.166)$$

The case of equality holds in (5.166) if and only if there exists a scalar $\lambda \in \mathbb{K}$ so that

$$\alpha \cdot \frac{\langle a, x \rangle}{\|x\|^2} x = a + \lambda b. \quad (5.167)$$

PROOF   We follow the proof by Dragomir [71].

Using Schwarz's inequality, we have that

$$\left| \left\langle \alpha \cdot \frac{\langle a, x \rangle}{\|x\|^2} x - a, b \right\rangle \right|^2 \leq \left\| \alpha \cdot \frac{\langle a, x \rangle}{\|x\|^2} x - a \right\|^2 \|b\|^2, \tag{5.168}$$

and since

$$\left\| \alpha \cdot \frac{\langle a, x \rangle}{\|x\|^2} x - a \right\|^2 = |\alpha|^2 \frac{|\langle a, x \rangle|^2}{\|x\|^2} - 2 \frac{|\langle a, x \rangle|^2}{\|x\|^2} \operatorname{Re} \alpha + \|a\|^2$$

$$= \frac{|\alpha - 1|^2 |\langle a, x \rangle|^2 + \|x\|^2 \|a\|^2 - |\langle a, x \rangle|^2}{\|x\|^2}$$

and

$$\left\langle \alpha \cdot \frac{\langle a, x \rangle}{\|x\|^2} x - a, b \right\rangle = \alpha \left[\frac{\langle a, x \rangle \langle x, b \rangle}{\|x\|^2} - \frac{\langle a, b \rangle}{\alpha} \right],$$

hence by (5.166) we deduce the desired inequality (5.166).

The case of equality is obvious from the above considerations related to the Schwarz inequality (5.162). □

Using the continuity property of the modulus, namely, $\|z| - |u\| \leq |z - u|$, $z, u \in \mathbb{K}$, we have:

$$\left| \frac{|\langle a, x \rangle \langle x, b \rangle|}{\|x\|^2} - \frac{|\langle a, b \rangle|}{|\alpha|} \right| \leq \left| \frac{\langle a, x \rangle \langle x, b \rangle}{\|x\|^2} - \frac{\langle a, b \rangle}{\alpha} \right|. \tag{5.169}$$

Therefore, by (5.166) and (5.169), one may deduce the following double inequality,

$$\frac{1}{|\alpha|} \left[|\langle a, b \rangle| - \frac{\|b\|}{\|x\|} \left[\left(|\alpha - 1|^2 |\langle x, a \rangle|^2 + \|x\|^2 \|a\|^2 - |\langle a, x \rangle|^2 \right)^{\frac{1}{2}} \right] \right] \tag{5.170}$$

$$\leq \frac{|\langle a, x \rangle \langle x, b \rangle|}{\|x\|^2}$$

$$\leq \frac{1}{|\alpha|} \left[|\langle a, b \rangle| + \frac{\|b\|}{\|x\|} \right] \left[\left(|\alpha - 1|^2 |\langle x, a \rangle|^2 + \|x\|^2 \|a\|^2 - |\langle x, a \rangle|^2 \right)^{\frac{1}{2}} \right],$$

for each $\alpha \in \mathbb{K} \setminus \{0\}$, $a, b, x \in H$, and $x \neq 0$.

It is obvious that we may obtain various particular inequalities from (5.166). We mention a class of these in the following, which is related to Buzano's result (5.161) [71].

Let $a, b, x \in H$, $x \neq 0$ and $\eta \in \mathbb{K}$ with $|\eta| = 1$, $\operatorname{Re} \eta \neq -1$. Then we have the inequality:

$$\left| \frac{\langle a, x \rangle \langle x, b \rangle}{\|x\|^2} - \frac{\langle a, b \rangle}{1 + \eta} \right| \leq \frac{\|a\| \|b\|}{\sqrt{2}\sqrt{1 + \operatorname{Re} \eta}}, \tag{5.171}$$

and, in particular, for $\eta = 1$, *the inequality:*

$$\left| \frac{\langle a, x \rangle \langle x, b \rangle}{\|x\|^2} - \frac{\langle a, b \rangle}{2} \right| \leq \frac{\|a\| \|b\|}{2}. \tag{5.172}$$

Using the continuity property of the modulus, we get from (5.171) that

$$\frac{|\langle a, x \rangle \langle x, b \rangle|}{\|x\|^2} \leq \frac{|\langle a, b \rangle| + \|a\| \|b\|}{\sqrt{2}\sqrt{1 + \operatorname{Re}\eta}}, \qquad |\eta| = 1, \quad \operatorname{Re}\eta \neq -1,$$

which provides, as the best possible inequality, the above result due to Buzano (5.161).

If the space is real, then the inequality (5.166) is equivalent to

$$\frac{\langle a, b \rangle}{\alpha} - \frac{\|b\|}{|\alpha| \|x\|} \left[(\alpha - 1)^2 \langle a, x \rangle^2 + \|x\|^2 \|a\|^2 - \langle a, x \rangle^2 \right]^{\frac{1}{2}} \tag{5.173}$$

$$\leq \frac{\langle a, x \rangle \langle x, b \rangle}{\|x\|^2}$$

$$\leq \frac{\langle a, b \rangle}{\alpha} + \frac{\|b\|}{|\alpha| \|x\|} \left[(\alpha - 1)^2 \langle a, x \rangle^2 + \|x\|^2 \|a\|^2 - \langle a, x \rangle^2 \right]^{\frac{1}{2}}$$

for any $\alpha \in \mathbb{R} \setminus \{0\}$ and $a, b, x \in H$, $x \neq 0$.

If in (5.173) we take $\alpha = 2$, then we get

$$\frac{1}{2} \left[\langle a, b \rangle - \|a\| \|b\| \right] \|x\|^2 \leq \langle a, x \rangle \langle x, b \rangle \tag{5.174}$$

$$\leq \frac{1}{2} \left[\langle a, b \rangle + \|a\| \|y\| \right] \|x\|^2,$$

which apparently, as mentioned by Precupanu [154], has been obtained independently of Buzano, by Richard [156].

Pečarić [149] gave a simple direct proof of (5.174) without mentioning the work of either Buzano or Richard, but tracked down the result, in a particular form, to an earlier paper due to Blatter [10].

Obviously, the following refinement of Buzano's result may be stated [71]:

Let $(H; \langle \cdot, \cdot \rangle)$ *be a real or complex inner product space and* $a, b, x \in H$. *Then*

$$|\langle a, x \rangle \langle x, b \rangle| \leq \left| \langle a, x \rangle \langle x, b \rangle - \frac{1}{2} \langle a, b \rangle \|x\|^2 \right| + \frac{1}{2} |\langle a, b \rangle| \|x\|^2 \tag{5.175}$$

$$\leq \frac{1}{2} \left[\|a\| \|b\| + |\langle a, b \rangle| \right] \|x\|^2.$$

PROOF The first inequality in (5.175) follows by the triangle inequality for the modulus $|\cdot|$. The second inequality is merely (5.172), in which we have added the same quantity to both sides. ⬜

For $\alpha = 1$, we deduce from (5.166) the following inequality:

$$\left| \frac{\langle a, x \rangle \langle x, b \rangle}{\|x\|^2} - \langle a, b \rangle \right| \le \frac{\|b\|}{\|x\|} \left[\|x\|^2 \|a\|^2 - |\langle a, x \rangle|^2 \right]^{\frac{1}{2}} \qquad (5.176)$$

for any $a, b, x \in H$ with $x \ne 0$.

If the space is real, then (5.176) is equivalent to

$$\langle a, b \rangle - \frac{\|b\|}{\|x\|} \left[\|x\|^2 \|a\|^2 - |\langle a, x \rangle|^2 \right]^{\frac{1}{2}} \qquad (5.177)$$

$$\le \frac{\langle a, x \rangle \langle x, b \rangle}{\|x\|^2}$$

$$\le \frac{\|b\|}{\|x\|} \left[\|x\|^2 \|a\|^2 - |\langle a, x \rangle|^2 \right]^{\frac{1}{2}} + \langle a, b \rangle,$$

which is similar to Richard's inequality (5.174).

Comments

The following refinement of Kurepa's result may be stated [71].

Let $(H; \langle \cdot, \cdot \rangle)$ be a real inner product space and $(H_{\mathbb{C}}, \langle \cdot, \cdot \rangle_{\mathbb{C}})$ be its complexification. Then for any $e \in H$ and $w \in H_{\mathbb{C}}$, one has the inequality

$$|\langle w, e \rangle_{\mathbb{C}}|^2 \le \left| \langle w, e \rangle_{\mathbb{C}}^2 - \frac{1}{2} \langle w, \bar{w} \rangle_{\mathbb{C}} \|e\|^2 \right| + \frac{1}{2} |\langle w, \bar{w} \rangle_{\mathbb{C}}| \|e\|^2 \qquad (5.178)$$

$$\le \frac{1}{2} \|e\|^2 \left[\|w\|_{\mathbb{C}}^2 + |\langle w, \bar{w} \rangle_{\mathbb{C}}| \right].$$

PROOF   We follow the proof by Dragomir [71].

If we apply (5.175) for $(H_{\mathbb{C}}, \langle \cdot, \cdot \rangle_{\mathbb{C}})$ and $x = e \in H$, $a = w$, and $b = \bar{w}$, then we have

$$|\langle w, e \rangle_{\mathbb{C}} \langle e, \bar{w} \rangle_{\mathbb{C}}| \qquad (5.179)$$

$$\le \left| \langle w, e \rangle_{\mathbb{C}} \langle e, \bar{w} \rangle_{\mathbb{C}} - \frac{1}{2} \langle w, \bar{w} \rangle_{\mathbb{C}} \|e\|^2 \right| + \frac{1}{2} |\langle w, \bar{w} \rangle_{\mathbb{C}}| \|e\|^2$$

$$\le \frac{1}{2} \|e\|^2 \left[\|w\|_{\mathbb{C}} \|\bar{w}\|_{\mathbb{C}} + |\langle w, \bar{w} \rangle_{\mathbb{C}}| \right].$$

Now, if we assume that $w = (x, y) \in H_{\mathbb{C}}$, then, by the definition of $\langle \cdot, \cdot \rangle_{\mathbb{C}}$ we have

$$\langle w, e \rangle_{\mathbb{C}} = \langle (x, y), (e, 0) \rangle_{\mathbb{C}}$$
$$= \langle x, e \rangle + \langle y, 0 \rangle + i \left[\langle y, e \rangle - \langle x, 0 \rangle \right]$$
$$= \langle e, x \rangle + i \langle e, y \rangle,$$

$$\langle e, \bar{w} \rangle_{\mathbb{C}} = \langle (e, 0), (x, -y) \rangle_{\mathbb{C}}$$
$$= \langle e, x \rangle + \langle 0, -y \rangle + i\left[\langle 0, x \rangle - \langle e, -y \rangle\right]$$
$$= \langle e, x \rangle + i\langle e, y \rangle = \langle w, e \rangle_{\mathbb{C}},$$

and

$$\|\bar{w}\|_{\mathbb{C}}^2 = \|x\|^2 + \|y\|^2 = \|w\|_{\mathbb{C}}^2.$$

Therefore, by (5.179), we deduce the desired result (5.178). ⬜

Denote by $\ell_\rho^2(\mathbb{C})$ the Hilbert space of all complex sequences $z = (z_i)_{i \in \mathbb{N}}$ with the property that for $\rho_i \geq 0$ with $\sum_{i=1}^\infty \rho_i = 1$ we have $\sum_{i=1}^\infty \rho_i |z_i|^2 < \infty$. If $a = (a_i)_{i \in \mathbb{N}}$ is a sequence of real numbers such that $a \in \ell_\rho^2(\mathbb{C})$, then for any $z \in \ell_\rho^2(\mathbb{C})$ we have the inequality:

$$\left| \sum_{i=1}^\infty \rho_i a_i z_i \right|^2 \tag{5.180}$$

$$\leq \left| \left(\sum_{i=1}^\infty \rho_i a_i z_i \right)^2 - \frac{1}{2} \sum_{i=1}^\infty \rho_i a_i^2 \sum_{i=1}^\infty \rho_i z_i^2 \right| + \frac{1}{2} \sum_{i=1}^\infty \rho_i a_i^2 \left| \sum_{i=1}^\infty \rho_i z_i^2 \right|$$

$$\leq \frac{1}{2} \sum_{i=1}^\infty \rho_i a_i^2 \left[\sum_{i=1}^\infty \rho_i |z_i|^2 + \left| \sum_{i=1}^\infty \rho_i z_i^2 \right| \right].$$

Similarly, if by $L_\rho^2(S, \Sigma, \mu)$ we understand the Hilbert space of all complex-valued functions $f : S \to \mathbb{C}$ with the property that for the μ-measurable function $\rho \geq 0$ with $\int_S \rho(t) \, d\mu(t) = 1$ we have

$$\int_S \rho(t) |f(t)|^2 \, d\mu(t) < \infty,$$

then for a real function $a \in L_\rho^2(S, \Sigma, \mu)$ and any $f \in L_\rho^2(S, \Sigma, \mu)$, we have the inequalities

$$\left| \int_S \rho(t) a(t) f(t) \, d\mu(t) \right|^2 \tag{5.181}$$

$$\leq \left| \left(\int_S \rho(t) a(t) f(t) \, d\mu(t) \right)^2 \right.$$

$$\left. - \frac{1}{2} \int_S \rho(t) f^2(t) \, d\mu(t) \int_S \rho(t) a^2(t) \, d\mu(t) \right|$$

$$+ \frac{1}{2} \left| \int_S \rho(t) f^2(t) \, d\mu(t) \right| \int_S \rho(t) a^2(t) \, d\mu(t)$$

$$\leq \frac{1}{2} \int_S \rho(t) a^2(t) \, d\mu(t) \left[\int_S \rho(t) |f(t)|^2 \, d\mu(t) + \left| \int_S \rho(t) f^2(t) \, d\mu(t) \right| \right].$$

5.12 A Generalisation of Buzano's Inequality

Buzano [14] obtained the following extension of the celebrated Schwarz inequality in a real or complex inner product space $(H; \langle \cdot, \cdot \rangle)$:

$$|\langle a, x \rangle \langle x, b \rangle| \leq \frac{1}{2} \left[\|a\| \, \|b\| + |\langle a, b \rangle| \right] \|x\|^2, \tag{5.182}$$

for any $a, b, x \in H$.

It is clear that when $a = b$, the above inequality becomes the Schwarz inequality, that is,

$$|\langle a, x \rangle|^2 \leq \|a\|^2 \, \|x\|^2, \quad a, x \in H; \tag{5.183}$$

in which equality holds if and only if there exists a scalar $\lambda \in \mathbb{K}$ $(\mathbb{R}, \mathbb{C})$ so that $x = \lambda a$.

As noted by Precupanu [154], independently of Buzano [14], Richard [156] obtained the following similar inequality which holds in real inner product spaces:

$$\frac{1}{2} \|x\|^2 \left[\langle a, b \rangle - \|a\| \, \|b\| \right] \leq \langle a, x \rangle \langle x, b \rangle \tag{5.184}$$

$$\leq \frac{1}{2} \|x\|^2 \left[\langle a, b \rangle + \|a\| \, \|b\| \right].$$

We say that the finite family $\{e_i\}_{i \in I}$ (I is finite) of vectors is *orthonormal* if $\langle e_i, e_j \rangle = 0$ if $i, j \in I$ with $i \neq j$ and $\|e_i\| = 1$ for each $i \in I$. The following result may be stated [58]:

Let $(H; \langle \cdot, \cdot \rangle)$ be an inner product space over the real or complex number field $\mathbb{K}$ and $\{e_i\}_{i \in I}$ a finite orthonormal family in H. Then for any $a, b \in H$, one has the inequality

$$\left| \sum_{i \in I} \langle a, e_i \rangle \langle e_i, b \rangle - \frac{1}{2} \langle a, b \rangle \right| \leq \frac{1}{2} \|a\| \, \|b\|. \tag{5.185}$$

The case of equality holds in (5.185) if and only if

$$\sum_{i \in I} \langle a, e_i \rangle e_i = \frac{1}{2} a + \left(\sum_{i \in I} \langle a, e_i \rangle \langle e_i, b \rangle - \frac{1}{2} \langle a, b \rangle \right) \cdot \frac{b}{\|b\|^2}. \tag{5.186}$$

PROOF We follow the proof by Dragomir [58].

It is well known that, for $e \neq 0$ and $f \in H$, the following identity holds:

$$\frac{\|f\|^2 \, \|e\|^2 - |\langle f, e \rangle|^2}{\|e\|^2} = \left\| f - \frac{\langle f, e \rangle e}{\|e\|^2} \right\|^2. \tag{5.187}$$

Therefore, in Schwarz's inequality,

$$|\langle f, e \rangle|^2 \leq \|f\|^2 \, \|e\|^2, \quad f, e \in H; \tag{5.188}$$

the case of equality, for $e \neq 0$, holds if and only if

$$f = \frac{\langle f, e \rangle \, e}{\|e\|^2}.$$

Let $f := 2 \sum_{i \in I} \langle a, e_i \rangle \, e_i - a$ and $e := b$. Then, by Schwarz's inequality (5.188), we may state that

$$\left| \left\langle 2 \sum_{i \in I} \langle a, e_i \rangle \, e_i - a, b \right\rangle \right|^2 \leq \left\| 2 \sum_{i \in I} \langle a, e_i \rangle \, e_i - a \right\|^2 \, \|b\|^2 \tag{5.189}$$

with equality, for $b \neq 0$, if and only if

$$2 \sum_{i \in I} \langle a, e_i \rangle \, e_i - a = \left\langle 2 \sum_{i \in I} \langle a, e_i \rangle \, e_i - a, b \right\rangle \frac{b}{\|b\|^2}. \tag{5.190}$$

Since

$$\left\langle 2 \sum_{i \in I} \langle a, e_i \rangle \, e_i - a, b \right\rangle = 2 \sum_{i \in I} \langle a, e_i \rangle \, \langle e_i, b \rangle - \langle a, b \rangle$$

and

$$\left\| 2 \sum_{i \in I} \langle a, e_i \rangle \, e_i - a \right\|^2$$

$$= 4 \left\| \sum_{i \in I} \langle a, e_i \rangle \, e_i \right\|^2 - 4 \operatorname{Re} \left\langle \sum_{i \in I} \langle a, e_i \rangle \, e_i, a \right\rangle + \|a\|^2$$

$$= 4 \sum_{i \in I} |\langle a, e_i \rangle|^2 - 4 \sum_{i \in I} |\langle a, e_i \rangle|^2 + \|a\|^2$$

$$= \|a\|^2,$$

hence by (5.189) we deduce the desired inequality (5.185).

Finally, as (5.186) is equivalent to

$$\sum_{i \in I} \langle a, e_i \rangle \, e_i - \frac{a}{2} = \left(\sum_{i \in I} \langle a, e_i \rangle \, \langle e_i, b \rangle - \frac{1}{2} \langle a, b \rangle \right) \frac{b}{\|b\|^2},$$

the equality thus holds in (5.185) if and only if (5.186) is valid. $\qquad\qquad \Box$

Comments

Recall the following result, known as Bessel's inequality,

$$\sum_{i \in I} |\langle x, e_i \rangle|^2 \leq \|x\|^2, \quad x \in H, \tag{5.191}$$

where, as above, $\{e_i\}_{i \in I}$ is a finite orthonormal family in the inner product space $(H; \langle \cdot, \cdot \rangle)$.

If one chooses $a = b = x$ in (5.185), then one gets the inequality

$$\left| \sum_{i \in I} |\langle x, e_i \rangle|^2 - \frac{1}{2} \|x\|^2 \right| \leq \frac{1}{2} \|x\|^2,$$

which is obviously equivalent to Bessel's inequality (5.191). Therefore, the inequality (5.185) may be regarded as a generalisation of Bessel's inequality as well.

Utilising the Bessel and Cauchy-Bunyakovsky-Schwarz inequalities, one may state that

$$\left| \sum_{i \in I} \langle a, e_i \rangle \langle e_i, b \rangle \right| \leq \left[\sum_{i \in I} |\langle a, e_i \rangle|^2 \sum_{i \in I} |\langle b, e_i \rangle|^2 \right]^{\frac{1}{2}} \leq \|a\| \|b\|. \tag{5.192}$$

A different refinement of the inequality between the first and last terms in (5.192) is incorporated in the following [58]:

With the above assumption, we have

$$\left| \sum_{i \in I} \langle a, e_i \rangle \langle e_i, b \rangle \right| \leq \left| \sum_{i \in I} \langle a, e_i \rangle \langle e_i, b \rangle - \frac{1}{2} \langle a, b \rangle \right| + \frac{1}{2} |\langle a, b \rangle| \tag{5.193}$$

$$\leq \frac{1}{2} \left[\|a\| \|b\| + |\langle a, b \rangle| \right]$$

$$\leq \|a\| \|b\|.$$

If the space $(H; \langle \cdot, \cdot \rangle)$ is real, then, obviously, (5.185) is equivalent to

$$\frac{1}{2} \left(\langle a, b \rangle - \|a\| \|b\| \right) \leq \sum_{i \in I} \langle a, e_i \rangle \langle e_i, b \rangle \leq \frac{1}{2} \left[\|a\| \|b\| + \langle a, b \rangle \right]. \tag{5.194}$$

It is obvious that if the family is comprised of only a single element $e = \frac{x}{\|x\|}$, $x \in H$, $x \neq 0$, then from (5.193) we recapture the refinement of Buzano's inequality incorporated in (5.182) while from (5.194) we deduce Richard's result from (5.184).

The following result is of interest as well [58]:

Let $\{e_i\}_{i \in I}$ be a finite orthonormal family in $(H; \langle \cdot, \cdot \rangle)$. If $x, y \in H \backslash \{0\}$ are such that there exists the constants $m_i, n_i, M_i, N_i \in \mathbb{R}$, $i \in I$ such that:

$$-1 \leq m_i \leq \frac{\text{Re} \langle x, e_i \rangle}{\|x\|} \cdot \frac{\text{Re} \langle y, e_i \rangle}{\|y\|} \leq M_i \leq 1, \quad i \in I \tag{5.195}$$

and

$$-1 \le n_i \le \frac{\text{Im}\,\langle x, e_i \rangle}{\|x\|} \cdot \frac{\text{Im}\,\langle y, e_i \rangle}{\|y\|} \le N_i \le 1, \quad i \in I, \tag{5.196}$$

then

$$2 \sum_{i \in I} (m_i + n_i) - 1 \le \frac{\text{Re}\,\langle x, y \rangle}{\|x\|\,\|y\|} \le 1 + 2 \sum_{i \in I} (M_i + N_i). \tag{5.197}$$

PROOF We follow the proof from Dragomir [58].

By using (5.185) and the fact that for any complex number z, $|z| \ge |\text{Re}\,z|$, we have

$$\left| \sum_{i \in I} \text{Re}\,[\langle x, e_i \rangle \langle e_i, y \rangle] - \frac{1}{2} \text{Re}\,\langle x, y \rangle \right| \tag{5.198}$$

$$\le \left| \sum_{i \in I} \langle x, e_i \rangle \langle e_i, y \rangle - \frac{1}{2} \langle x, y \rangle \right|$$

$$\le \frac{1}{2} \|x\|\,\|y\|\,.$$

Since

$$\text{Re}\,[\langle x, e_i \rangle \langle e_i, y \rangle] = \text{Re}\,\langle x, e_i \rangle\,\text{Re}\,\langle y, e_i \rangle + \text{Im}\,\langle x, e_i \rangle\,\text{Im}\,\langle y, e_i \rangle,$$

we have the following from (5.198):

$$-\frac{1}{2}\|x\|\,\|y\| + \frac{1}{2}\text{Re}\,\langle x, y \rangle \tag{5.199}$$

$$\le \sum_{i \in I} \text{Re}\,\langle x, e_i \rangle\,\text{Re}\,\langle y, e_i \rangle + \sum_{i \in I} \text{Im}\,\langle x, e_i \rangle\,\text{Im}\,\langle y, e_i \rangle$$

$$\le \frac{1}{2}\|x\|\,\|y\| + \frac{1}{2}\text{Re}\,\langle x, y \rangle\,.$$

Utilising the assumptions (5.195) and (5.196), we have

$$\sum_{i \in I} m_i \le \sum_{i \in I} \frac{\text{Re}\,\langle x, e_i \rangle\,\text{Re}\,\langle y, e_i \rangle}{\|x\|\,\|y\|} \le \sum_{i \in I} M_i \tag{5.200}$$

and

$$\sum_{i \in I} n_i \le \sum_{i \in I} \frac{\text{Im}\,\langle x, e_i \rangle\,\text{Im}\,\langle y, e_i \rangle}{\|x\|\,\|y\|} \le \sum_{i \in I} N_i. \tag{5.201}$$

Finally, by making use of (5.199)–(5.201), we deduce the desired result (5.197).
$\Box$

By Schwarz's inequality, is it obvious that, in general,

$$-1 \leq \frac{\mathrm{Re}\,\langle x, y \rangle}{\|x\|\,\|y\|} \leq 1.$$

Consequently, the left inequality in (5.197) is of interest when $\sum_{i \in I} (m_i + n_i) > 0$, while the right inequality in (5.197) is of interest when $\sum_{i \in I} (M_i + N_i) < 0$.

5.13   Generalisations of Precupanu's Inequality

In 1976, Precupanu [154] obtained the following result related to the Schwarz inequality in a real inner product space $(H; \langle \cdot, \cdot \rangle)$:

For any $a \in H$, $x, y \in H \backslash \{0\}$, we have the inequality:

$$
\begin{aligned}
&\frac{-\|a\|\,\|b\| + \langle a, b \rangle}{2} \\
&\leq \frac{\langle x, a \rangle \langle x, b \rangle}{\|x\|^2} + \frac{\langle y, a \rangle \langle y, b \rangle}{\|y\|^2} - 2 \cdot \frac{\langle x, a \rangle \langle y, b \rangle \langle x, y \rangle}{\|x\|^2 \|y\|^2} \\
&\leq \frac{\|a\|\,\|b\| + \langle a, b \rangle}{2}.
\end{aligned}
\tag{5.202}
$$

In the right-hand side or in the left-hand side of (5.202) we have equality if and only if there are $\lambda, \mu \in \mathbb{R}$ such that

$$\lambda \frac{\langle x, a \rangle}{\|x\|^2} \cdot x + \mu \frac{\langle y, b \rangle}{\|y\|^2} \cdot y = \frac{1}{2}\left(\lambda a + \mu b\right). \tag{5.203}$$

Note for instance that Precupanu [154], if $y \perp b$, i.e., $\langle y, b \rangle = 0$, then by (5.202) one may deduce:

$$\frac{-\|a\|\,\|b\| + \langle a, b \rangle}{2} \|x\|^2 \leq \langle x, a \rangle \langle x, b \rangle \leq \frac{\|a\|\,\|b\| + \langle a, b \rangle}{2} \|x\|^2 \tag{5.204}$$

for any $a, b, x \in H$, which is an inequality that has been obtained previously by Richard [156]. The case of equality in the right-hand side or in the left-hand side of (5.204) holds if and only if there are $\lambda, \mu \in \mathbb{R}$ with

$$2\lambda \langle x, a \rangle x = (\lambda a + \mu b) \|x\|^2. \tag{5.205}$$

For $a = b$, we may obtain from (5.202) the following inequality [154]:

$$0 \leq \frac{\langle x, a \rangle^2}{\|x\|^2} + \frac{\langle y, a \rangle^2}{\|y\|^2} - 2 \cdot \frac{\langle x, a \rangle \langle y, a \rangle \langle x, y \rangle}{\|x\|^2 \|y\|^2} \leq \|a\|^2. \tag{5.206}$$

This inequality implies [154]:

$$\frac{\langle x, y\rangle}{\|x\|\,\|y\|} \geq \frac{1}{2}\left[\frac{\langle x, a\rangle}{\|x\|\,\|a\|} + \frac{\langle y, a\rangle}{\|y\|\,\|a\|}\right]^2 - \frac{3}{2}. \tag{5.207}$$

Moore [143] pointed out the following reverse of the Schwarz inequality,

$$|\langle y, z\rangle| \leq \|y\|\,\|z\|, \qquad y, z \in H, \tag{5.208}$$

where some information about a third vector x is known:

Let $(H; \langle \cdot, \cdot \rangle)$ be an inner product space over the real field $\mathbb{R}$ and $x, y, z \in H$ such that:

$$|\langle x, y\rangle| \geq (1-\varepsilon)\,\|x\|\,\|y\|, \qquad |\langle x, z\rangle| \geq (1-\varepsilon)\,\|x\|\,\|z\|, \tag{5.209}$$

where ε is a positive real number, reasonably small. Then

$$|\langle y, z\rangle| \geq \max\left\{1 - \varepsilon - \sqrt{2\varepsilon}, 1 - 4\varepsilon, 0\right\}\|y\|\,\|z\|. \tag{5.210}$$

By utilising Richard's inequality (5.204), which is written in the following equivalent form,

$$2 \cdot \frac{\langle x, a\rangle\,\langle x, b\rangle}{\|x\|^2} - \|a\|\,\|b\| \leq \langle a, b\rangle \leq 2 \cdot \frac{\langle x, a\rangle\,\langle x, b\rangle}{\|x\|^2} + \|a\|\,\|b\| \tag{5.211}$$

for any $a, b \in H$ and $a \in H \backslash \{0\}$, Precupanu has obtained the following Moore's type result:

Let $(H; \langle \cdot, \cdot \rangle)$ be a real inner product space. If $a, b, x \in H$ and $0 < \varepsilon_1 < \varepsilon_2$ are such that:

$$\varepsilon_1 \|x\|\,\|a\| \leq \langle x, a\rangle \leq \varepsilon_2 \|x\|\,\|a\|, \tag{5.212}$$
$$\varepsilon_1 \|x\|\,\|b\| \leq \langle x, b\rangle \leq \varepsilon_2 \|x\|\,\|b\|,$$

then

$$\left(2\varepsilon_1^2 - 1\right)\|a\|\,\|b\| \leq \langle a, b\rangle \leq \left(2\varepsilon_1^2 + 1\right)\|a\|\,\|b\|. \tag{5.213}$$

We remark that the right inequality is always satisfied, since by Schwarz's inequality we have $\langle a, b\rangle \leq \|a\|\,\|b\|$. The left inequality may be useful when one assumes that $\varepsilon_1 \in (0, 1]$. In that case, from (5.213) we obtain

$$-\|a\|\,\|b\| \leq \left(2\varepsilon_1^2 - 1\right)\|a\|\,\|b\| \leq \langle a, b\rangle \tag{5.214}$$

provided $\varepsilon_1 \|x\|\,\|a\| \leq \langle x, a\rangle$ and $\varepsilon_1 \|x\|\,\|b\| \leq \langle x, b\rangle$, which is a refinement of Schwarz's inequality

$$-\|a\|\,\|b\| \leq \langle a, b\rangle.$$

In the complex case, independently of Richard, Buzano [14] obtained the following inequality,

$$|\langle x, a \rangle \langle x, b \rangle| \leq \frac{\|a\| \, \|b\| + |\langle a, b \rangle|}{2} \cdot \|x\|^2, \tag{5.215}$$

provided x, a, b are vectors in the complex inner product space $(H; \langle \cdot, \cdot \rangle)$.

In the same paper [154], Precupanu, without mentioning Buzano's name in relation to the inequality (5.215), observed that, on utilising (5.215), one may obtain the following result of the Moore type [143]:

Let $(H; \langle \cdot, \cdot \rangle)$ be a (real or) complex inner product space. If $x, a, b \in H$ are such that

$$|\langle x, a \rangle| \geq (1 - \varepsilon) \, \|x\| \, \|a\|, \qquad |\langle x, b \rangle| \geq (1 - \varepsilon) \, \|x\| \, \|b\|, \tag{5.216}$$

then

$$|\langle a, b \rangle| \geq \left(1 - 4\varepsilon + 2\varepsilon^2\right) \|a\| \, \|b\|. \tag{5.217}$$

Note that the above result is useful when, for $\varepsilon \in (0, 1]$, the quantity $1 - 4\varepsilon + 2\varepsilon^2 > 0$, i.e., $\varepsilon \in \left(0, 1 - \frac{\sqrt{2}}{2}\right]$.

When the space is real, inequality (5.217) provides a better lower bound for $|\langle a, b \rangle|$ than the second bound in Moore's result (5.210). However, it is not known if the first bound in (5.210) remains valid for the case of complex spaces. From Moore's original proof [143], apparently, the fact that the space $(H; \langle \cdot, \cdot \rangle)$ is real plays an essential role.

Before we point out some new results for orthonormal families of vectors in real or complex inner product spaces, we state the following result that complements the Moore type results outlined above for real spaces [56]:

Let $(H; \langle \cdot, \cdot \rangle)$ be a real inner product space and $a, b, x, y \in H \backslash \{0\}$.
(i) If there exist $\delta_1, \delta_2 \in (0, 1]$ such that

$$\frac{\langle x, a \rangle}{\|x\| \, \|a\|} \geq \delta_1, \qquad \frac{\langle y, a \rangle}{\|y\| \, \|a\|} \geq \delta_2$$

and $\delta_1 + \delta_2 \geq 1$, then

$$\frac{\langle x, y \rangle}{\|x\| \, \|y\|} \geq \frac{1}{2} \left(\delta_1 + \delta_2\right)^2 - \frac{3}{2} \qquad (\geq -1). \tag{5.218}$$

(ii) If there exist $\mu_1 \, (\mu_2) \in R$ such that

$$\mu_1 \, \|a\| \, \|b\| \leq \frac{\langle x, a \rangle \langle x, b \rangle}{\|x\|^2} \quad (\leq \mu_2 \, \|a\| \, \|b\|)$$

and $1 \geq \mu_1 \geq 0 \, (-1 \leq \mu_2 \leq 0)$, then

$$[-1 \leq] \, 2\mu_1 - 1 \leq \frac{\langle a, b \rangle}{\|a\| \, \|b\|} \quad (\leq 2\mu_2 + 1 \, [\leq 1]). \tag{5.219}$$

The proof is obvious by the inequalities (5.207) and (5.211). We omit the details.

The following result concerning inequalities for orthonormal families may be stated [56]:

Let $\{e_i\}_{i\in I}$ and $\{f_j\}_{j\in J}$ be two finite families of orthonormal vectors in $(H; \langle \cdot, \cdot \rangle)$. For any $x, y \in H \setminus \{0\}$ one has the inequality

$$\left| \sum_{i\in I} \langle x, e_i \rangle \langle e_i, y \rangle + \sum_{j\in J} \langle x, f_j \rangle \langle f_j, y \rangle \right.$$

$$\left. - 2 \sum_{i\in I, j\in J} \langle x, e_i \rangle \langle f_j, y \rangle \langle e_i, f_j \rangle - \frac{1}{2} \langle x, y \rangle \right| \le \frac{1}{2} \|x\| \|y\|. \quad (5.220)$$

Equality holds in (5.220) if and only if there exists a $\lambda \in \mathbb{K}$ such that

$$x - \lambda y = 2 \left(\sum_{i\in I} \langle x, e_i \rangle e_i - \lambda \sum_{j\in J} \langle y, f_j \rangle f_j \right). \quad (5.221)$$

PROOF We follow the proof by Dragomir [56].

We know that if $u, v \in H$, $v \ne 0$, then

$$\left\| u - \frac{\langle u, v \rangle}{\|v\|^2} \cdot v \right\|^2 = \frac{\|u\|^2 \|v\|^2 - |\langle u, v \rangle|^2}{\|v\|^2}, \quad (5.222)$$

showing that, in Schwarz's inequality

$$|\langle u, v \rangle|^2 \le \|u\|^2 \|v\|^2, \quad (5.223)$$

the case of equality for $v \ne 0$ holds if and only if

$$u = \frac{\langle u, v \rangle}{\|v\|^2} \cdot v, \quad (5.224)$$

so that there exists a $\lambda \in \mathbb{R}$ such that $u = \lambda v$.

Now, let $u := 2 \sum_{i\in I} \langle x, e_i \rangle e_i - x$ and $v := 2 \sum_{j\in J} \langle y, f_j \rangle f_j - y$. Observe that

$$\|u\|^2 = \left\| 2 \sum_{i\in I} \langle x, e_i \rangle e_i \right\|^2 - 4 \operatorname{Re} \left\langle \sum_{i\in I} \langle x, e_i \rangle e_i, x \right\rangle + \|x\|^2$$

$$= 4 \sum_{i\in I} |\langle x, e_i \rangle|^2 - 4 \sum_{i\in I} |\langle x, e_i \rangle|^2 + \|x\|^2 = \|x\|^2,$$

and, similarly

$$\|v\|^2 = \|y\|^2 .$$

Also,

$$\langle u, v \rangle = 4 \sum_{i \in I, j \in J} \langle x, e_i \rangle \langle f_j, y \rangle \langle e_i, f_j \rangle + \langle x, y \rangle$$

$$- 2 \sum_{i \in I} \langle x, e_i \rangle \langle e_i, y \rangle - 2 \sum_{j \in J} \langle x, f_j \rangle \langle f_j, y \rangle .$$

Therefore, by Schwarz's inequality (5.223) we deduce the desired inequality (5.220). By (5.224), the case of equality holds in (5.220) if and only if there exists a $\lambda \in \mathbb{K}$ such that

$$2 \sum_{i \in I} \langle x, e_i \rangle e_i - x = \lambda \left(2 \sum_{j \in J} \langle y, f_j \rangle f_j - y \right),$$

which is equivalent to (5.221). □

If in (5.221) we choose $x = y$, then we get the inequality

$$\left| \sum_{i \in I} |\langle x, e_i \rangle|^2 + \sum_{j \in J} |\langle x, f_j \rangle|^2 - 2 \sum_{i \in I, j \in J} \langle x, e_i \rangle \langle f_j, x \rangle \langle e_i, f_j \rangle - \frac{1}{2} \|x\|^2 \right|$$

$$\leq \frac{1}{2} \|x\|^2 \quad (5.225)$$

for any $x \in H$.

If in the above result we assume that $I = J$ and $f_i = e_i$, $i \in I$, then we get from (5.220) the Schwarz inequality $|\langle x, y \rangle| \leq \|x\| \|y\|$.

If $I \cap J = \varnothing$, $I \cup J = K$, $g_k = e_k$, $k \in I$, $g_k = f_k$, $k \in J$, and $\{g_k\}_{k \in K}$ is orthonormal, then from (5.220) we get

$$\left| \sum_{k \in K} \langle x, g_k \rangle \langle g_k, y \rangle - \frac{1}{2} \langle x, y \rangle \right| \leq \frac{1}{2} \|x\| \|y\|, \qquad x, y \in H \qquad (5.226)$$

which has been obtained earlier by Dragomir [58].

If I and J reduce to one element, namely, $e_1 = \frac{e}{\|e\|}$, $f_1 = \frac{f}{\|f\|}$ with $e, f \neq 0$, then from (5.220) we get

$$\left| \frac{\langle x, e \rangle \langle e, y \rangle}{\|e\|^2} + \frac{\langle x, f \rangle \langle f, y \rangle}{\|f\|^2} - 2 \cdot \frac{\langle x, e \rangle \langle f, y \rangle \langle e, f \rangle}{\|e\|^2 \|f\|^2} - \frac{1}{2} \langle x, y \rangle \right|$$

$$\leq \frac{1}{2} \|x\| \|y\|, \qquad x, y \in H, \quad (5.227)$$

which is the corresponding complex version of Precupanu's inequality (5.202). If in (5.227) we assume that $x = y$, then we get

$$\left| \frac{|\langle x, e \rangle|^2}{\|e\|^2} + \frac{|\langle x, f \rangle|^2}{\|f\|^2} - 2 \cdot \frac{\langle x, e \rangle \langle f, e \rangle \langle e, f \rangle}{\|e\|^2 \|f\|^2} - \frac{1}{2} \|x\|^2 \right| \leq \frac{1}{2} \|x\|^2. \quad (5.228)$$

The following result may be stated [56]:

With the assumptions leading to the result (5.220), we have:

$$\left| \sum_{i \in I} \langle x, e_i \rangle \langle e_i, y \rangle + \sum_{j \in J} \langle x, f_j \rangle \langle f_j, y \rangle \right. \quad (5.229)$$

$$\left. - 2 \sum_{i \in I, j \in J} \langle x, e_i \rangle \langle f_j, y \rangle \langle e_i, f_j \rangle \right|$$

$$\leq \frac{1}{2} |\langle x, y \rangle| + \left| \sum_{i \in I} \langle x, e_i \rangle \langle e_i, y \rangle + \sum_{j \in J} \langle x, f_j \rangle \langle f_j, y \rangle \right.$$

$$\left. - 2 \sum_{i \in I, j \in J} \langle x, e_i \rangle \langle f_j, y \rangle \langle e_i, f_j \rangle - \frac{1}{2} |\langle x, y \rangle| \right|$$

$$\leq \frac{1}{2} \left[|\langle x, y \rangle| + \|x\| \|y\| \right].$$

PROOF The first inequality follows by the triangle inequality for the modulus. The second inequality follows by (5.220) on adding the quantity $\frac{1}{2} |\langle x, y \rangle|$ on both sides. ◻

Comments
(a) If we choose $x = y$ in (5.229), then we get:

$$\left| \sum_{i \in I} |\langle x, e_i \rangle|^2 + \sum_{j \in J} |\langle x, f_j \rangle|^2 - 2 \sum_{i \in I, j \in J} \langle x, e_i \rangle \langle f_j, x \rangle \langle e_i, f_j \rangle \right| \quad (5.230)$$

$$\leq \left| \sum_{i \in I} |\langle x, e_i \rangle|^2 + \sum_{j \in J} |\langle x, f_j \rangle|^2 \right.$$

$$\left. - 2 \sum_{i \in I, j \in J} \langle x, e_i \rangle \langle f_j, x \rangle \langle e_i, f_j \rangle - \frac{1}{2} \|x\|^2 \right| + \frac{1}{2} \|x\|^2$$

$$\leq \|x\|^2.$$

We observe that (5.230) will generate Bessel's inequality if $\{e_i\}_{i\in I}$, $\{f_j\}_{j\in J}$ are disjoint parts of a larger orthonormal family.

(b) From (5.227) one can obtain:

$$\left| \frac{\langle x, e \rangle \langle e, y \rangle}{\|e\|^2} + \frac{\langle x, f \rangle \langle f, y \rangle}{\|f\|^2} - 2 \cdot \frac{\langle x, e \rangle \langle f, y \rangle \langle e, f \rangle}{\|e\|^2 \|f\|^2} \right|$$

$$\leq \frac{1}{2} \left[\|x\| \|y\| + |\langle x, y \rangle| \right] \qquad (5.231)$$

and in particular

$$\left| \frac{|\langle x, e \rangle|^2}{\|e\|^2} + \frac{|\langle x, f \rangle|^2}{\|f\|^2} - 2 \cdot \frac{\langle x, e \rangle \langle f, e \rangle \langle e, f \rangle}{\|e\|^2 \|f\|^2} \right| \leq \|x\|^2, \qquad (5.232)$$

for any $x, y \in H$.

The case of real inner products will provide a natural generalisation for Precupanu's inequality (5.202) [56]:

Let $(H; \langle \cdot, \cdot \rangle)$ be a real inner product space and $\{e_i\}_{i\in I}$, $\{f_j\}_{j\in J}$ be two finite families of orthonormal vectors in $(H; \langle \cdot, \cdot \rangle)$. For any $x, y \in H \backslash \{0\}$ one has the double inequality

$$\frac{1}{2} \left[|\langle x, y \rangle| - \|x\| \|y\| \right] \leq \sum_{i\in I} \langle x, e_i \rangle \langle y, e_i \rangle + \sum_{j\in J} \langle x, f_j \rangle \langle y, f_j \rangle \qquad (5.233)$$

$$- 2 \sum_{i\in I, j\in J} \langle x, e_i \rangle \langle y, f_j \rangle \langle e_i, f_j \rangle$$

$$\leq \frac{1}{2} \left[\|x\| \|y\| + |\langle x, y \rangle| \right].$$

In particular, we have

$$0 \leq \sum_{i\in I} \langle x, e_i \rangle^2 + \sum_{j\in J} \langle x, f_j \rangle^2 - 2 \sum_{i\in I, j\in J} \langle x, e_i \rangle \langle x, f_j \rangle \langle e_i, f_j \rangle \qquad (5.234)$$

$$\leq \|x\|^2,$$

for any $x \in H$.

5.14   The Dunkl-Williams Inequality

If a, b are nonnull vectors in the real or complex inner product space $(H; \langle \cdot, \cdot \rangle)$, then

$$\|a - b\| \geq \frac{1}{2} \left(\|a\| + \|b\| \right) \left\| \frac{a}{\|a\|} - \frac{b}{\|b\|} \right\|. \qquad (5.235)$$

Equality holds in (5.235) if and only if either $\|a\| = \|b\|$ *or* $\|a\| + \|b\| = \|a - b\|$.

PROOF We follow the proof by Dunkl and Williams [116].
Observe, by the properties of inner products, that

$$\left\| \frac{a}{\|a\|} - \frac{b}{\|b\|} \right\|^2 = \left\langle \frac{a}{\|a\|} - \frac{b}{\|b\|}, \frac{a}{\|a\|} - \frac{b}{\|b\|} \right\rangle$$

$$= 2 - 2\operatorname{Re}\left\langle \frac{a}{\|a\|}, \frac{b}{\|b\|} \right\rangle$$

$$= \frac{1}{\|a\|\,\|b\|}\left(2\,\|a\|\,\|b\| - 2\operatorname{Re}\langle a, b\rangle \right)$$

$$= \frac{1}{\|a\|\,\|b\|}\left[\|a - b\|^2 - (\|a\| - \|b\|)^2 \right].$$

Hence

$$\|a - b\|^2 - \frac{1}{2}\left(\|a\| + \|b\| \right)^2 \left\| \frac{a}{\|a\|} - \frac{b}{\|b\|} \right\|^2$$

$$= \frac{(\|a\| - \|b\|)^2}{4\,\|a\|\,\|b\|}\left[(\|a\| + \|b\|)^2 - \|a - b\|^2 \right].$$

Inequality (5.235) follows directly from the above inequality by the use of the triangle inequality. ☐

Comments
One may observe that in an inner product space $(H; \langle \cdot, \cdot \rangle)$, for $x, y \neq 0$, the following two statements are equivalent:

(i) $\left\| \frac{x}{\|x\|} - \frac{y}{\|y\|} \right\| \leq (\geq) r$;

(ii) The following reverse (improvement) of Schwarz's inequality holds:

$$\|x\|\,\|y\| - \operatorname{Re}\langle x, y\rangle \leq (\geq) \frac{1}{2}r^2\,\|x\|\,\|y\|. \tag{5.236}$$

Then, by utilising the Dunkl-Williams inequality, we have

$$\|x\|\,\|y\| - \operatorname{Re}\langle x, y\rangle \leq 2 \cdot \frac{\|x - y\|^2}{(\|x\| + \|y\|)^2}\,\|x\|\,\|y\|.$$

Now, consider the inequality

$$\frac{2\,\|x\|\,\|y\|}{(\|x\| + \|y\|)^2} \leq \varepsilon \qquad (\varepsilon > 0). \tag{5.237}$$

This is equivalent to:

$$\varepsilon \|x\|^2 - 2(1-\varepsilon)\|x\|\|y\| + \varepsilon \|y\|^2 \geq 0. \tag{5.238}$$

If we divide (5.238) by $\|y\|^2 > 0$ and denote $\frac{\|x\|}{\|y\|} = t$, then we get

$$\varepsilon t^2 - 2(1-\varepsilon)t + \varepsilon \geq 0.$$

We have

$$\Delta = 4(1-\varepsilon)^2 - 4\varepsilon^2 = 4(1-2\varepsilon).$$

Since inequality (5.237) is of interest for small ε, we assume that $\varepsilon \in \left(0, \frac{1}{2}\right]$. Then (5.237) holds iff

$$1 - \varepsilon - \sqrt{1-2\varepsilon} \leq \frac{\|x\|}{\|y\|} \leq -\varepsilon - \sqrt{1-2\varepsilon}. \tag{5.239}$$

Therefore, if x, y satisfy (5.239), then we have the following reverse of the Schwarz inequality:

$$\|x\|\|y\| - \operatorname{Re}\langle x, y\rangle \leq \varepsilon \|x-y\|^2,$$

where $\varepsilon \in \left(0, \frac{1}{2}\right]$.

5.15   The Grüss Inequality in Inner Product Spaces

The following elementary inequality holds [79]:

Let a, x, A be vectors in the inner product space $(H, \langle \cdot, \cdot \rangle)$ over $\mathbb{K}$ ($\mathbb{K} = \mathbb{R}, \mathbb{C}$) with $a \neq A$. Then

$$\operatorname{Re}\langle A - x, x - a\rangle \geq 0$$

if and only if

$$\left\|x - \frac{a+A}{2}\right\| \leq \frac{1}{2}\|A - a\|.$$

PROOF   Define

$$I_1 := \operatorname{Re}\langle A - x, x - a\rangle, \qquad I_2 := \frac{1}{4}\|A - a\|^2 - \left\|x - \frac{a+A}{2}\right\|^2.$$

A simple calculation shows that

$$I_1 = I_2 = \operatorname{Re}\left[\langle x, a\rangle + \langle A, x\rangle\right] - \operatorname{Re}\langle A, a\rangle - \|x\|^2.$$

Thus, $I_1 \geq 0$ iff $I_2 \geq 0$, showing the required equivalence. ☐

The following particular case is obvious:

Let $x, e \in H$ with $\|e\| = 1$ and $\delta, \Delta \in K$ with $\delta \neq \Delta$. Then

$$\mathrm{Re}\,\langle \Delta e - x, x - \delta e \rangle \geq 0$$

iff

$$\left\| x - \frac{\delta + \Delta}{2} \cdot e \right\| \leq \frac{1}{2} |\Delta - \delta|.$$

If $H = \mathbb{C}$, then

$$\mathrm{Re}\,[(A - x)(\bar{x} - \bar{a})] \geq 0$$

if and only if

$$\left| x - \frac{a + A}{2} \right| \leq \frac{1}{2} |A - a|,$$

where $a, x, A \in \mathbb{C}$. If $H = \mathbb{R}$ and $A > a$, then $a \leq x \leq A$ if and only if $\left| x - \frac{a+A}{2} \right| \leq \frac{1}{2} |A - a|$.

The following representation of the difference in Schwarz's inequality is of interest [79]:

Let $x, e \in H$ with $\|e\| = 1$. Then one has the following representation:

$$\|x\|^2 - |\langle x, e \rangle|^2 = \inf_{\lambda \in \mathbb{K}} \|x - \lambda e\|^2 \geq 0. \tag{5.240}$$

PROOF   Observe, for any $\lambda \in \mathbb{K}$, that

$$\langle x - \lambda e, x - \langle x, e \rangle e \rangle = \|x\|^2 - |\langle x, e \rangle|^2 - \lambda \left[\langle e, x \rangle - \langle e, x \rangle \|e\|^2 \right]$$

$$= \|x\|^2 - |\langle x, e \rangle|^2.$$

Using Schwarz's inequality, we have

$$\left[\|x\|^2 - |\langle x, e \rangle|^2 \right]^2 = |\langle x - \lambda e, x - \langle x, e \rangle e \rangle|^2$$

$$\leq \|x - \lambda e\|^2 \|x - \langle x, e \rangle e\|^2$$

$$= \|x - \lambda e\|^2 \left[\|x\|^2 - |\langle x, e \rangle|^2 \right],$$

giving the bound

$$\|x\|^2 - |\langle x, e \rangle|^2 \leq \|x - \lambda e\|^2, \qquad \lambda \in \mathbb{K}. \tag{5.241}$$

Taking the infimum in (5.241) over $\lambda \in \mathbb{K}$, we deduce

$$\|x\|^2 - |\langle x, e \rangle|^2 \leq \inf_{\lambda \in \mathbb{K}} \|x - \lambda e\|^2.$$

Since for $\lambda_0 = \langle x, e \rangle$ we get $\|x - \lambda_0 e\|^2 = \|x\|^2 - |\langle x, e \rangle|^2$, then the representation (5.240) is proved. $\square$

The following Grüss type inequality in inner product spaces, with a different proof than the one from Dragomir [47], may be stated [79]:

Let $(H, \langle \cdot, \cdot \rangle)$ be an inner product space over K and $e \in H, \|e\| = 1$. If $\varphi, \gamma, \Phi, \Gamma$ are real or complex numbers and x, y are vectors in H such that the conditions

$$\text{Re} \langle \Phi e - x, x - \varphi e \rangle \geq 0, \quad \text{Re} \langle \Gamma e - x, x - \gamma e \rangle \geq 0$$

hold, or, equivalently, the following assumptions

$$\left\| x - \frac{\varphi + \Phi}{2} \cdot e \right\| \leq \frac{1}{2} |\Phi - \varphi|, \quad \left\| y - \frac{\gamma + \Gamma}{2} \cdot e \right\| \leq \frac{1}{2} |\Gamma - \gamma| \qquad (5.242)$$

are valid, then one has the inequality

$$|\langle x, y \rangle - \langle x, e \rangle \langle e, y \rangle| \leq \frac{1}{4} |\Phi - \varphi| \cdot |\Gamma - \gamma|. \qquad (5.243)$$

The constant $\frac{1}{4}$ is best possible.

PROOF It can be easily shown (see for example the proof of Theorem 1 from Dragomir [47]) that

$$|\langle x, y \rangle - \langle x, e \rangle \langle e, y \rangle| \leq \left[\|x\|^2 - |\langle x, e \rangle|^2 \right]^{\frac{1}{2}} \left[\|y\|^2 - |\langle y, e \rangle|^2 \right]^{\frac{1}{2}} \qquad (5.244)$$

for any $x, y \in H$ and $e \in H, \|e\| = 1$. On using the representation (5.240) and the conditions (5.242), we have that

$$\left[\|x\|^2 - |\langle x, e \rangle|^2 \right]^{\frac{1}{2}} = \inf_{\lambda \in \mathbb{K}} \|x - \lambda e\| \leq \left\| x - \frac{\varphi + \Phi}{2} \cdot e \right\| \leq \frac{1}{2} |\Phi - \varphi|$$

and

$$\left[\|y\|^2 - |\langle y, e \rangle|^2 \right]^{\frac{1}{2}} = \inf_{\lambda \in \mathbb{K}} \|y - \lambda e\| \leq \left\| y - \frac{\gamma + \Gamma}{2} \cdot e \right\| \leq \frac{1}{2} |\Gamma - \gamma|.$$

By (5.244), the desired inequality (5.243) is obtained. $\square$

Comments

Some particular cases of interest for integrable functions with real or complex values and the corresponding discrete versions are listed below.

Let $f, g : [a, b] \to \mathbb{K}$ $(\mathbb{K} = \mathbb{R}, \mathbb{C})$ be Lebesgue integrable and such that

$$\text{Re} \left[(\Phi - f(x)) \left(\overline{f(x)} - \overline{\varphi} \right) \right] \geq 0, \quad \text{Re} \left[(\Gamma - g(x)) \left(\overline{g(x)} - \overline{\gamma} \right) \right] \geq 0$$

for a.e. $x \in [a,b]$, *where* $\varphi, \gamma, \Phi, \Gamma$ *are real or complex numbers and* $\bar{z}$ *denotes the complex conjugate of* z. *Then we have the inequality*

$$\left| \frac{1}{b-a} \int_a^b f(x) \overline{g(x)} dx - \frac{1}{b-a} \int_a^b f(x) dx \cdot \frac{1}{b-a} \int_a^b \overline{g(x)} dx \right|$$
$$\leq \frac{1}{4} |\Phi - \varphi| \cdot |\Gamma - \gamma|.$$

The constant $\frac{1}{4}$ *is best possible.*

The discrete case is embodied in:

Let $x, y \in \mathbb{K}^n$ *and* $\varphi, \gamma, \Phi, \Gamma$ *be real or complex numbers such that*

$$\text{Re}\left[(\Phi - x_i)(\overline{x_i} - \overline{\varphi})\right] \geq 0 \qquad \text{and} \qquad \text{Re}\left[(\Gamma - y_i)(\overline{y_i} - \overline{\gamma})\right] \geq 0$$

for each $i \in \{1, \ldots, n\}$. *Then we have the inequality*

$$\left| \frac{1}{n} \sum_{i=1}^n x_i \overline{y_i} - \frac{1}{n} \sum_{i=1}^n x_i \cdot \frac{1}{n} \sum_{i=1}^n \overline{y_i} \right| \leq \frac{1}{4} |\Phi - \varphi| \cdot |\Gamma - \gamma|.$$

The constant $\frac{1}{4}$ *is best possible.*

5.16 A Refinement of the Grüss Inequality in Inner Product Spaces

The following result gives an improvement to the Grüss inequality in inner product spaces [79]:

Let $(H, \langle \cdot, \cdot \rangle)$ *be an inner product space over* $\mathbb{K}$ *and* $e \in H, \|e\| = 1$. *If* $\varphi, \gamma, \Phi, \Gamma$ *are real or complex numbers and* x, y *are vectors in* H *such that the conditions*

$$\text{Re}\,\langle \Phi e - x, x - \varphi e \rangle \geq 0, \qquad \text{Re}\,\langle \Gamma e - x, x - \gamma e \rangle \geq 0,$$

or, equivalently,

$$\left\| x - \frac{\varphi + \Phi}{2} \cdot e \right\| \leq \frac{1}{2} |\Phi - \varphi|, \qquad \left\| y - \frac{\gamma + \Gamma}{2} \cdot e \right\| \leq \frac{1}{2} |\Gamma - \gamma|$$

hold, then we have the inequality

$$|\langle x, y \rangle - \langle x, e \rangle \langle e, y \rangle| \tag{5.245}$$
$$\leq \frac{1}{4} |\Phi - \varphi| \cdot |\Gamma - \gamma|$$
$$- [\text{Re}\,\langle \Phi e - x, x - \varphi e \rangle]^{\frac{1}{2}} [\text{Re}\,\langle \Gamma e - y, y - \gamma e \rangle]^{\frac{1}{2}}$$
$$\leq \left(\frac{1}{4} |\Phi - \varphi| \cdot |\Gamma - \gamma| \right).$$

The constant $\frac{1}{4}$ is best possible.

PROOF   As in Dragomir [47], we have

$$|\langle x, y \rangle - \langle x, e \rangle \langle e, y \rangle|^2 \le \left[\|x\|^2 - |\langle x, e \rangle|^2 \right] \left[\|y\|^2 - |\langle y, e \rangle|^2 \right], \qquad (5.246)$$

$$\|x\|^2 - |\langle x, e \rangle|^2 = \mathrm{Re} \left[(\Phi - \langle x, e \rangle) \left(\overline{\langle x, e \rangle} - \overline{\varphi} \right) \right] - \mathrm{Re} \langle \Phi e - x, x - \varphi e \rangle, \qquad (5.247)$$

and

$$\|y\|^2 - |\langle y, e \rangle|^2 = \mathrm{Re} \left[(\Gamma - \langle y, e \rangle) \left(\overline{\langle y, e \rangle} - \overline{\gamma} \right) \right] - \mathrm{Re} \langle \Gamma e - x, x - \gamma e \rangle. \quad (5.248)$$

Using the elementary inequality

$$4 \, \mathrm{Re} \left(a\bar{b} \right) \le |a + b|^2 ; \quad a, b \in \mathbb{K} \ (\mathbb{K} = \mathbb{R}, \mathbb{C}),$$

we may state that

$$\mathrm{Re} \left[(\Phi - \langle x, e \rangle) \left(\overline{\langle x, e \rangle} - \overline{\varphi} \right) \right] \le \frac{1}{4} |\Phi - \varphi|^2, \qquad (5.249)$$

and

$$\mathrm{Re} \left[(\Gamma - \langle y, e \rangle) \left(\overline{\langle y, e \rangle} - \overline{\gamma} \right) \right] \le \frac{1}{4} |\Gamma - \gamma|^2. \qquad (5.250)$$

Consequently, by (5.246)–(5.250) we may state that

$$|\langle x, y \rangle - \langle x, e \rangle \langle e, y \rangle|^2$$
$$\le \left[\frac{1}{4} |\Phi - \varphi|^2 - \left([\mathrm{Re} \langle \Phi e - x, x - \varphi e \rangle]^{\frac{1}{2}} \right)^2 \right]$$
$$\times \left[\frac{1}{4} |\Gamma - \gamma|^2 - \left([\mathrm{Re} \langle \Gamma e - y, y - \gamma e \rangle]^{\frac{1}{2}} \right)^2 \right]. \quad (5.251)$$

Finally, using the elementary inequality for positive real numbers,

$$\left(m^2 - n^2 \right) \left(p^2 - q^2 \right) \le (mp - nq)^2,$$

we have

$$\left[\frac{1}{4} |\Phi - \varphi|^2 - \left([\mathrm{Re} \langle \Phi e - x, x - \varphi e \rangle]^{\frac{1}{2}} \right)^2 \right]$$
$$\times \left[\frac{1}{4} |\Gamma - \gamma|^2 - \left([\mathrm{Re} \langle \Gamma e - y, y - \gamma e \rangle]^{\frac{1}{2}} \right)^2 \right]$$
$$\le \left(\frac{1}{4} |\Phi - \varphi| \cdot |\Gamma - \gamma| - [\mathrm{Re} \langle \Phi e - x, x - \varphi e \rangle]^{\frac{1}{2}} [\mathrm{Re} \langle \Gamma e - y, y - \gamma e \rangle]^{\frac{1}{2}} \right)^2,$$

giving the desired inequality (5.245).

The fact that $\frac{1}{4}$ is the best constant can be proven in a similar manner to the one in the Grüss inequality (see, for instance, Dragomir [47]) and we omit the details. □

Comments

The following companion of the Grüss inequality in inner product spaces holds [79]:

Let $(H, \langle \cdot, \cdot \rangle)$ be an inner product space over $\mathbb{K}$ and $e \in H$, with $\|e\| = 1$. If $\gamma, \Gamma \in \mathbb{K}$ and $x, y \in H$ are such that

$$\mathrm{Re}\left\langle \Gamma e - \frac{x+y}{2}, \frac{x+y}{2} - \gamma e \right\rangle \geq 0, \qquad (5.252)$$

or, equivalently,

$$\left\| \frac{x+y}{2} - \frac{\gamma + \Gamma}{2} \cdot e \right\| \leq \frac{1}{2} |\Gamma - \gamma|,$$

then we have the inequality

$$\mathrm{Re}\left[\langle x, y \rangle - \langle x, e \rangle \langle e, y \rangle \right] \leq \frac{1}{4} |\Gamma - \gamma|^2. \qquad (5.253)$$

The constant $\frac{1}{4}$ is best possible.

PROOF We start with the following inequality:

$$\mathrm{Re}\langle z, u \rangle \leq \frac{1}{4} \|z + u\|^2; \quad z, u \in H. \qquad (5.254)$$

Since

$$\langle x, y \rangle - \langle x, e \rangle \langle e, y \rangle = \langle x - \langle x, e \rangle e, y - \langle y, e \rangle e \rangle,$$

we may write the following, on using (5.254):

$$
\begin{aligned}
\mathrm{Re}\left[\langle x, y \rangle - \langle x, e \rangle \langle e, y \rangle \right] &= \mathrm{Re}\left[\langle x - \langle x, e \rangle e, y - \langle y, e \rangle e \rangle \right] && (5.255) \\
&\leq \frac{1}{4} \| x - \langle x, e \rangle e + y - \langle y, e \rangle e \|^2 \\
&= \left\| \frac{x+y}{2} - \left\langle \frac{x+y}{2}, e \right\rangle \cdot e \right\|^2 \\
&= \left\| \frac{x+y}{2} \right\|^2 - \left| \left\langle \frac{x+y}{2}, e \right\rangle \right|^2.
\end{aligned}
$$

If we apply Grüss' inequality in inner product spaces for, say, $a = b = \frac{x+y}{2}$, we get

$$\left\| \frac{x+y}{2} \right\|^2 - \left| \left\langle \frac{x+y}{2}, e \right\rangle \right|^2 \leq \frac{1}{4} |\Gamma - \gamma|^2. \qquad (5.256)$$

By making use of (5.255) and (5.256) we deduce (5.253).

The fact that $\frac{1}{4}$ is the best possible constant in (5.253) follows by choosing $x = y$ in (5.252), whence we obtain

$$\operatorname{Re}\langle \Gamma e - x, x - \gamma e\rangle \geq 0,$$

implying that $0 \leq \|x\|^2 - |\langle x, e\rangle|^2 \leq \frac{1}{4}|\Gamma - \gamma|^2$, for which, by Grüss' inequality in inner product spaces, we know that the constant $\frac{1}{4}$ is best possible. □

The following result might be of interest if one wanted to evaluate the absolute value of

$$\operatorname{Re}\left[\langle x, y\rangle - \langle x, e\rangle\langle e, y\rangle\right].$$

Let $(H, \langle \cdot, \cdot\rangle)$ be an inner product space over $\mathbb{K}$ and $e \in H, \|e\| = 1$. If $\gamma, \Gamma \in \mathbb{K}$ and $x, y \in H$ are such that

$$\operatorname{Re}\left\langle \Gamma e - \frac{x \pm y}{2}, \frac{x \pm y}{2} - \gamma e\right\rangle \geq 0,$$

or, equivalently,

$$\left\|\frac{x \pm y}{2} - \frac{\gamma + \Gamma}{2} \cdot e\right\| \leq \frac{1}{2}|\Gamma - \gamma|,$$

holds, then we have the inequality

$$|\operatorname{Re}\left[\langle x, y\rangle - \langle x, e\rangle\langle e, y\rangle\right]| \leq \frac{1}{4}|\Gamma - \gamma|^2. \tag{5.257}$$

If the inner product space H is real, then (for $m, M \in \mathbb{R}, M > m$)

$$\left\langle Me - \frac{x \pm y}{2}, \frac{x \pm y}{2} - me\right\rangle \geq 0,$$

or, equivalently,

$$\left\|\frac{x \pm y}{2} - \frac{m + M}{2} \cdot e\right\| \leq \frac{1}{2}(M - m),$$

which implies that

$$|\langle x, y\rangle - \langle x, e\rangle\langle e, y\rangle| \leq \frac{1}{4}(M - m)^2. \tag{5.258}$$

The constant $\frac{1}{4}$ is best possible in both inequalities (5.257) and (5.258).

PROOF We only remark that, if

$$\operatorname{Re}\left\langle \Gamma e - \frac{x - y}{2}, \frac{x - y}{2} - \gamma e\right\rangle \geq 0$$

holds, then by (5.245), we get

$$\mathrm{Re}\left[-\langle x, y\rangle + \langle x, e\rangle \langle e, y\rangle\right] \leq \frac{1}{4}\left|\Gamma - \gamma\right|^2,$$

showing that

$$\mathrm{Re}\left[\langle x, y\rangle - \langle x, e\rangle \langle e, y\rangle\right] \geq -\frac{1}{4}\left|\Gamma - \gamma\right|^2. \tag{5.259}$$

Making use of (5.253) and (5.259), we deduce the desired result (5.257). ☐

Chapter 6

Inequalities in Normed Linear Spaces and for Functionals

The concept of a normed, and subsequently a Banach space has become over the last 70 years, a turning point in functional analysis with countless applications in mathematics and its use for modeling in science. There is no field of analysis which cannot be viewed from this perspective. Further, the unifying power of this concept provides a rich background of results and techniques for mathematicians to use in order to solve their specific problems.

In this context, both the discrete and integral versions of the triangle inequality are of crucial importance. It is therefore a natural problem to find sufficient conditions for the vectors or functions involved in order to produce reverse inequalities in both multiplicative and additive forms.

In this chapter some classical as well as recent results providing reverses of the generalised triangle inequalities are presented. The results of Diaz and Metcalf [45], together with more recent developments, are surveyed. The cases for bounded linear functionals are also provided.

Last but not least, the abstract functional versions of Jensen's and Hermite-Hadamard's inequalities for convex functions and related results are presented. They are accompanied by numerous remarks and comments that will lead the reader to further work in both research or in the use of the results for applications.

6.1 A Multiplicative Reverse for the Continuous Triangle Inequality

Let $f : [a, b] \to \mathbb{K}$, $\mathbb{K} = \mathbb{C}$ or $\mathbb{R}$ be a Lebesgue integrable function. The following inequality, which is the continuous version of the *triangle inequality*,

$$\left| \int_a^b f(x)\,dx \right| \leq \int_a^b |f(x)|\,dx, \tag{6.1}$$

plays a fundamental role in mathematical analysis and its applications.

The first reverse inequality for (6.1) was obtained by Karamata in his book [126] from 1949 (see Mitrinović, Pečarić, and Fink [141, p. 492]). It can be stated as

$$\cos\theta \int_a^b |f(x)|\, dx \le \left| \int_a^b f(x)\, dx \right| \tag{6.2}$$

provided

$$-\theta \le \arg f(x) \le \theta, \quad x \in [a, b]$$

for given $\theta \in \left(0, \frac{\pi}{2}\right)$.

We consider now the case of vector-valued functions in a Hilbert space.

We recall that $f \in L([a, b]; H)$, the space of Bochner integrable functions with values in a Hilbert space H, if and only if $f : [a, b] \to H$ is Bochner measurable on $[a, b]$ and the Lebesgue integral $\int_a^b \|f(t)\|\, dt$ is finite.

The following result holds [75]:

If $f \in L([a, b]; H)$ is such that there exists a constant $K \ge 1$ and a vector $e \in H$, $\|e\| = 1$ with

$$\|f(t)\| \le K \operatorname{Re} \langle f(t), e \rangle \quad \text{for a.e. } t \in [a, b], \tag{6.3}$$

then we have the inequality

$$\int_a^b \|f(t)\|\, dt \le K \left\| \int_a^b f(t)\, dt \right\|. \tag{6.4}$$

Equality holds in (6.4) if and only if

$$\int_a^b f(t)\, dt = \frac{1}{K} \left(\int_a^b \|f(t)\|\, dt \right) e. \tag{6.5}$$

PROOF By the Schwarz inequality in inner product spaces, we have

$$\left\| \int_a^b f(t)\, dt \right\| = \left\| \int_a^b f(t)\, dt \right\| \|e\| \tag{6.6}$$

$$\ge \left| \left\langle \int_a^b f(t)\, dt, e \right\rangle \right| \ge \left| \operatorname{Re} \left\langle \int_a^b f(t)\, dt, e \right\rangle \right|$$

$$\ge \operatorname{Re} \left\langle \int_a^b f(t)\, dt, e \right\rangle = \int_a^b \operatorname{Re} \langle f(t), e \rangle\, dt.$$

From the condition (6.3), on integrating over $[a, b]$, we deduce

$$\int_a^b \operatorname{Re} \langle f(t), e \rangle\, dt \ge \frac{1}{K} \int_a^b \|f(t)\|\, dt. \tag{6.7}$$

Thus, on making use of (6.6) and (6.7), we obtain the desired inequality (6.4).

If (6.5) holds true, then

$$K \left\| \int_a^b f(t)\, dt \right\| = \|e\| \int_a^b \|f(t)\|\, dt = \int_a^b \|f(t)\|\, dt,$$

showing that (6.4) is identically true.

If we assume that equality holds in (6.4), then by the argument provided at the beginning of our proof, we must have equality in each of the inequalities from (6.6) and (6.7).

Observe that in Schwarz's inequality $\|x\|\, \|y\| \geq \mathrm{Re}\,\langle x, y\rangle$, $x, y \in H$, the case of equality holds if and only if there exists a positive scalar μ such that $x = \mu e$. Therefore, equality holds in the first inequality in (6.6) iff $\int_a^b f(t)\, dt = \lambda e$, with $\lambda \geq 0$.

If we assume that a strict inequality holds in (6.3) for t in a nonzero measure subset of $[a, b]$, then $\int_a^b \|f(t)\|\, dt < K \int_a^b \mathrm{Re}\,\langle f(t), e\rangle\, dt$, and by (6.6) we deduce a strict inequality in (6.4), which contradicts the assumption. Thus, we must have $\|f(t)\| = K\,\mathrm{Re}\,\langle f(t), e\rangle$ for a.e. $t \in [a, b]$.

If we integrate this equality, we deduce

$$\int_a^b \|f(t)\|\, dt = K \int_a^b \mathrm{Re}\,\langle f(t), e\rangle\, dt = K\,\mathrm{Re}\,\left\langle \int_a^b f(t)\, dt, e\right\rangle$$

$$= K\,\mathrm{Re}\,\langle \lambda e, e\rangle = \lambda K,$$

giving

$$\lambda = \frac{1}{K}\int_a^b \|f(t)\|\, dt,$$

and thus the equality (6.5) is necessary.

This completes the proof.   ∎

Comments

A more appropriate result from an application point of view is stated in the following [75]:

Let e be a unit vector in the Hilbert space $(H; \langle \cdot, \cdot\rangle)$, $\rho \in (0, 1)$, and $f \in L([a, b]; H)$ so that

$$\|f(t) - e\| \leq \rho \quad \text{for a.e. } t \in [a, b]. \tag{6.8}$$

Then we have the inequality

$$\sqrt{1 - \rho^2}\int_a^b \|f(t)\|\, dt \leq \left\| \int_a^b f(t)\, dt \right\|, \tag{6.9}$$

with equality if and only if

$$\int_a^b f(t)\,dt = \sqrt{1 - \rho^2}\left(\int_a^b \|f(t)\|\,dt\right)e. \qquad (6.10)$$

PROOF From (6.8) we have

$$\|f(t)\|^2 - 2\operatorname{Re}\langle f(t),e\rangle + 1 \le \rho^2,$$

giving

$$\|f(t)\|^2 + 1 - \rho^2 \le 2\operatorname{Re}\langle f(t),e\rangle$$

for a.e. $t \in [a,b]$.

Dividing by $\sqrt{1-\rho^2} > 0$, we deduce

$$\frac{\|f(t)\|^2}{\sqrt{1-\rho^2}} + \sqrt{1-\rho^2} \le \frac{2\operatorname{Re}\langle f(t),e\rangle}{\sqrt{1-\rho^2}} \qquad (6.11)$$

for a.e. $t \in [a,b]$.

On the other hand, by the elementary inequality

$$\frac{p}{\alpha} + q\alpha \ge 2\sqrt{pq}, \quad p,q \ge 0, \ \alpha > 0$$

we have

$$2\|f(t)\| \le \frac{\|f(t)\|^2}{\sqrt{1-\rho^2}} + \sqrt{1-\rho^2} \qquad (6.12)$$

for each $t \in [a,b]$.

On making use of (6.11) and (6.12), we deduce

$$\|f(t)\| \le \frac{1}{\sqrt{1-\rho^2}}\operatorname{Re}\langle f(t),e\rangle$$

for a.e. $t \in [a,b]$.

By applying the inequality (6.4) for $K = \frac{1}{\sqrt{1-\rho^2}}$, we deduce the desired inequality (6.9). □

In the same spirit, we also have the following result [75]:

Let e be a unit vector in H and $M \ge m > 0$. If $f \in L([a,b];H)$ is such that

$$\operatorname{Re}\langle Me - f(t), f(t) - me\rangle \ge 0 \qquad (6.13)$$

or, equivalently,

$$\left\|f(t) - \frac{M+m}{2}e\right\| \le \frac{1}{2}(M-m) \qquad (6.14)$$

for a.e. $t \in [a, b]$, *then we have the inequality*

$$\frac{2\sqrt{mM}}{M+m} \int_a^b \|f(t)\| \, dt \leq \left\| \int_a^b f(t) \, dt \right\| \tag{6.15}$$

or, equivalently,

$$(0 \leq) \int_a^b \|f(t)\| \, dt - \left\| \int_a^b f(t) \, dt \right\| \tag{6.16}$$

$$\leq \frac{\left(\sqrt{M} - \sqrt{m}\right)^2}{M+m} \left\| \int_a^b f(t) \, dt \right\|.$$

Equality holds in (6.15) (or in the second part of Equation 6.16) if and only if

$$\int_a^b f(t) \, dt = \frac{2\sqrt{mM}}{M+m} \left(\int_a^b \|f(t)\| \, dt \right) e. \tag{6.17}$$

PROOF   First, we remark that if $x, z, Z \in H$, then the following statements are equivalent:

(i) $\operatorname{Re} \langle Z - x, x - z \rangle \geq 0$

and

(ii) $\left\| x - \frac{Z+z}{2} \right\| \leq \frac{1}{2} \|Z - z\|$.

By using this fact, we may realise that (6.11) and (6.12) are equivalent. Now, from (6.11), we obtain

$$\|f(t)\|^2 + mM \leq (M + m) \operatorname{Re} \langle f(t), e \rangle$$

for a.e. $t \in [a, b]$. Dividing this inequality with $\sqrt{mM} > 0$, we deduce the following inequality that will be used in the sequel:

$$\frac{\|f(t)\|^2}{\sqrt{mM}} + \sqrt{mM} \leq \frac{M + m}{\sqrt{mM}} \operatorname{Re} \langle f(t), e \rangle \tag{6.18}$$

which holds for a.e. $t \in [a, b]$.
On the other hand,

$$2 \|f(t)\| \leq \frac{\|f(t)\|^2}{\sqrt{mM}} + \sqrt{mM} \tag{6.19}$$

for any $t \in [a, b]$.

By utilising (6.18) and (6.19), we may conclude the following inequality,

$$\|f(t)\| \le \frac{M+m}{2\sqrt{mM}} \operatorname{Re}\langle f(t), e\rangle,$$

for a.e. $t \in [a, b]$.

Applying inequality (6.4) for the constant $K := \frac{m+M}{2\sqrt{mM}} \ge 1$, we deduce the desired result. $\square$

6.2 Additive Reverses for the Continuous Triangle Inequality

The following result concerning an additive reverse for the continuous triangle inequality of vector-valued functions in Hilbert spaces holds [52]:

If $f \in L([a, b]; H)$ is such that there exists a vector $e \in H$, $\|e\| = 1$ and $k : [a, b] \to [0, \infty)$, a Lebesgue integrable function with

$$\|f(t)\| - \operatorname{Re}\langle f(t), e\rangle \le k(t) \quad \text{for a.e. } t \in [a, b], \tag{6.20}$$

then we have the inequality:

$$(0 \le) \int_a^b \|f(t)\| \, dt - \left\| \int_a^b f(t) \, dt \right\| \le \int_a^b k(t) \, dt. \tag{6.21}$$

Equality holds in (6.21) if and only if

$$\int_a^b \|f(t)\| \, dt \ge \int_a^b k(t) \, dt \tag{6.22}$$

and

$$\int_a^b f(t) \, dt = \left(\int_a^b \|f(t)\| \, dt - \int_a^b k(t) \, dt \right) e. \tag{6.23}$$

PROOF If we integrate the inequality (6.20), then we get

$$\int_a^b \|f(t)\| \, dt \le \operatorname{Re}\left\langle \int_a^b f(t) \, dt, e \right\rangle + \int_a^b k(t) \, dt. \tag{6.24}$$

By Schwarz's inequality for e and $\int_a^b f(t)\, dt$, we have

$$\mathrm{Re} \left\langle \int_a^b f(t)\, dt, e \right\rangle \tag{6.25}$$

$$\leq \left| \mathrm{Re} \left\langle \int_a^b f(t)\, dt, e \right\rangle \right| \leq \left| \left\langle \int_a^b f(t)\, dt, e \right\rangle \right|$$

$$\leq \left\| \int_a^b f(t)\, dt \right\| \|e\| = \left\| \int_a^b f(t)\, dt \right\|.$$

We deduce the desired inequality (6.21) on making use of (6.24) and (6.25).

If (6.22) and (6.23) hold true, then

$$\left\| \int_a^b f(t)\, dt \right\| = \left| \int_a^b \|f(t)\|\, dt - \int_a^b k(t)\, dt \right| \|e\|$$

$$= \int_a^b \|f(t)\|\, dt - \int_a^b k(t)\, dt$$

and equality holds in (6.21).

Conversely, if equality holds in (6.21), then (6.22) is valid and we need only to prove (6.23).

If $\|f(t)\| - \mathrm{Re}\langle f(t), e\rangle < k(t)$ for t in a nonzero measure subset of $[a, b]$, then (6.24) holds as a strict inequality, implying that (6.21) also holds as a strict inequality. Therefore, if we assume that equality holds in (6.21), then we must have

$$\|f(t)\| = \mathrm{Re}\langle f(t), e\rangle + k(t) \quad \text{for a.e. } t \in [a, b]. \tag{6.26}$$

It is well known that equality holds in Schwarz's inequality $\|x\|\,\|y\| \geq \mathrm{Re}\langle x, y\rangle$ iff there exists a $\lambda \geq 0$ such that $x = \lambda y$. Therefore, if we assume that equality holds in all of (6.25), then there exists a $\lambda \geq 0$ such that

$$\int_a^b f(t)\, dt = \lambda e. \tag{6.27}$$

Integrating (6.26) on $[a, b]$, we deduce

$$\int_a^b \|f(t)\|\, dt = \mathrm{Re}\left\langle \int_a^b f(t)\, dt, e \right\rangle + \int_a^b k(t)\, dt,$$

and thus, by (6.27) we get

$$\int_a^b \|f(t)\|\, dt = \lambda \|e\|^2 + \int_a^b k(t)\, dt,$$

giving $\lambda = \int_a^b \|f(t)\|\, dt - \int_a^b k(t)\, dt$.

Using (6.27), we deduce (6.23) and the result is completely proved. ⬚

Comments

The following particular case may be useful for applications [52]:

If $f \in L([a,b];H)$ is such that there exists a vector $e \in H$, $\|e\| = 1$ and $\rho \in (0,1)$ such that

$$\|f(t) - e\| \leq \rho \quad \text{for a.e. } t \in [a,b],\tag{6.28}$$

then we have the inequality

$$(0 \leq) \int_a^b \|f(t)\|\, dt - \left\|\int_a^b f(t)\, dt\right\| \tag{6.29}$$

$$\leq \frac{\rho^2}{\sqrt{1-\rho^2}\left(1 + \sqrt{1-\rho^2}\right)} \operatorname{Re}\left\langle \int_a^b f(t)\, dt, e\right\rangle$$

$$\left(\leq \frac{\rho^2}{\sqrt{1-\rho^2}\left(1 + \sqrt{1-\rho^2}\right)} \left\|\int_a^b f(t)\, dt\right\|\right).$$

Equality holds in (6.29) *if and only if*

$$\int_a^b \|f(t)\|\, dt \geq \frac{\rho^2}{\sqrt{1-\rho^2}\left(1 + \sqrt{1-\rho^2}\right)} \operatorname{Re}\left\langle \int_a^b f(t)\, dt, e\right\rangle \tag{6.30}$$

and

$$\int_a^b f(t)\, dt$$

$$= \left(\int_a^b \|f(t)\|\, dt - \frac{\rho^2}{\sqrt{1-\rho^2}\left(1 + \sqrt{1-\rho^2}\right)} \operatorname{Re}\left\langle \int_a^b f(t)\, dt, e\right\rangle\right) e. \tag{6.31}$$

PROOF   First, note that (6.22) is equivalent to

$$\|f(t)\|^2 + 1 - \rho^2 \leq 2\operatorname{Re}\langle f(t), e\rangle,$$

giving

$$\frac{\|f(t)\|^2}{\sqrt{1-\rho^2}} + \sqrt{1-\rho^2} \leq \frac{2\operatorname{Re}\langle f(t), e\rangle}{\sqrt{1-\rho^2}}$$

for a.e. $t \in [a,b]$.

Since

$$2 \left\| f\left(t\right) \right\| \leq \frac{\left\| f\left(t\right) \right\|^{2}}{\sqrt{1 - \rho^{2}}} + \sqrt{1 - \rho^{2}}$$

for any $t \in [a, b]$, we deduce the inequality

$$\left\| f\left(t\right) \right\| \leq \frac{\operatorname{Re} \left\langle f\left(t\right), e \right\rangle}{\sqrt{1 - \rho^{2}}} \quad \text{for a.e. } t \in [a, b],$$

which is clearly equivalent to

$$\left\| f\left(t\right) \right\| - \operatorname{Re} \left\langle f\left(t\right), e \right\rangle \leq \frac{\rho^{2}}{\sqrt{1 - \rho^{2}} \left(1 + \sqrt{1 - \rho^{2}} \right)} \operatorname{Re} \left\langle f\left(t\right), e \right\rangle$$

for a.e. $t \in [a, b]$.

Applying inequality (6.21) for $k\left(t\right) := \frac{\rho^{2}}{\sqrt{1 - \rho^{2}} \left(1 + \sqrt{1 - \rho^{2}} \right)} \operatorname{Re} \left\langle f\left(t\right), e \right\rangle$, we deduce the desired result. $\square$

In the same spirit, we also have the following result [52]:

If $f \in L\left([a, b]; H\right)$ is such that there exists a vector $e \in H$, $\left\| e \right\| = 1$ and $M \geq m > 0$ such that either

$$\operatorname{Re} \left\langle Me - f\left(t\right), f\left(t\right) - me \right\rangle \geq 0 \tag{6.32}$$

or, equivalently,

$$\left\| f\left(t\right) - \frac{M + m}{2} e \right\| \leq \frac{1}{2} \left(M - m\right) \tag{6.33}$$

for a.e. $t \in [a, b]$, then we have the inequality

$$\left(0 \leq\right) \int_{a}^{b} \left\| f\left(t\right) \right\| dt - \left\| \int_{a}^{b} f\left(t\right) dt \right\| \tag{6.34}$$

$$\leq \frac{\left(\sqrt{M} - \sqrt{m} \right)^{2}}{2\sqrt{mM}} \operatorname{Re} \left\langle \int_{a}^{b} f\left(t\right) dt, e \right\rangle$$

$$\left(\leq \frac{\left(\sqrt{M} - \sqrt{m} \right)^{2}}{2\sqrt{mM}} \left\| \int_{a}^{b} f\left(t\right) dt \right\| \right).$$

Equality holds in (6.34) if and only if

$$\int_{a}^{b} \left\| f\left(t\right) \right\| dt \geq \frac{\left(\sqrt{M} - \sqrt{m} \right)^{2}}{2\sqrt{mM}} \operatorname{Re} \left\langle \int_{a}^{b} f\left(t\right) dt, e \right\rangle$$

and

$$\int_a^b f(t)\, dt = \left(\int_a^b \|f(t)\|\, dt - \frac{\left(\sqrt{M} - \sqrt{m}\right)^2}{2\sqrt{mM}} \operatorname{Re} \left\langle \int_a^b f(t)\, dt, e \right\rangle \right) e.$$

PROOF   Observe that (6.32) is clearly equivalent to

$$\|f(t)\|^2 + mM \leq (M + m) \operatorname{Re} \langle f(t), e \rangle$$

for a.e. $t \in [a, b]$, giving the inequality

$$\frac{\|f(t)\|^2}{\sqrt{mM}} + \sqrt{mM} \leq \frac{M + m}{\sqrt{mM}} \operatorname{Re} \langle f(t), e \rangle$$

for a.e. $t \in [a, b]$.

Since

$$2\|f(t)\| \leq \frac{\|f(t)\|^2}{\sqrt{mM}} + \sqrt{mM}$$

for any $t \in [a, b]$, hence we deduce the inequality

$$\|f(t)\| \leq \frac{M + m}{\sqrt{mM}} \operatorname{Re} \langle f(t), e \rangle \quad \text{for a.e. } t \in [a, b],$$

which is clearly equivalent to

$$\|f(t)\| - \operatorname{Re} \langle f(t), e \rangle \leq \frac{\left(\sqrt{M} - \sqrt{m}\right)^2}{2\sqrt{mM}} \operatorname{Re} \langle f(t), e \rangle$$

for a.e. $t \in [a, b]$.

Finally, by applying the inequality (6.21), we obtain the desired result.   ▯

We can state now (see also Dragomir [52]) the following fact:

If $f \in L([a, b]; H)$ and $r \in L_2([a, b]; H)$, $e \in H$, $\|e\| = 1$ are such that

$$\|f(t) - e\| \leq r(t) \quad \text{for a.e. } t \in [a, b], \tag{6.35}$$

then we have the inequality

$$(0 \leq) \int_a^b \|f(t)\|\, dt - \left\| \int_a^b f(t)\, dt \right\| \leq \frac{1}{2} \int_a^b r^2(t)\, dt. \tag{6.36}$$

Equality holds in (6.36) if and only if

$$\int_a^b \|f(t)\|\, dt \geq \frac{1}{2} \int_a^b r^2(t)\, dt$$

and

$$\int_a^b f(t)\, dt = \left(\int_a^b \|f(t)\|\, dt - \frac{1}{2} \int_a^b r^2(t)\, dt \right) e.$$

PROOF   The condition (6.35) is obviously equivalent to

$$\|f(t)\|^2 + 1 \le 2\, \mathrm{Re}\, \langle f(t), e \rangle + r^2(t)$$

for a.e. $t \in [a, b]$.

Using the elementary inequality

$$2\,\|f(t)\| \le \|f(t)\|^2 + 1, \quad t \in [a, b],$$

we deduce

$$\|f(t)\| - \mathrm{Re}\, \langle f(t), e \rangle \le \frac{1}{2} r^2(t)$$

for a.e. $t \in [a, b]$.

Applying inequality (6.21) for $k(t) := \frac{1}{2} r^2(t)$, $t \in [a, b]$, we deduce the desired result. □

Finally, we may state and prove the following result as well [52]:

If $f \in L([a, b]; H)$, $e \in H$, $\|e\| = 1$, and $M, m : [a, b] \to [0, \infty)$ with $M \ge m$ for a.e. on $[a, b]$ are such that $\frac{(M-m)^2}{M+m} \in L[a, b]$ and either

$$\left\| f(t) - \frac{M(t) + m(t)}{2} e \right\| \le \frac{1}{2} [M(t) - m(t)] \tag{6.37}$$

or, equivalently,

$$\mathrm{Re}\, \langle M(t)e - f(t), f(t) - m(t)e \rangle \ge 0 \tag{6.38}$$

for a.e. $t \in [a, b]$, then we have the inequality

$$(0 \le) \int_a^b \|f(t)\|\, dt - \left\| \int_a^b f(t)\, dt \right\| \le \frac{1}{4} \int_a^b \frac{[M(t) - m(t)]^2}{M(t) + m(t)}\, dt. \tag{6.39}$$

Equality holds in (6.39) if and only if

$$\int_a^b \|f(t)\|\, dt \ge \frac{1}{4} \int_a^b \frac{[M(t) - m(t)]^2}{M(t) + m(t)}\, dt$$

and

$$\int_a^b f(t)\, dt = \left(\int_a^b \|f(t)\|\, dt - \frac{1}{4} \int_a^b \frac{[M(t) - m(t)]^2}{M(t) + m(t)}\, dt \right) e.$$

PROOF   The condition (6.37) is equivalent to

$$\|f(t)\|^2 + \left(\frac{M(t)+m(t)}{2}\right)^2$$

$$\leq 2\left(\frac{M(t)+m(t)}{2}\right)\operatorname{Re}\langle f(t),e\rangle + \frac{1}{4}\left[M(t)-m(t)\right]^2$$

for a.e. $t \in [a,b]$, and since

$$2\left(\frac{M(t)+m(t)}{2}\right)\|f(t)\| \leq \|f(t)\|^2 + \left(\frac{M(t)+m(t)}{2}\right)^2, \quad t \in [a,b],$$

hence

$$\|f(t)\| - \operatorname{Re}\langle f(t),e\rangle \leq \frac{1}{4}\frac{\left[M(t)-m(t)\right]^2}{M(t)+m(t)}$$

for a.e. $t \in [a,b]$.

Now, applying the inequality (6.21) for $k(t) := \frac{1}{4}\frac{[M(t)-m(t)]^2}{M(t)+m(t)}$, $t \in [a,b]$, we deduce the desired inequality.   □

6.3   Reverses of the Discrete Triangle Inequality in Normed Spaces

Diaz and Metcalf [45] established the following reverse of the generalised triangle inequality in real or complex normed linear spaces:

Let $\mathbb{K}$ be the field of real or complex numbers. If $F : X \to \mathbb{K}$ is a linear functional of a unit norm defined on the normed linear space X endowed with the norm $\|\cdot\|$ and the vectors $x_1, \ldots, x_n$ satisfy the condition

$$0 \leq r \leq \operatorname{Re} F(x_i), \qquad i \in \{1,\ldots,n\}, \tag{6.40}$$

then

$$r\sum_{i=1}^{n}\|x_i\| \leq \left\|\sum_{i=1}^{n}x_i\right\|. \tag{6.41}$$

Equality holds if and only if both

$$F\left(\sum_{i=1}^{n}x_i\right) = r\sum_{i=1}^{n}\|x_i\| \tag{6.42}$$

and

$$F\left(\sum_{i=1}^{n}x_i\right) = \left\|\sum_{i=1}^{n}x_i\right\|. \tag{6.43}$$

If $X = H$, $(H; \langle \cdot, \cdot \rangle)$ is an inner product space and $F(x) = \langle x, e \rangle$, $\|e\| = 1$, then the condition (6.40) may be replaced with the simpler assumption

$$0 \le r \|x_i\| \le \operatorname{Re} \langle x_i, e \rangle, \qquad i = 1, \ldots, n, \tag{6.44}$$

which implies the reverse of the generalised triangle inequality (6.41). In this case equality holds in (6.41) if and only if [45]

$$\sum_{i=1}^{n} x_i = r \left(\sum_{i=1}^{n} \|x_i\| \right) e. \tag{6.45}$$

Let $F_1, \ldots, F_m$ be linear functionals on X, each of unit norm. As in Diaz and Metcalf [45], consider the real number c defined by

$$c = \sup_{x \ne 0} \left[\frac{\sum_{k=1}^{m} |F_k(x)|^2}{\|x\|^2} \right];$$

it then follows that $1 \le c \le m$. Suppose the vectors $x_1, \ldots, x_n$, whenever $x_i \ne 0$, satisfy

$$0 \le r_k \|x_i\| \le \operatorname{Re} F_k(x_i), \qquad i = 1, \ldots, n, \ k = 1, \ldots, m. \tag{6.46}$$

Then one has the following reverse of the generalised triangle inequality [45]:

$$\left(\frac{\sum_{k=1}^{m} r_k^2}{c} \right)^{\frac{1}{2}} \sum_{i=1}^{n} \|x_i\| \le \left\| \sum_{i=1}^{n} x_i \right\|. \tag{6.47}$$

Equality holds if and only if

$$F_k \left(\sum_{i=1}^{n} x_i \right) = r_k \sum_{i=1}^{n} \|x_i\|, \qquad k = 1, \ldots, m \tag{6.48}$$

and

$$\sum_{k=1}^{m} \left[F_k \left(\sum_{i=1}^{n} x_i \right) \right]^2 = c \left\| \sum_{i=1}^{n} x_i \right\|^2. \tag{6.49}$$

If $X = H$, an inner product space, then, for $F_k(x) = \langle x, e_k \rangle$, where $\{e_k\}_{k=\overline{1,n}}$ is an orthonormal family in H, the condition (6.46) may be replaced by

$$0 \le r_k \|x_i\| \le \operatorname{Re} \langle x_i, e_k \rangle, \qquad i = 1, \ldots, n, \ k = 1, \ldots, m. \tag{6.50}$$

Hence, the following reverse of the generalised triangle inequality holds:

$$\left(\sum_{k=1}^{m} r_k^2 \right)^{\frac{1}{2}} \sum_{i=1}^{n} \|x_i\| \le \left\| \sum_{i=1}^{n} x_i \right\|, \tag{6.51}$$

where equality results if and only if

$$\sum_{i=1}^{n} x_i = \left(\sum_{i=1}^{n} \|x_i\| \right) \sum_{k=1}^{m} r_k e_k. \tag{6.52}$$

The following result may be stated [51]:

Let $(X, \|\cdot\|)$ *be a normed linear space over the real or complex number field* $\mathbb{K}$ *and* $F_k : X \to \mathbb{K}$, $k \in \{1, \dots, m\}$ *continuous linear functionals on* X. *If* $x_i \in X \backslash \{0\}$, $i \in \{1, \dots, n\}$ *are such that there exists the constants* $r_k \geq 0$, $k \in \{1, \dots, m\}$ *with* $\sum_{k=1}^{m} r_k > 0$ *and*

$$\operatorname{Re} F_k (x_i) \geq r_k \|x_i\| \tag{6.53}$$

for each $i \in \{1, \dots, n\}$ *and* $k \in \{1, \dots, m\}$, *then*

$$\sum_{i=1}^{n} \|x_i\| \leq \frac{\left\| \sum_{k=1}^{m} F_k \right\|}{\sum_{k=1}^{m} r_k} \left\| \sum_{i=1}^{n} x_i \right\|. \tag{6.54}$$

The case of equality holds in (6.54) *if both*

$$\left(\sum_{k=1}^{m} F_k \right) \left(\sum_{i=1}^{n} x_i \right) = \left(\sum_{k=1}^{m} r_k \right) \sum_{i=1}^{n} \|x_i\| \tag{6.55}$$

and

$$\left(\sum_{k=1}^{m} F_k \right) \left(\sum_{i=1}^{n} x_i \right) = \left\| \sum_{k=1}^{m} F_k \right\| \left\| \sum_{i=1}^{n} x_i \right\|. \tag{6.56}$$

PROOF   Utilising the hypothesis (6.53) and the properties of the modulus, we have

$$I := \left| \left(\sum_{k=1}^{m} F_k \right) \left(\sum_{i=1}^{n} x_i \right) \right| \geq \left| \operatorname{Re} \left[\left(\sum_{k=1}^{m} F_k \right) \left(\sum_{i=1}^{n} x_i \right) \right] \right| \tag{6.57}$$

$$\geq \sum_{k=1}^{m} \operatorname{Re} F_k \left(\sum_{i=1}^{n} x_i \right) = \sum_{k=1}^{m} \sum_{i=1}^{n} \operatorname{Re} F_k (x_i)$$

$$\geq \left(\sum_{k=1}^{m} r_k \right) \sum_{i=1}^{n} \|x_i\|.$$

On the other hand, by the continuity property of F_k, $k \in \{1, \dots, m\}$ we obviously have

$$I = \left| \left(\sum_{k=1}^{m} F_k \right) \left(\sum_{i=1}^{n} x_i \right) \right| \leq \left\| \sum_{k=1}^{m} F_k \right\| \left\| \sum_{i=1}^{n} x_i \right\|. \tag{6.58}$$

We deduce the desired inequality (6.54) on making use of (6.57) and (6.58).

Now, if (6.55) and (6.56) are valid, then obviously the case of equality holds true in the inequality (6.54).

Conversely, if equality holds in (6.54), then it must hold in all the inequalities used to prove (6.54). Therefore, we have

$$\operatorname{Re} F_k\left(x_i\right) = r_k \left\|x_i\right\| \tag{6.59}$$

for each $i \in \{1, \ldots, n\}$, $k \in \{1, \ldots, m\}$;

$$\sum_{k=1}^{m} \operatorname{Im} F_k\left(\sum_{i=1}^{n} x_i\right) = 0 \tag{6.60}$$

and

$$\sum_{k=1}^{m} \operatorname{Re} F_k\left(\sum_{i=1}^{n} x_i\right) = \left\|\sum_{k=1}^{m} F_k\right\| \left\|\sum_{i=1}^{n} x_i\right\|. \tag{6.61}$$

Note that, from (6.59), by summation over i and k, we get

$$\operatorname{Re}\left[\left(\sum_{k=1}^{m} F_k\right)\left(\sum_{i=1}^{n} x_i\right)\right] = \left(\sum_{k=1}^{m} r_k\right) \sum_{i=1}^{n} \left\|x_i\right\|. \tag{6.62}$$

Since (6.60) and (6.62) imply (6.55), while (6.11) and (6.62) imply (6.56), the proof is thus complete. □

If the norms $\left\|F_k\right\|$, $k \in \{1, \ldots, m\}$ are easier to find, then, from (6.54) one may get the (coarser) inequality that might be more useful in practice:

$$\sum_{i=1}^{n} \left\|x_i\right\| \leq \frac{\sum_{k=1}^{m} \left\|F_k\right\|}{\sum_{k=1}^{m} r_k} \left\|\sum_{i=1}^{n} x_i\right\|. \tag{6.63}$$

Comments

The case of inner product spaces, in which we may provide a simpler condition for equality, is of interest in applications [51].

Let $(H; \langle \cdot, \cdot \rangle)$ be an inner product space over the real or complex number field $\mathbb{K}$, e_k, $x_i \in H \backslash \{0\}$, $k \in \{1, \ldots, m\}$, $i \in \{1, \ldots, n\}$. If $r_k \geq 0$, $k \in \{1, \ldots, m\}$ with $\sum_{k=1}^{m} r_k > 0$ satisfy

$$\operatorname{Re} \langle x_i, e_k \rangle \geq r_k \left\|x_i\right\| \tag{6.64}$$

for each $i \in \{1, \ldots, n\}$ and $k \in \{1, \ldots, m\}$, then

$$\sum_{i=1}^{n} \left\|x_i\right\| \leq \frac{\left\|\sum_{k=1}^{m} e_k\right\|}{\sum_{k=1}^{m} r_k} \left\|\sum_{i=1}^{n} x_i\right\|. \tag{6.65}$$

Equality holds in (6.65) if and only if

$$\sum_{i=1}^{n} x_i = \frac{\sum_{k=1}^{m} r_k}{\left\|\sum_{k=1}^{m} e_k\right\|^2} \left(\sum_{i=1}^{n} \|x_i\|\right) \sum_{k=1}^{m} e_k. \tag{6.66}$$

PROOF By the properties of inner product and by (6.64), we have

$$\left| \left\langle \sum_{i=1}^{n} x_i, \sum_{k=1}^{m} e_k \right\rangle \right| \tag{6.67}$$

$$\geq \left| \sum_{k=1}^{m} \mathrm{Re} \left\langle \sum_{i=1}^{n} x_i, e_k \right\rangle \right| \geq \sum_{k=1}^{m} \mathrm{Re} \left\langle \sum_{i=1}^{n} x_i, e_k \right\rangle$$

$$= \sum_{k=1}^{m} \sum_{i=1}^{n} \mathrm{Re} \left\langle x_i, e_k \right\rangle \geq \left(\sum_{k=1}^{m} r_k \right) \sum_{i=1}^{n} \|x_i\| > 0.$$

Observe also that, by (6.67), $\sum_{k=1}^{m} e_k \neq 0$.

On utilising Schwarz's inequality in the inner product space $(H; \langle \cdot, \cdot \rangle)$ for $\sum_{i=1}^{n} x_i$, $\sum_{k=1}^{m} e_k$, we have

$$\left\| \sum_{i=1}^{n} x_i \right\| \left\| \sum_{k=1}^{m} e_k \right\| \geq \left| \left\langle \sum_{i=1}^{n} x_i, \sum_{k=1}^{m} e_k \right\rangle \right|. \tag{6.68}$$

Making use of (6.67) and (6.68), we can conclude that (5.129) holds.

Now, if (6.66) holds true, then, by taking the norm we have

$$\left\| \sum_{i=1}^{n} x_i \right\| = \frac{\left(\sum_{k=1}^{m} r_k\right) \sum_{i=1}^{n} \|x_i\|}{\left\|\sum_{k=1}^{m} e_k\right\|^2} \left\| \sum_{k=1}^{m} e_k \right\|$$

$$= \frac{\left(\sum_{k=1}^{m} r_k\right)}{\left\|\sum_{k=1}^{m} e_k\right\|} \sum_{i=1}^{n} \|x_i\|,$$

so that the case of equality holds in (6.65).

Conversely, if the case of equality holds in (6.65), then it must hold in all the inequalities used to prove (6.65). Therefore, we have

$$\mathrm{Re} \langle x_i, e_k \rangle = r_k \|x_i\| \tag{6.69}$$

for each $i \in \{1, \ldots, n\}$ and $k \in \{1, \ldots, m\}$,

$$\left\| \sum_{i=1}^{n} x_i \right\| \left\| \sum_{k=1}^{m} e_k \right\| = \left| \left\langle \sum_{i=1}^{n} x_i, \sum_{k=1}^{m} e_k \right\rangle \right| \tag{6.70}$$

and

$$\mathrm{Im} \left\langle \sum_{i=1}^{n} x_i, \sum_{k=1}^{m} e_k \right\rangle = 0. \tag{6.71}$$

From (6.69), on summing over i and k, we get

$$\mathrm{Re}\left\langle \sum_{i=1}^{n} x_i, \sum_{k=1}^{m} e_k \right\rangle = \left(\sum_{k=1}^{m} r_k\right) \sum_{i=1}^{n} \|x_i\|. \tag{6.72}$$

By (6.71) and (6.72), we have

$$\left\langle \sum_{i=1}^{n} x_i, \sum_{k=1}^{m} e_k \right\rangle = \left(\sum_{k=1}^{m} r_k\right) \sum_{i=1}^{n} \|x_i\|. \tag{6.73}$$

On the other hand, from the following identity in inner product spaces,

$$\left\| u - \frac{\langle u, v \rangle v}{\|v\|^2} \right\|^2 = \frac{\|u\|^2 \|v\|^2 - |\langle u, v \rangle|^2}{\|v\|^2}, \quad v \neq 0, \tag{6.74}$$

the relation (6.70) holds if and only if

$$\sum_{i=1}^{n} x_i = \frac{\langle \sum_{i=1}^{n} x_i, \sum_{k=1}^{m} e_k \rangle}{\left\| \sum_{k=1}^{m} e_k \right\|^2} \sum_{k=1}^{m} e_k. \tag{6.75}$$

Finally, on utilising (6.73) and (6.75), we deduce that the condition (6.66) is necessary for the equality case in (6.65). □

Before we give some particular results, we need to state the following elementary inequality that has been basically obtained by Dragomir [53]. For the sake of completeness, we provide a short proof here as well.

Let $(H; \langle \cdot, \cdot \rangle)$ *be an inner product space over the real or complex number field* $\mathbb{K}$ *and* $x, a \in H$, $r > 0$ *such that:*

$$\|x - a\| \leq r < \|a\|. \tag{6.76}$$

Then we have the inequality

$$\|x\| \left(\|a\|^2 - r^2\right)^{\frac{1}{2}} \leq \mathrm{Re}\,\langle x, a \rangle \tag{6.77}$$

or, equivalently,

$$\|x\|^2 \|a\|^2 - [\mathrm{Re}\,\langle x, a \rangle]^2 \leq r^2 \|x\|^2. \tag{6.78}$$

Equality holds in (6.77) (or in (6.78)) if and only if

$$\|x - a\| = r \quad \text{and} \quad \|x\|^2 + r^2 = \|a\|^2. \tag{6.79}$$

PROOF From the first part of (6.76) we have

$$\|x\|^2 + \|a\|^2 - r^2 \leq 2\,\mathrm{Re}\,\langle x, a \rangle. \tag{6.80}$$

By the second part of (6.76) we have $\left(\|a\|^2 - r^2 \right)^{\frac{1}{2}} > 0$, therefore, by (6.80), we may state that

$$0 < \frac{\|x\|^2}{\left(\|a\|^2 - r^2 \right)^{\frac{1}{2}}} + \left(\|a\|^2 - r^2 \right)^{\frac{1}{2}} \leq \frac{2 \operatorname{Re} \langle x, a \rangle}{\left(\|a\|^2 - r^2 \right)^{\frac{1}{2}}}. \tag{6.81}$$

Utilising the elementary inequality

$$\frac{1}{\alpha} q + \alpha p \geq 2\sqrt{pq}, \quad \alpha > 0, \ p > 0, \ q \geq 0$$

with equality if and only if $\alpha = \sqrt{\frac{q}{p}}$, we may state (for $\alpha = \left(\|a\|^2 - r^2 \right)^{\frac{1}{2}}$, $p = 1$, $q = \|x\|^2$) that

$$2 \|x\| \leq \frac{\|x\|^2}{\left(\|a\|^2 - r^2 \right)^{\frac{1}{2}}} + \left(\|a\|^2 - r^2 \right)^{\frac{1}{2}}. \tag{6.82}$$

Inequality (6.77) follows now by (6.81) and (6.82).

From the above argument, it is clear that equality holds in (6.77) if and only if it holds in (6.81) and (6.82). However, equality holds in (6.81) if and only if $\|x - a\| = r$ and in (6.82) if and only if $\left(\|a\|^2 - r^2 \right)^{\frac{1}{2}} = \|x\|$.

The proof is thus completed. $\qquad\qquad\square$

We may now state the following result [51]:

Let $(H; \langle \cdot, \cdot \rangle)$ be an inner product space over the real or complex number field $\mathbb{K}$, e_k, $x_i \in H \backslash \{0\}$, $k \in \{1, \ldots, m\}$, $i \in \{1, \ldots, n\}$. If $\rho_k \geq 0$, $k \in \{1, \ldots, m\}$ with

$$\|x_i - e_k\| \leq \rho_k < \|e_k\| \tag{6.83}$$

for each $i \in \{1, \ldots, n\}$ and $k \in \{1, \ldots, m\}$, then

$$\sum_{i=1}^{n} \|x_i\| \leq \frac{\left\| \sum_{k=1}^{m} e_k \right\|}{\sum_{k=1}^{m} \left(\|e_k\|^2 - \rho_k^2 \right)^{\frac{1}{2}}} \left\| \sum_{i=1}^{n} x_i \right\|. \tag{6.84}$$

Equality holds in (6.84) if and only if

$$\sum_{i=1}^{n} x_i = \frac{\sum_{k=1}^{m} \left(\|e_k\|^2 - \rho_k^2 \right)^{\frac{1}{2}}}{\left\| \sum_{k=1}^{m} e_k \right\|^2} \left(\sum_{i=1}^{n} \|x_i\| \right) \sum_{k=1}^{m} e_k.$$

PROOF   Utilising (6.78), we have from (6.83) that

$$\|x_i\| \left(\|e_k\|^2 - \rho_k^2 \right)^{\frac{1}{2}} \leq \operatorname{Re} \langle x_i, e_k \rangle$$

for each $k \in \{1, \ldots, m\}$ and $i \in \{1, \ldots, n\}$.

Applying (6.65) for

$$r_k := \left(\|e_k\|^2 - \rho_k^2 \right)^{\frac{1}{2}}, \quad k \in \{1, \ldots, m\},$$

we deduce the desired result. □

If $\{e_k\}_{k \in \{1, \ldots, m\}}$ are orthogonal, then (6.84) becomes

$$\sum_{i=1}^{n} \|x_i\| \le \frac{\left(\sum_{k=1}^{m} \|e_k\|^2 \right)^{\frac{1}{2}}}{\sum_{k=1}^{m} \left(\|e_k\|^2 - \rho_k^2 \right)^{\frac{1}{2}}} \left\| \sum_{i=1}^{n} x_i \right\| \tag{6.85}$$

with equality if and only if

$$\sum_{i=1}^{n} x_i = \frac{\sum_{k=1}^{m} \left(\|e_k\|^2 - \rho_k^2 \right)^{\frac{1}{2}}}{\sum_{k=1}^{m} \|e_k\|^2} \left(\sum_{i=1}^{n} \|x_i\| \right) \sum_{k=1}^{m} e_k.$$

Moreover, if $\{e_k\}_{k \in \{1, \ldots, m\}}$ is assumed to be orthonormal and

$$\|x_i - e_k\| \le \rho_k \text{ for } k \in \{1, \ldots, m\}, \ i \in \{1, \ldots, n\}$$

where $\rho_k \in [0, 1)$ for $k \in \{1, \ldots, m\}$, then

$$\sum_{i=1}^{n} \|x_i\| \le \frac{\sqrt{m}}{\sum_{k=1}^{m} (1 - \rho_k^2)^{\frac{1}{2}}} \left\| \sum_{i=1}^{n} x_i \right\| \tag{6.86}$$

with equality if and only if

$$\sum_{i=1}^{n} x_i = \frac{\sum_{k=1}^{m} (1 - \rho_k^2)^{\frac{1}{2}}}{m} \left(\sum_{i=1}^{n} \|x_i\| \right) \sum_{k=1}^{m} e_k.$$

The following elementary inequality may be stated as well [77]:

Let $(H; \langle \cdot, \cdot \rangle)$ *be an inner product space over the real or complex number field* $\mathbb{K}$, $x, y \in H$ *and* $M \ge m > 0$. *If*

$$\operatorname{Re} \langle My - x, x - my \rangle \ge 0 \tag{6.87}$$

or, equivalently,

$$\left\| x - \frac{m + M}{2} y \right\| \le \frac{1}{2} (M - m) \|y\|, \tag{6.88}$$

then

$$\|x\| \, \|y\| \le \frac{1}{2} \cdot \frac{M + m}{\sqrt{mM}} \operatorname{Re} \langle x, y \rangle. \tag{6.89}$$

Equality holds in (6.89) if and only if the case of equality holds in (6.87) and

$$\|x\| = \sqrt{mM}\, \|y\|.\qquad(6.90)$$

PROOF Obviously,

$$\mathrm{Re}\,\langle My - x, x - my\rangle = (M + m)\,\mathrm{Re}\,\langle x, y\rangle - \|x\|^2 - mM\,\|y\|^2.$$

Then (6.87) is clearly equivalent to

$$\frac{\|x\|^2}{\sqrt{mM}} + \sqrt{mM}\,\|y\|^2 \le \frac{M + m}{\sqrt{mM}}\,\mathrm{Re}\,\langle x, y\rangle.\qquad(6.91)$$

Since, obviously,

$$2\,\|x\|\,\|y\| \le \frac{\|x\|^2}{\sqrt{mM}} + \sqrt{mM}\,\|y\|^2,\qquad(6.92)$$

with equality iff $\|x\| = \sqrt{mM}\,\|y\|$. Hence (6.91) and (6.92) imply (6.89). The case of equality is obvious and we omit the details. $\qquad\Box$

Comments

Finally, we may state the following result (see Dragomir [51]):

Let $(H; \langle\cdot,\cdot\rangle)$ be an inner product space over the real or complex number field $\mathbb{K}$, e_k, $x_i \in H\setminus\{0\}$, $k \in \{1,\ldots,m\}$, $i \in \{1,\ldots,n\}$. If $M_k > \mu_k > 0$, $k \in \{1,\ldots,m\}$ are such that either

$$\mathrm{Re}\,\langle M_k e_k - x_i, x_i - \mu_k e_k\rangle \ge 0\qquad(6.93)$$

or, equivalently,

$$\left\| x_i - \frac{M_k + \mu_k}{2} e_k \right\| \le \frac{1}{2}\,(M_k - \mu_k)\,\|e_k\|$$

for each $k \in \{1,\ldots,m\}$ and $i \in \{1,\ldots,n\}$, then

$$\sum_{i=1}^{n} \|x_i\| \le \frac{\|\sum_{k=1}^{m} e_k\|}{\sum_{k=1}^{m} \frac{2\cdot\sqrt{\mu_k M_k}}{\mu_k + M_k}\,\|e_k\|}\,\left\| \sum_{i=1}^{n} x_i \right\|.\qquad(6.94)$$

Equality holds in (6.94) if and only if

$$\sum_{i=1}^{n} x_i = \frac{\sum_{k=1}^{m} \frac{2\cdot\sqrt{\mu_k M_k}}{\mu_k + M_k}\,\|e_k\|}{\|\sum_{k=1}^{m} e_k\|^2}\,\sum_{i=1}^{n} \|x_i\| \sum_{k=1}^{m} e_k.$$

PROOF Utilising (6.89) and (6.93), we deduce

$$\frac{2\cdot\sqrt{\mu_k M_k}}{\mu_k + M_k}\,\|x_i\|\,\|e_k\| \le \mathrm{Re}\,\langle x_i, e_k\rangle$$

for each $k \in \{1, \ldots, m\}$ and $i \in \{1, \ldots, n\}$.

Applying (6.89) for

$$r_k := \frac{2 \cdot \sqrt{\mu_k M_k}}{\mu_k + M_k} \|e_k\|, \quad k \in \{1, \ldots, m\},$$

we deduce the desired result. ☐

6.4 Other Multiplicative Reverses for a Finite Sequence of Functionals

Assume that F_k, $k \in \{1, \ldots, m\}$ are bounded linear functionals defined on the normed linear space X.

For $p \in [1, \infty)$, define [76]

$$c_p := \sup_{x \neq 0} \left[\frac{\sum_{k=1}^m |F_k(x)|^p}{\|x\|^p} \right]^{\frac{1}{p}} \tag{c_p}$$

and for $p = \infty$,

$$c_\infty := \sup_{x \neq 0} \left[\max_{1 \leq k \leq m} \left\{ \frac{|F_k(x)|}{\|x\|} \right\} \right]. \tag{c_∞}$$

Since $|F_k(x)| \leq \|F_k\| \|x\|$ for any $x \in X$, where $\|F_k\|$ is the norm of the functional F_k, we have

$$c_p \leq \left(\sum_{k=1}^m \|F_k\|^p \right)^{\frac{1}{p}}, \quad p \geq 1$$

and

$$c_\infty \leq \max_{1 \leq k \leq m} \|F_k\|.$$

We may now state and prove a new reverse inequality for the generalised triangle inequality in normed linear spaces [76]:

Let $(X, \|\cdot\|)$ be a normed linear space over the real or complex number field $\mathbb{K}$ and $F_k : X \to \mathbb{K}$, $k \in \{1, \ldots, m\}$ continuous linear functionals on X. If $x_i \in X \setminus \{0\}$, $i \in \{1, \ldots, n\}$ are such that there exist constants $r_k \geq 0$, $k \in \{1, \ldots, m\}$ with $\sum_{k=1}^m r_k > 0$ and

$$\mathrm{Re}\, F_k(x_i) \geq r_k \|x_i\| \tag{6.95}$$

for each $i \in \{1, \ldots, n\}$ and $k \in \{1, \ldots, m\}$, then we have the inequalities

$$(1 \leq) \frac{\sum_{i=1}^n \|x_i\|}{\|\sum_{i=1}^n x_i\|} \leq \frac{c_\infty}{\max_{1 \leq k \leq m} \{r_k\}} \left(\leq \frac{\max_{1 \leq k \leq m} \|F_k\|}{\max_{1 \leq k \leq m} \{r_k\}} \right). \tag{6.96}$$

Equality holds in (6.96) if and only if

$$\text{Re}\left[F_k\left(\sum_{i=1}^{n} x_i\right)\right] = r_k \sum_{i=1}^{n} \|x_i\| \quad \text{for each } k \in \{1,\ldots,m\} \tag{6.97}$$

and

$$\max_{1\le k\le m} \text{Re}\left[F_k\left(\sum_{i=1}^{n} x_i\right)\right] = c_\infty \left\|\sum_{i=1}^{n} x_i\right\|. \tag{6.98}$$

PROOF Since, by the definition of c_∞, we have

$$c_\infty \|x\| \ge \max_{1\le k\le m} |F_k(x)|, \quad \text{for any } x \in X,$$

we can state, for $x = \sum_{i=1}^{n} x_i$, that

$$c_\infty \left\|\sum_{i=1}^{n} x_i\right\| \ge \max_{1\le k\le m}\left|F_k\left(\sum_{i=1}^{n} x_i\right)\right| \ge \max_{1\le k\le m}\left[\left|\text{Re}\,F_k\left(\sum_{i=1}^{n} x_i\right)\right|\right] \tag{6.99}$$

$$\ge \max_{1\le k\le m}\left[\text{Re}\sum_{i=1}^{n} F_k(x_i)\right] = \max_{1\le k\le m}\left[\sum_{i=1}^{n}\text{Re}\,F_k(x_i)\right].$$

Utilising the hypothesis (6.95) we obviously have

$$\max_{1\le k\le m}\left[\sum_{i=1}^{n}\text{Re}\,F_k(x_i)\right] \ge \max_{1\le k\le m}\{r_k\}\cdot\sum_{i=1}^{n}\|x_i\|.$$

Also, $\sum_{i=1}^{n} x_i \ne 0$, since from the initial assumptions, not all r_k and x_i with $k \in \{1,\ldots,m\}$ and $i \in \{1,\ldots,n\}$ are allowed to be zero. Hence the desired inequality (6.96) is obtained.

Now, if (6.97) is valid, then, taking the maximum over $k \in \{1,\ldots,m\}$ in this equality, we get

$$\max_{1\le k\le m} \text{Re}\left[F_k\left(\sum_{i=1}^{n} x_i\right)\right] = \max_{1\le k\le m}\{r_k\}\left\|\sum_{i=1}^{n} x_i\right\|,$$

which, together with (6.98), provides the equality case in (6.96).

Now, if equality holds in (6.96), it must hold in all the inequalities used to prove (6.96); therefore, we have

$$\text{Re}\,F_k(x_i) = r_k \|x_i\| \quad \text{for each } i \in \{1,\ldots,n\} \text{ and } k \in \{1,\ldots,m\} \tag{6.100}$$

and, from (6.99),

$$c_\infty \left\|\sum_{i=1}^{n} x_i\right\| = \max_{1\le k\le m} \text{Re}\left[F_k\left(\sum_{i=1}^{n} x_i\right)\right],$$

which is (6.98).

From (6.100), on summing over $i \in \{1, \ldots, n\}$, we get (6.97), and the inequality is proved. ☐

The following result in normed spaces also holds:

Let $(X, \|\cdot\|)$ be a normed linear space over the real or complex number field $\mathbb{K}$ and $F_k : X \to \mathbb{K}$, $k \in \{1, \ldots, m\}$ continuous linear functionals on X. If $x_i \in X \setminus \{0\}$, $i \in \{1, \ldots, n\}$ are such that there exist the constants $r_k \geq 0$, $k \in \{1, \ldots, m\}$ with $\sum_{k=1}^{m} r_k > 0$ and

$$\operatorname{Re} F_k(x_i) \geq r_k \|x_i\| \tag{6.101}$$

for each $i \in \{1, \ldots, n\}$ and $k \in \{1, \ldots, m\}$, then we have the inequality

$$(1 \leq) \frac{\sum_{i=1}^{n} \|x_i\|}{\left\| \sum_{i=1}^{n} x_i \right\|} \leq \frac{c_p}{\left(\sum_{k=1}^{m} r_k^p \right)^{\frac{1}{p}}} \left(\leq \frac{\sum_{k=1}^{m} \|F_k\|^p}{\sum_{k=1}^{m} r_k^p} \right)^{\frac{1}{p}}, \tag{6.102}$$

where $p \geq 1$.

Equality holds in (6.102) if and only if

$$\operatorname{Re} \left[F_k \left(\sum_{i=1}^{n} x_i \right) \right] = r_k \sum_{i=1}^{n} \|x_i\| \quad \text{for each } k \in \{1, \ldots, m\} \tag{6.103}$$

and

$$\sum_{k=1}^{m} \left[\operatorname{Re} F_k \left(\sum_{i=1}^{n} x_i \right) \right]^p = c_p^p \left\| \sum_{i=1}^{n} x_i \right\|^p. \tag{6.104}$$

PROOF By the definition of c_p, $p \geq 1$, we have

$$c_p^p \|x\|^p \geq \sum_{k=1}^{m} |F_k(x)|^p \quad \text{for any } x \in X,$$

implying that

$$c_p^p \left\| \sum_{i=1}^{n} x_i \right\|^p \geq \sum_{k=1}^{m} \left| F_k \left(\sum_{i=1}^{n} x_i \right) \right|^p \geq \sum_{k=1}^{m} \left| \operatorname{Re} F_k \left(\sum_{i=1}^{n} x_i \right) \right|^p \tag{6.105}$$

$$\geq \sum_{k=1}^{m} \left[\operatorname{Re} F_k \left(\sum_{i=1}^{n} x_i \right) \right]^p = \sum_{k=1}^{m} \left[\sum_{i=1}^{n} \operatorname{Re} F_k(x_i) \right]^p.$$

Utilising the hypothesis (6.101), we have that

$$\sum_{k=1}^{m} \left[\sum_{i=1}^{n} \operatorname{Re} F_k(x_i) \right]^p \geq \sum_{k=1}^{m} \left[\sum_{i=1}^{n} r_k \|x_i\| \right]^p = \sum_{k=1}^{m} r_k^p \left(\sum_{i=1}^{n} \|x_i\| \right)^p. \tag{6.106}$$

Making use of (6.105) and (6.106), we deduce

$$c_p^p \left\| \sum_{i=1}^{n} x_i \right\|^p \geq \left(\sum_{k=1}^{m} r_k^p \right) \left(\sum_{i=1}^{n} \|x_i\| \right)^p,$$

which implies the desired inequality (6.102).

If (6.103) holds true, then, by taking the power p and summing over $k \in \{1, \ldots, m\}$, we deduce

$$\sum_{k=1}^{m} \left[\operatorname{Re} \left[F_k \left(\sum_{i=1}^{n} x_i \right) \right] \right]^p = \sum_{k=1}^{m} r_k^p \left(\sum_{i=1}^{n} \|x_i\| \right)^p,$$

which, together with (6.104), shows that equality holds true in (6.102).

Conversely, if equality holds in (6.102), then it must hold in all inequalities needed to prove (6.102); therefore, we must have:

$$\operatorname{Re} F_k(x_i) = r_k \|x_i\| \quad \text{for each } i \in \{1, \ldots, n\} \text{ and } k \in \{1, \ldots, m\} \quad (6.107)$$

and, from (6.105),

$$c_p^p \left\| \sum_{i=1}^{n} x_i \right\|^p = \sum_{k=1}^{m} \left[\operatorname{Re} F_k \left(\sum_{i=1}^{n} x_i \right) \right]^p,$$

which is exactly (6.104).

From (6.107), on summing over i from 1 to n, we deduce (6.103). This completes the proof.   □

Comments

Similar results may be stated in the case of inner product spaces where one could choose the following bounded linear functionals $F_k(x) = \langle x, e_k \rangle$, where $\{e_k\}_{k=1,\ldots,m}$ are vectors in the inner product space $(H, \langle \cdot, \cdot \rangle)$. The details are omitted.

6.5   The Diaz-Metcalf Inequality for Semi-Inner Products

In 1961, Lumer [132] introduced the following concept.

Let X be a linear space over the real or complex number field $\mathbb{K}$. The mapping $[\cdot, \cdot] : X \times X \to \mathbb{K}$ is called a semi-inner product on X, if the following properties are satisfied (see also Dragomir [77, p. 17]):

(i) $[x + y, z] = [x, z] + [y, z]$ *for all* $x, y, z \in X$;

(ii) $[\lambda x, y] = \lambda [x, y]$ *for all* $x, y \in X$ *and* $\lambda \in \mathbb{K}$;

(iii) $[x, x] \geq 0$ *for all* $x \in X$ *and* $[x, x] = 0$ *implies* $x = 0$;

(iv) $|[x, y]|^2 \leq [x, x][y, y]$ *for all* $x, y \in X$;

(v) $[x, \lambda y] = \bar{\lambda} [x, y]$ *for all* $x, y \in X$ *and* $\lambda \in \mathbb{K}$.

It is well known that the mapping $X \ni x \longmapsto [x, x]^{\frac{1}{2}} \in \mathbb{R}$ is a norm on X and, for any $y \in X$, the functional $X \ni x \xmapsto{\varphi_y} [x, y] \in \mathbb{K}$ is a continuous linear functional on X endowed with the norm $\|\cdot\|$ generated by $[\cdot, \cdot]$. Moreover, one has $\|\varphi_y\| = \|y\|$ (see, for instance, Dragomir [77, p. 17]).

Let $(X, \|\cdot\|)$ be a real or complex normed space. If $J : X \to_2 X^*$ is the *normalised duality mapping* defined on X, i.e., we recall that (see, for instance, Dragomir [77, p. 1])

$$J(x) = \{\varphi \in X^* | \varphi(x) = \|\varphi\| \|x\|, \ \|\varphi\| = \|x\|\}, \quad x \in X,$$

then we may state the following representation result (see, for instance Dragomir [77, p. 18]):

Each semi-inner product $[\cdot, \cdot] : X \times X \to \mathbb{K}$ *that generates the norm* $\|\cdot\|$ *of the normed linear space* $(X, \|\cdot\|)$ *over the real or complex number field* $\mathbb{K}$ *is of the form*

$$[x, y] = \left\langle \tilde{J}(y), x \right\rangle \text{ for any } x, y \in X,$$

where $\tilde{J}$ *is a selection of the normalised duality mapping and* $\langle \varphi, x \rangle := \varphi(x)$ *for* $\varphi \in X^*$ *and* $x \in X$.

Utilising the concept of semi-inner products, we can state the following particular case of the Diaz-Metcalf inequality:

Let $(X, \|\cdot\|)$ *be a normed linear space,* $[\cdot, \cdot] : X \times X \to \mathbb{K}$ *a semi-inner product generating the norm* $\|\cdot\|$ *and* $e \in X$, *with* $\|e\| = 1$. *If* $x_i \in X$, $i \in \{1, \ldots, n\}$, *and* $r \geq 0$ *such that*

$$r \|x_i\| \leq \text{Re}\,[x_i, e] \quad \text{for each} \ i \in \{1, \ldots, n\}, \tag{6.108}$$

then we have the inequality

$$r \sum_{i=1}^{n} \|x_i\| \leq \left\| \sum_{i=1}^{n} x_i \right\|. \tag{6.109}$$

Equality holds in (6.109) *if and only if both*

$$\left[\sum_{i=1}^{n} x_i, e \right] = r \sum_{i=1}^{n} \|x_i\| \tag{6.110}$$

and

$$\left[\sum_{i=1}^{n} x_i, e \right] = \left\| \sum_{i=1}^{n} x_i \right\|. \tag{6.111}$$

The proof is obvious from the Diaz-Metcalf theorem [45, Theorem 3] applied for the continuous linear functional $F_e(x) = [x, e]$, $x \in X$.

Comments
Before we provide a simpler necessary and sufficient condition of equality in (6.109), we need to recall the concept of strictly convex normed spaces and a classical characterisation of these spaces.

A normed linear space $(X, \|\cdot\|)$ *is said to be strictly convex if for every* x, y *from* X *with* $x \neq y$ *and* $\|x\| = \|y\| = 1$ *we have* $\|\lambda x + (1 - \lambda) y\| < 1$ *for all* $\lambda \in (0, 1)$.

The following characterisation of strictly convex spaces is useful in what follows (see Berkson [8], Gudder and Strawther [120], or Dragomir [77, p. 21]):

Let $(X, \|\cdot\|)$ *be a normed linear space over* $\mathbb{K}$ *equipped with a semi-inner product* $[\cdot, \cdot]$ *which generates its norm. The following statements are equivalent:*

(i) $(X, \|\cdot\|)$ *is strictly convex;*

(ii) *For every* $x, y \in X$, $x, y \neq 0$ *with* $[x, y] = \|x\| \|y\|$, *there exists a* $\lambda > 0$ *such that* $x = \lambda y$.

The following result may be stated:

Let $(X, \|\cdot\|)$ *be a strictly convex normed linear space, equipped with a semi-inner product* $[\cdot, \cdot]$ *which generates the norm and* e, x_i $(i \in \{1, \dots, n\})$ *as above. Then the case of equality holds in (6.109) if and only if*

$$\sum_{i=1}^{n} x_i = r \left(\sum_{i=1}^{n} \|x_i\| \right) e. \tag{6.112}$$

PROOF   If (6.112) holds true, then

$$\left\| \sum_{i=1}^{n} x_i \right\| = r \left(\sum_{i=1}^{n} \|x_i\| \right) \|e\| = r \sum_{i=1}^{n} \|x_i\|,$$

which is the equality case in (6.109).

Conversely, if equality holds in (6.109), then (6.110) and (6.111) hold true. By utilising the criterion (i) and (ii) above, we conclude that there exists a $\mu > 0$ such that

$$\sum_{i=1}^{n} x_i = \mu e. \tag{6.113}$$

By substituting this in (6.110) we get

$$\mu \|e\|^2 = r \sum_{i=1}^{n} \|x_i\|,$$

giving

$$\mu = r \sum_{i=1}^{n} \|x_i\|. \tag{6.114}$$

Finally, by (6.113) and (6.114) we deduce (6.112) and the statement is proved.
□

6.6 Multiplicative Reverses of the Continuous Triangle Inequality

Let $(X, \|\cdot\|)$ be a Banach space over the real or complex number field. Then one has the following reverse of the continuous triangle inequality [75]:

Let F be a continuous linear functional of unit norm on X. Suppose that the function $f : [a, b] \to X$ is Bochner integrable on $[a, b]$ and there exists an $r \geq 0$ such that

$$r \|f(t)\| \leq \operatorname{Re} F(f(t)) \quad \text{for a.e. } t \in [a, b]. \tag{6.115}$$

Then

$$r \int_a^b \|f(t)\| \, dt \leq \left\| \int_a^b f(t) \, dt \right\|, \tag{6.116}$$

and equality holds in (6.116) if and only if both

$$F\left(\int_a^b f(t) \, dt \right) = r \int_a^b \|f(t)\| \, dt \tag{6.117}$$

and

$$F\left(\int_a^b f(t) \, dt \right) = \left\| \int_a^b f(t) \, dt \right\|. \tag{6.118}$$

PROOF   Since the norm of F is one, then

$$|F(x)| \leq \|x\| \quad \text{for any } x \in X.$$

Applying this inequality for the vector $\int_a^b f(t)\, dt$, we get

$$\left\| \int_a^b f(t)\, dt \right\| \geq \left| F\left(\int_a^b f(t)\, dt \right) \right| \tag{6.119}$$

$$\geq \left| \operatorname{Re} F\left(\int_a^b f(t)\, dt \right) \right| = \left| \int_a^b \operatorname{Re} F(f(t))\, dt \right|.$$

Now, by integration of (6.115), we obtain

$$\int_a^b \operatorname{Re} F(f(t))\, dt \geq r \int_a^b \|f(t)\|\, dt, \tag{6.120}$$

and by (6.119) and (6.120) we deduce the desired inequality (6.115).

Obviously, if (6.117) and (6.118) are true, then equality holds in (6.116).

Conversely, if equality holds in (6.116), then it must hold in all the inequalities used before in proving this inequality. Therefore, we must have

$$r \|f(t)\| = \operatorname{Re} F(f(t)) \quad \text{for a.e. } t \in [a, b], \tag{6.121}$$

$$\operatorname{Im} F\left(\int_a^b f(t)\, dt \right) = 0, \tag{6.122}$$

and

$$\left\| \int_a^b f(t)\, dt \right\| = \operatorname{Re} F\left(\int_a^b f(t)\, dt \right). \tag{6.123}$$

Integrating (6.121) on $[a, b]$, we get

$$r \int_a^b \|f(t)\|\, dt = \operatorname{Re} F\left(\int_a^b f(t)\, dt \right). \tag{6.124}$$

By utilising (6.124) and (6.122), we deduce (6.117), while (6.123) and (6.124) would imply (6.118). The statement is thus proved. ▯

Comments

Let $(X, \|\cdot\|)$ be a Banach space equipped with a semi-inner product $[\cdot, \cdot]$ that generates the norm $\|\cdot\|$ and $e \in X$, with $\|e\| = 1$. Suppose that the function $f : [a, b] \to X$ is Bochner integrable on $[a, b]$ and there exists an $r \geq 0$ such that

$$r \|f(t)\| \leq \operatorname{Re}[f(t), e] \quad \text{for a.e. } t \in [a, b]. \tag{6.125}$$

Then

$$r \int_a^b \|f(t)\|\, dt \leq \left\| \int_a^b f(t)\, dt \right\| \tag{6.126}$$

and equality holds in (6.126) if and only if both

$$\left[\int_a^b f(t)\,dt, e \right] = r \int_a^b \|f(t)\|\,dt \qquad (6.127)$$

and

$$\left[\int_a^b f(t)\,dt, e \right] = \left\| \int_a^b f(t)\,dt \right\|. \qquad (6.128)$$

The proof follows from (6.116) for the continuous linear functional $F(x) = [x, e]$, $x \in X$, and we omit the details.

The following particular case which provides more information for the equality case may be stated [75]:

Let $(X, \|\cdot\|)$ be a strictly convex Banach space, equipped with a semi-inner product $[\cdot, \cdot]$ which generates the norm $\|\cdot\|$ and $e \in X$, with $\|e\| = 1$. If $f : [a, b] \to X$ is Bochner integrable on $[a, b]$ and there exists a $r \geq 0$ such that (6.125) holds true, then (6.126) is valid. Equality holds in (6.126) if and only if

$$\int_a^b f(t)\,dt = r \left(\int_a^b \|f(t)\|\,dt \right) e. \qquad (6.129)$$

PROOF   If (6.129) holds true, then, obviously

$$\left\| \int_a^b f(t)\,dt \right\| = r \left(\int_a^b \|f(t)\|\,dt \right) \|e\| = r \int_a^b \|f(t)\|\,dt,$$

which is the equality case in (6.126).

Conversely, if equality holds in (6.126), then we must have (6.127) and (6.128). Utilising (6.128) we can conclude that there exists a $\mu > 0$ such that

$$\int_a^b f(t)\,dt = \mu e. \qquad (6.130)$$

Replacing this in (6.127), we get

$$\mu \|e\|^2 = r \int_a^b \|f(t)\|\,dt,$$

giving

$$\mu = r \int_a^b \|f(t)\|\,dt. \qquad (6.131)$$

Utilising (6.130) and (6.131), we deduce (6.129) and the proof is completed.
□

It may be noted that the above proof is similar to that leading to (6.113) since the result is its integral counterpart.

6.7   Reverses in Terms of a Finite Sequence of Functionals

The following result provides a reverse for the continuous triangle inequality in terms of m functionals [75]:

Let $(X, \|\cdot\|)$ be a Banach space over the real or complex number field $\mathbb{K}$ and $F_k : X \rightarrow \mathbb{K}$, $k \in \{1, \ldots, m\}$ continuous linear functionals on X. If $f : [a, b] \rightarrow X$ is a Bochner integrable function on $[a, b]$ and there exists $r_k \geq 0$, $k \in \{1, \ldots, m\}$ with $\sum_{k=1}^m r_k > 0$ and

$$r_k \|f(t)\| \leq \operatorname{Re} F_k(f(t)) \tag{6.132}$$

for each $k \in \{1, \ldots, m\}$ and a.e. $t \in [a, b]$, then

$$\int_a^b \|f(t)\| \, dt \leq \frac{\left\|\sum_{k=1}^m F_k\right\|}{\sum_{k=1}^m r_k} \left\|\int_a^b f(t) \, dt\right\|. \tag{6.133}$$

The case of equality holds in (6.133) if both

$$\left(\sum_{k=1}^m F_k\right)\left(\int_a^b f(t) \, dt\right) = \left(\sum_{k=1}^m r_k\right)\int_a^b \|f(t)\| \, dt \tag{6.134}$$

and

$$\left(\sum_{k=1}^m F_k\right)\left(\int_a^b f(t) \, dt\right) = \left\|\sum_{k=1}^m F_k\right\| \left\|\int_a^b f(t) \, dt\right\|. \tag{6.135}$$

PROOF   Utilising the hypothesis (6.132), we have

$$
\begin{aligned}
I &:= \left|\sum_{k=1}^m F_k\left(\int_a^b f(t) \, dt\right)\right| \geq \left|\operatorname{Re}\left[\sum_{k=1}^m F_k\left(\int_a^b f(t) \, dt\right)\right]\right| \\
&\geq \operatorname{Re}\left[\sum_{k=1}^m F_k\left(\int_a^b f(t) \, dt\right)\right] = \sum_{k=1}^m\left(\int_a^b \operatorname{Re} F_k f(t) \, dt\right) \\
&\geq \left(\sum_{k=1}^m r_k\right) \cdot \int_a^b \|f(t)\| \, dt.
\end{aligned}
\tag{6.136}
$$

On the other hand, by the continuity property of F_k, $k \in \{1, \ldots, m\}$, we obviously have

$$I = \left|\left(\sum_{k=1}^m F_k\right)\left(\int_a^b f(t) \, dt\right)\right| \leq \left\|\sum_{k=1}^m F_k\right\| \left\|\int_a^b f(t) \, dt\right\|. \tag{6.137}$$

Making use of (6.136) and (6.137), we deduce (6.133).

Now, obviously, if (6.134) and (6.135) are valid, then equality holds in (6.133).

Conversely, if equality holds in (6.133), then it must hold in all the inequalities used to prove (6.133); therefore we have

$$r_k \, \|f \, (t)\| = \operatorname{Re} F_k \, (f \, (t)) \qquad (6.138)$$

for each $k \in \{1, \ldots, m\}$ and a.e. $t \in [a, b]$,

$$\operatorname{Im} \left(\sum_{k=1}^{m} F_k \right) \left(\int_a^b f \, (t) \, dt \right) = 0, \qquad (6.139)$$

$$\operatorname{Re} \left(\sum_{k=1}^{m} F_k \right) \left(\int_a^b f \, (t) \, dt \right) = \left\| \sum_{k=1}^{m} F_k \right\| \left\| \int_a^b f \, (t) \, dt \right\|. \qquad (6.140)$$

Note that, by (6.138), on integrating on $[a, b]$ and summing over $k \in \{1, \ldots, m\}$, we get

$$\operatorname{Re} \left(\sum_{k=1}^{m} F_k \right) \left(\int_a^b f \, (t) \, dt \right) = \left(\sum_{k=1}^{m} r_k \right) \int_a^b \|f \, (t)\| \, dt. \qquad (6.141)$$

Now, (6.139) and (6.141) imply (6.134), while (6.139) and (6.140) imply (6.135). Therefore the result is proved. □

The following new results may be stated as well [76]:

Let $(X, \|\cdot\|)$ be a Banach space over the real or complex number field K and $F_k : X \rightarrow K$, $k \in \{1, \ldots, m\}$ continuous linear functionals on X. Also, assume that $f : [a, b] \rightarrow X$ is a Bochner integrable function on $[a, b]$ and there exists $r_k \geq 0$, $k \in \{1, \ldots, m\}$ with $\sum_{k=1}^{m} r_k > 0$ and

$$r_k \, \|f \, (t)\| \leq \operatorname{Re} F_k \, (f \, (t))$$

for each $k \in \{1, \ldots, m\}$ and a.e. $t \in [a, b]$.

(i) If c_∞ is defined by (c_∞) on page 339, then we have the inequality

$$(1 \leq) \, \frac{\int_a^b \|f \, (t)\| \, dt}{\left\| \int_a^b f \, (t) \, dt \right\|} \leq \frac{c_\infty}{\max_{1 \leq k \leq m} \{r_k\}} \left(\leq \frac{\max_{1 \leq k \leq m} \|F_k\|}{\max_{1 \leq k \leq m} \{r_k\}} \right). \qquad (6.142)$$

Equality holds if and only if

$$\operatorname{Re} (F_k) \left(\int_a^b f \, (t) \, dt \right) = r_k \int_a^b \|f \, (t)\| \, dt$$

for each $k \in \{1, \ldots, m\}$ and

$$\max_{1 \leq k \leq m} \left[\operatorname{Re} (F_k) \left(\int_a^b f(t) \, dt \right) \right] = c_\infty \int_a^b \| f(t) \| \, dt.$$

(ii) *If $c_p, p \geq 1$ is defined by (c_p) on page 339, then we have the inequality*

$$(1 \leq) \frac{\int_a^b \| f(t) \| \, dt}{\left\| \int_a^b f(t) \, dt \right\|} \leq \frac{c_p}{\left(\sum_{k=1}^m r_k^p \right)^{\frac{1}{p}}} \left(\leq \frac{\sum_{k=1}^m \| F_k \|^p}{\sum_{k=1}^m r_k^p} \right)^{\frac{1}{p}}.$$

Equality holds if and only if

$$\operatorname{Re} (F_k) \left(\int_a^b f(t) \, dt \right) = r_k \int_a^b \| f(t) \| \, dt$$

for each $k \in \{1, \ldots, m\}$ and

$$\sum_{k=1}^m \left[\operatorname{Re} F_k \left(\int_a^b f(t) \, dt \right) \right]^p = c_p^p \left\| \int_a^b f(t) \, dt \right\|^p$$

where $p \geq 1$.

The proof is similar to that immediately above and we omit the details.

The case of Hilbert spaces, which provide a simpler condition for equality, is of interest for applications [75]:

Let $(X, \|\cdot\|)$ *be a Hilbert space over the real or complex number field $\mathbb{K}$ and $e_k \in H \backslash \{0\}$, $k \in \{1, \ldots, m\}$. If $f : [a, b] \to H$ is a Bochner integrable function and $r_k \geq 0$, $k \in \{1, \ldots, m\}$ and $\sum_{k=1}^m r_k > 0$ satisfy*

$$r_k \| f(t) \| \leq \operatorname{Re} \langle f(t), e_k \rangle \tag{6.143}$$

for each $k \in \{1, \ldots, m\}$ and for a.e. $t \in [a, b]$, then

$$\int_a^b \| f(t) \| \, dt \leq \frac{\left\| \sum_{k=1}^m e_k \right\|}{\sum_{k=1}^m r_k} \left\| \int_a^b f(t) \, dt \right\|. \tag{6.144}$$

Equality holds in (6.144) for $f \neq 0$ a.e. on $[a, b]$ if and only if

$$\int_a^b f(t) \, dt = \frac{\left(\sum_{k=1}^m r_k \right) \int_a^b \| f(t) \| \, dt}{\left\| \sum_{k=1}^m e_k \right\|^2} \sum_{k=1}^m e_k. \tag{6.145}$$

PROOF   Utilising the hypothesis (6.143) and the modulus properties, we have

$$\left| \left\langle \int_a^b f(t)\, dt, \sum_{k=1}^m e_k \right\rangle \right| \geq \left| \sum_{k=1}^m \operatorname{Re} \left\langle \int_a^b f(t)\, dt, e_k \right\rangle \right| \tag{6.146}$$

$$\geq \sum_{k=1}^m \operatorname{Re} \left\langle \int_a^b f(t)\, dt, e_k \right\rangle$$

$$= \sum_{k=1}^m \int_a^b \operatorname{Re} \left\langle f(t), e_k \right\rangle dt$$

$$\geq \left(\sum_{k=1}^m r_k \right) \int_a^b \| f(t) \|\, dt.$$

By Schwarz's inequality in Hilbert spaces applied for $\int_a^b f(t)\, dt$ and $\sum_{k=1}^m e_k$, we have

$$\left\| \int_a^b f(t)\, dt \right\| \left\| \sum_{k=1}^m e_k \right\| \geq \left| \left\langle \int_a^b f(t)\, dt, \sum_{k=1}^m e_k \right\rangle \right|. \tag{6.147}$$

Making use of (6.146) and (6.147), we deduce (6.144).

Now, if $f \neq 0$ a.e. on $[a,b]$, then $\int_a^b \| f(t) \|\, dt \neq 0$. Note also that $\sum_{k=1}^m e_k \neq 0$ from (6.146). If (6.145) is valid, then taking the norm we have

$$\left\| \int_a^b f(t)\, dt \right\| = \frac{\left(\sum_{k=1}^m r_k \right) \int_a^b \| f(t) \|\, dt}{\left\| \sum_{k=1}^m e_k \right\|^2} \left\| \sum_{k=1}^m e_k \right\|$$

$$= \frac{\sum_{k=1}^m r_k}{\left\| \sum_{k=1}^m e_k \right\|} \int_a^b \| f(t) \|\, dt,$$

i.e., the case of equality holds true in (6.144).

Conversely, if equality holds in (6.144), then it must hold in all the inequalities used to prove (6.144); therefore we have

$$\operatorname{Re} \left\langle f(t), e_k \right\rangle = r_k \| f(t) \| \tag{6.148}$$

for each $k \in \{1, \ldots, m\}$ and a.e. $t \in [a,b]$,

$$\left\| \int_a^b f(t)\, dt \right\| \left\| \sum_{k=1}^m e_k \right\| = \left| \left\langle \int_a^b f(t)\, dt, \sum_{k=1}^m e_k \right\rangle \right|, \tag{6.149}$$

and

$$\operatorname{Im} \left\langle \int_a^b f(t)\, dt, \sum_{k=1}^m e_k \right\rangle = 0. \tag{6.150}$$

From (6.148), on integrating on $[a, b]$ and summing over k from 1 to m, we get

$$\operatorname{Re}\left\langle \int_a^b f(t)\, dt, \sum_{k=1}^m e_k \right\rangle = \left(\sum_{k=1}^m r_k \right) \int_a^b \| f(t) \|\, dt, \tag{6.151}$$

and so by (6.150) and (6.151), we have

$$\left\langle \int_a^b f(t)\, dt, \sum_{k=1}^m e_k \right\rangle = \left(\sum_{k=1}^m r_k \right) \int_a^b \| f(t) \|\, dt. \tag{6.152}$$

On the other hand, by the use of the identity

$$\left\| u - \langle u, v \rangle \frac{v}{\|v\|^2} \right\|^2 = \frac{\|u\|^2 \|v\|^2 - |\langle u, v \rangle|^2}{\|v\|^2}, \quad u, v \in H, v \neq 0,$$

the relation (6.149) holds true if and only if

$$\int_a^b f(t)\, dt = \frac{\left\langle \int_a^b f(t)\, dt, \sum_{k=1}^m e_k \right\rangle}{\left\| \sum_{k=1}^m e_k \right\|} \sum_{k=1}^m e_k. \tag{6.153}$$

Finally, by (6.152) and (6.153) we deduce that (6.145) is also necessary for equality to hold in (6.144). The result is thus proven.   □

If $\{e_k\}_{k \in \{1,\dots,m\}}$ are orthogonal, then (6.144) can be replaced by

$$\int_a^b \| f(t) \|\, dt \leq \frac{\left(\sum_{k=1}^m \|e_k\|^2 \right)^{\frac{1}{2}}}{\sum_{k=1}^m r_k} \left\| \int_a^b f(t)\, dt \right\| \tag{6.154}$$

with equality if and only if

$$\int_a^b f(t)\, dt = \frac{\left(\sum_{k=1}^m r_k \right) \int_a^b \| f(t) \|\, dt}{\sum_{k=1}^m \|e_k\|^2} \sum_{k=1}^m e_k. \tag{6.155}$$

Moreover, if $\{e_k\}_{k \in \{1,\dots,m\}}$ are orthonormal, then (6.154) becomes

$$\int_a^b \| f(t) \|\, dt \leq \frac{\sqrt{m}}{\sum_{k=1}^m r_k} \left\| \int_a^b f(t)\, dt \right\| \tag{6.156}$$

with equality if and only if

$$\int_a^b f(t)\, dt = \frac{1}{m} \left(\sum_{k=1}^m r_k \right) \left(\int_a^b \| f(t) \|\, dt \right) \sum_{k=1}^m e_k. \tag{6.157}$$

The following particular case may be stated as well [75]:

Let $(H; \langle \cdot, \cdot \rangle)$ be a Hilbert space over the real or complex number field $\mathbb{K}$ and $e_k \in H \backslash \{0\}$, $k \in \{1, \ldots, m\}$. If $f : [a, b] \rightarrow H$ is a Bochner integrable function on $[a, b]$ and $\rho_k > 0$, $k \in \{1, \ldots, m\}$ with

$$\| f(t) - e_k \| \le \rho_k < \| e_k \| \tag{6.158}$$

for each $k \in \{1, \ldots, m\}$ and a.e. $t \in [a, b]$, then

$$\int_a^b \| f(t) \| \, dt \le \frac{\left\| \sum_{k=1}^m e_k \right\|}{\sum_{k=1}^m \left(\| e_k \|^2 - \rho_k^2 \right)^{\frac{1}{2}}} \left\| \int_a^b f(t) \, dt \right\|. \tag{6.159}$$

Equality holds in (6.159) if and only if

$$\int_a^b f(t) \, dt = \frac{\sum_{k=1}^m \left(\| e_k \|^2 - \rho_k^2 \right)^{\frac{1}{2}}}{\left\| \sum_{k=1}^m e_k \right\|^2} \left(\int_a^b \| f(t) \| \, dt \right) \sum_{k=1}^m e_k. \tag{6.160}$$

Comments

If $\{e_k\}_{k \in \{1, \ldots, m\}}$ are orthogonal, then (6.159) becomes

$$\int_a^b \| f(t) \| \, dt \le \frac{\left(\sum_{k=1}^m \| e_k \|^2 \right)^{\frac{1}{2}}}{\sum_{k=1}^m \left(\| e_k \|^2 - \rho_k^2 \right)^{\frac{1}{2}}} \left\| \int_a^b f(t) \, dt \right\| \tag{6.161}$$

with equality if and only if

$$\int_a^b f(t) \, dt = \frac{\sum_{k=1}^m \left(\| e_k \|^2 - \rho_k^2 \right)^{\frac{1}{2}}}{\sum_{k=1}^m \| e_k \|^2} \left(\int_a^b \| f(t) \| \, dt \right) \sum_{k=1}^m e_k. \tag{6.162}$$

Moreover, if $\{e_k\}_{k \in \{1, \ldots, m\}}$ is assumed to be orthonormal and

$$\| f(t) - e_k \| \le \rho_k \quad \text{for a.e. } t \in [a, b],$$

where $\rho_k \in [0, 1)$, $k \in \{1, \ldots, m\}$, then

$$\int_a^b \| f(t) \| \, dt \le \frac{\sqrt{m}}{\sum_{k=1}^m \left(1 - \rho_k^2 \right)^{\frac{1}{2}}} \left\| \int_a^b f(t) \, dt \right\| \tag{6.163}$$

with equality if and only if

$$\int_a^b f(t) \, dt = \frac{\sum_{k=1}^m \left(1 - \rho_k^2 \right)^{\frac{1}{2}}}{m} \left(\int_a^b \| f(t) \| \, dt \right) \sum_{k=1}^m e_k. \tag{6.164}$$

Finally, we may state the following special case as well [75]:

Let $(H; \langle \cdot, \cdot \rangle)$ be a Hilbert space over the real or complex number field $\mathbb{K}$ and $e_k \in H \setminus \{0\}$, $k \in \{1, \ldots, m\}$. If $f : [a, b] \to H$ is a Bochner integrable function on $[a, b]$ and $M_k \geq \mu_k > 0$, $k \in \{1, \ldots, m\}$ are such that either

$$\operatorname{Re} \langle M_k e_k - f(t), f(t) - \mu_k e_k \rangle \geq 0 \tag{6.165}$$

or, equivalently,

$$\left\| f(t) - \frac{M_k + \mu_k}{2} e_k \right\| \leq \frac{1}{2} (M_k - \mu_k) \|e_k\| \tag{6.166}$$

for each $k \in \{1, \ldots, m\}$ and a.e. $t \in [a, b]$, then

$$\int_a^b \|f(t)\| \, dt \leq \frac{\left\| \sum_{k=1}^m e_k \right\|}{\sum_{k=1}^m \frac{2 \cdot \sqrt{\mu_k M_k}}{\mu_k + M_k} \|e_k\|} \left\| \int_a^b f(t) \, dt \right\|. \tag{6.167}$$

Equality holds if and only if

$$\int_a^b f(t) \, dt = \frac{\sum_{k=1}^m \frac{2 \cdot \sqrt{\mu_k M_k}}{\mu_k + M_k} \|e_k\|}{\left\| \sum_{k=1}^m e_k \right\|^2} \left(\int_a^b \|f(t)\| \, dt \right) \cdot \sum_{k=1}^m e_k.$$

6.8 Generalisations of the Hermite-Hadamard Inequality for Isotonic Linear Functionals

In this section we shall give some generalisations of the Hermite-Hadamard inequality for isotonic linear functionals.

Let E be a nonempty set and let L be a linear class of real-valued functions $g : E \to \mathbb{R}$ having the properties:

L1: $f, g \in L$ implies $(af + bg) \in L$ for all $a, b \in \mathbb{R}$;

L2: $\mathbb{I} \in L$, that is, if $f(t) = 1$ $(t \in E)$ then $f \in L$.

We also consider isotonic linear functionals $A : L \to \mathbb{R}$. That is, we suppose:

A1: $A(af + bg) = aA(f) + bA(g)$ for $f, g \in L$, $a, b \in \mathbb{R}$;

A2: $f \in L, f(t) \geq 0$ on E implies $A(f) \geq 0$.

We note that common examples of such isotonic linear functionals A are given by

$$A(g) = \int_E g \, d\mu \text{ or } A(g) = \sum_{k \in E} p_k g_k,$$

where μ is a positive measure on E in the first case and E is a subset of the natural numbers $\mathbb{N}$, in the second $(p_k \geq 0, k \in E)$.

We shall use the following result which is well known in the literature as *Jessen's Inequality* (see, for example, Pečarić and Dragomir [152] or Dragomir [49]):

Let L satisfy properties L1 and L2 on a nonempty set E and suppose ϕ is a convex function on an interval $I \subseteq \mathbb{R}$. If A is any isotonic functional with $A(\mathbb{I}) = 1$, then, for all $g \in L$ such that $\phi(g) \in L$, we have $A(g) \in I$ and

$$\phi(A(g)) \leq A(\phi(g)). \tag{6.168}$$

The following result holds [152]:

Let X be a real linear space and C its convex subset. Then the following statements are equivalent for a mapping $F : X \to \mathbb{R}$:
(i) *f is convex on C;*
(ii) *for all $x, y \in C$ the mapping $g_{x,y} : [0,1] \to \mathbb{R}$, $g_{x,y}(t) := f(tx + (1-t)y)$ is convex on $[0,1]$.*

PROOF   Assume that (i) holds. Suppose $x, y \in C$ and let $t_1, t_2 \in [0,1]$, $\lambda_1, \lambda_2 \geq 0$ with $\lambda_1 + \lambda_2 = 1$. Then

$$\begin{aligned}
g_{x,y}(\lambda_1 t_1 + \lambda_2 t_2) &= f[(\lambda_1 t_1 + \lambda_2 t_2)x + (1 - \lambda_1 t_1 - \lambda_2 t_2)y] \\
&= f[(\lambda_1 t_1 + \lambda_2 t_2)x + [\lambda_1(1 - t_1) + \lambda_2(1 - t_2)]y] \\
&\leq \lambda_1 f(t_1 x + (1 - t_1)y) + \lambda_2 f(t_2 x + (1 - t_2)y).
\end{aligned}$$

That is, $g_{x,y}$ is convex on $[0,1]$.

Conversely, assume that (ii) is true. Let $x, y \in C$ and $\lambda_1, \lambda_2 \geq 0$ with $\lambda_1 + \lambda_2 = 1$. Then we have:

$$\begin{aligned}
f(\lambda_1 x + \lambda_2 y) &= f(\lambda_1 x + (1 - \lambda_1)y) = g_{x,y}(\lambda_1 \cdot 1 + \lambda_2 \cdot 0) \\
&\leq \lambda_1 g_{x,y}(1) + \lambda_2 g_{x,y}(0) = \lambda_1 f(x) + \lambda_2 f(y),
\end{aligned}$$

that is, f is convex on C. This completes the proof.   ▯

The following generalisation of Hermite-Hadamard's inequality for isotonic linear functionals holds [152]:

Let $f : C \subseteq X \to \mathbb{R}$ be a convex function on C, L and A satisfy conditions L1, L2 and A1, A2, and $h : E \to \mathbb{R}$, $0 \leq h(t) \leq 1$, $h \in L$ is such that $g_{x,y} \circ h \in L$ for x, y given in C. If $A(\mathbb{I}) = 1$, then we have the inequality

$$f(A(h)x + (1 - A(h))y) \leq A[f(hx + (\mathbb{I} - h)y)] \tag{6.169}$$
$$\leq A(h)f(x) + (1 - A(h))f(y).$$

PROOF Consider the mapping $g_{x,y} : [0, 1] \to \mathbb{R}$, $g_{x,y}(s) := f(sx + (1 - s)y)$. Then, by the above result, we have that $g_{x,y}$ is convex on $[0, 1]$. For all $t \in E$ we have:

$$g_{x,y}(h(t) \cdot 1 + (1 - h(t)) \cdot 0) \le h(t) g_{x,y}(1) + (1 - h(t)) g_{x,y}(0),$$

which implies that

$$A(g_{x,y}(h)) \le A(h) g_{x,y}(1) + (1 - A(h)) g_{x,y}(0),$$

that is,

$$A[f(hx + (\mathbb{I} - h)y)] \le A(h) f(x) + (1 - A(h)) f(y).$$

On the other hand, by Jessen's inequality, applied for $g_{x,y}$ we have:

$$g_{x,y}(A(h)) \le A(g_{x,y}(h)),$$

which gives:

$$f(A(h)x + (1 - A(h))y) \le A[f(hx + (\mathbb{I} - h)y)]$$

and the proof is completed. □

If $h : E \to [0, 1]$ is such that $A(h) = \frac{1}{2}$, we get from the inequality (6.169) that

$$f\left(\frac{x+y}{2}\right) \le A[f(hx + (\mathbb{I} - h)y)] \le \frac{f(x) + f(y)}{2}, \tag{6.170}$$

for all x, y in C.

Comments
(a) If $A = \int_0^1$, $E = [0, 1]$, $h(t) = t$, $C = [x, y] \subset \mathbb{R}$, then we recapture from (6.169) the classical inequality of Hermite-Hadamard, because

$$\int_0^1 f(tx + (1 - t)y)\, dt = \frac{1}{y - x} \int_x^y f(t)\, dt.$$

(b) If $A = \frac{2}{\pi} \int_0^{\frac{\pi}{2}}$, $E = \left[0, \frac{\pi}{2}\right]$, $h(t) = \sin^2 t$, $C \subseteq \mathbb{R}$, then, from (6.170) we get

$$f\left(\frac{x+y}{2}\right) \le \frac{2}{\pi} \int_0^{\frac{\pi}{2}} f(x \sin^2 t + y \cos^2 t)\, dt \le \frac{f(x) + f(y)}{2},$$

$x, y \in C$, which is a new inequality of Hadamard's type. This is as $\frac{2}{\pi} \int_0^{\frac{\pi}{2}} \sin^2 t\, dt = \frac{1}{2}$.

(c) If $A = \int_0^1$, $E = [0, 1]$, $h(t) = t$, and X is a normed linear space, then (6.170) implies that for $f(x) = \|x\|^p$, $x \in X$, $p \ge 1$:

$$\left\|\frac{x+y}{2}\right\|^p \le \int_0^1 \|tx + (1 - t)y\|^p\, dt \le \frac{\|x\|^p + \|y\|^p}{2}$$

for all $x, y \in X$.

(d) If $A = \frac{1}{n} \sum_{i=1}^{n}$, $E = \{1, \ldots, n\}$, $\sum_{i=1}^{n} t_i = \frac{n}{2}$, $C \subseteq \mathbb{R}$, $n \geq 1$, then from (6.170) we also have

$$f\left(\frac{x+y}{2}\right) \leq \frac{1}{n} \sum_{i=1}^{n} f\left(t_i x + (1 - t_i) y\right) \leq \frac{f(x) + f(y)}{2}$$

for all $x, y \in C$, which is a discrete variant of the Hermite-Hadamard inequality.

6.9   A Symmetric Generalisation

To give a symmetric generalisation of the Hermite-Hadamard inequality, we present the following result which is interesting in itself [49]:

Let X be a real linear space and C be its convex subset. If $f : C \to \mathbb{R}$ is convex on C, then for all x, y in C the mapping $g_{x,y} : [0, 1] \to \mathbb{R}$ given by

$$g_{x,y}(t) := \frac{1}{2} \left[f\left(tx + (1 - t) y\right) + f\left((1 - t) x + ty\right) \right]$$

is also convex on $[0, 1]$. In addition, we have the inequality

$$f\left(\frac{x+y}{2}\right) \leq g_{x,y}(t) \leq \frac{f(x) + f(y)}{2} \tag{6.171}$$

for all $x, y \in C$ and $t \in [0, 1]$.

PROOF   Suppose $x, y \in C$ and let $t_1, t_2 \in [0, 1]$, $\alpha, \beta \geq 0$, and $\alpha + \beta = 1$. Then

$$
\begin{aligned}
g_{x,y}(\alpha t_1 + \beta t_2) &= \frac{1}{2} \left[f\left((\alpha t_1 + \beta t_2) x + (1 - \alpha t_1 - \beta t_2) y\right) \right. \\
&\quad \left. + f\left((1 - \alpha t_1 - \beta t_2) x + (\alpha t_1 + \beta t_2) y\right) \right] \\
&= \frac{1}{2} \left(f\left[\alpha\left(t_1 x + (1 - t_1) y\right) + \beta\left(t_2 x + (1 - t_2) y\right)\right] \right. \\
&\quad \left. + f\left[\alpha\left((1 - t_1) + t_1 xy\right) + \beta\left((1 - t_2) x + t_2 y\right)\right] \right) \\
&\leq \frac{1}{2} \left(\alpha f\left[t_1 x + (1 - t_1) y\right] + \beta f\left[t_2 x + (1 - t_2) y\right] \right. \\
&\quad \left. + \alpha f\left[(1 - t_1) + t_1 xy\right] + \beta f\left[(1 - t_2) x + t_2 y\right] \right) \\
&= \alpha g_{x,y}(t_1) + \beta g_{x,y}(t_2),
\end{aligned}
$$

which shows that $g_{x,y}$ is convex on $[0, 1]$.

By the convexity of f we can state that

$$g_{x,y}(t) \geq f\left[\frac{1}{2}(tx + (1-t)y + (1-t)x + ty)\right] = f\left(\frac{x+y}{2}\right).$$

In addition,

$$g_{x,y}(t) \leq \frac{1}{2}[tf(x) + (1-t)f(y) + (1-t)f(x) + tf(y)] \leq \frac{f(x) + f(y)}{2}$$

for all t in $[0,1]$, which completes the proof. ⬜

By the inequality (6.171) we deduce the bounds

$$\sup_{t \in [0,1]} g_{x,y}(t) = \frac{f(x) + f(y)}{2} \quad \text{and} \quad \inf_{t \in [0,1]} g_{x,y}(t) = f\left(\frac{x+y}{2}\right)$$

for all x, y in C.

The following symmetric generalisation of the Hermite-Hadamard inequality holds [49]:

Let $f : C \subseteq X \to \mathbb{R}$ be a convex function on the convex set C, where L and A satisfy the conditions L1, L2 and A1, A2. Further, suppose that $h : E \to \mathbb{R}$, $0 \leq h(t) \leq 1$ $(t \in E)$, and $h \in L$ is such that $f(hx + (1-h)y)$, $f((1-h)x + hy)$ belong to L for x, y fixed in C. If $A(\mathbb{I}) = 1$, then we have the inequality:

$$f\left(\frac{x+y}{2}\right) \tag{6.172}$$
$$\leq \frac{1}{2}[f(A(h)x + (1 - A(h))y) + f((1 - A(h))x + A(h)y)]$$
$$\leq \frac{1}{2}(A[f(hx + (\mathbb{I} - h)y)] + A[f((\mathbb{I} - h)x + hy)])$$
$$\leq \frac{f(x) + f(y)}{2}.$$

PROOF Let us consider the mapping $g_{x,y} : [0,1] \to \mathbb{R}$ given above. Then, by the above result we know that $g_{x,y}$ is convex on $[0,1]$.
Applying Jensen's inequality to the mapping $g_{x,y}$ we get:

$$g_{x,y}(A(h)) \leq A(g_{x,y}(h)).$$

However,

$$g_{x,y}(A(h)) = \frac{1}{2}[f(A(h)x + (1 - A(h))y) + f((1 - A(h))x + A(h)y)]$$

and

$$A(g_{x,y}(h)) = \frac{1}{2}(A[f(hx + (\mathbb{I} - h)y)] + A[f((\mathbb{I} - h)x + hy)])$$

and thus proves the second inequality in (6.172).

To prove the first inequality in (6.172) we observe, by (6.171), that

$$f\left(\frac{x+y}{2}\right) \le g_{x,y}\left(A\left(h\right)\right) \text{ as } 0 \le A\left(h\right) \le 1,$$

which is exactly the desired outcome.

Finally, by the convexity of f, we observe that

$$\frac{1}{2}\left[f\left(hx + \left(\mathbb{I} - h\right)y\right) + f\left(\left(\mathbb{I} - h\right)x + hy\right)\right] \le \frac{f\left(x\right) + f\left(y\right)}{2}$$

on E.

By applying the functional A, since $A\left(\mathbb{I}\right) = 1$, we obtain the last part of (6.172). $\quad\square$

Note that, if we choose $A = \int_0^1, E = [0,1], h\left(t\right) = t, C = [x,y] \subset \mathbb{R}$, we recapture, by (6.172), the Hermite-Hadamard inequality for integrals. This is because

$$\int_0^1 f\left(tx + \left(1 - t\right)y\right)dt = \int_0^1 f\left(\left(1 - t\right)x + ty\right)dt = \frac{1}{y - x}\int_x^y f\left(t\right)dt.$$

Comments

(a) Let $h : [0,1] \to [0,1]$ be a Riemann integrable function on $[0,1]$ and $p \ge 1$. Then, for all x, y vectors in the normed space $\left(X; \|\cdot\|\right)$ we have the inequality:

$$\left\|\frac{x+y}{2}\right\|^p \le \frac{1}{2}\left[\left\|\left(1 - \int_0^1 h\left(t\right)dt\right)x + \left(\int_0^1 h\left(t\right)dt\right)y\right\|^p \right.$$
$$\left. + \left\|\left(\int_0^1 h\left(t\right)dt\right)x + \left(1 - \int_0^1 h\left(t\right)dt\right)y\right\|^p\right]$$
$$\le \frac{1}{2}\left[\int_0^1 \left\|\left(h\left(t\right)\right)x + \left(1 - h\left(t\right)\right)y\right\|^p dt\right.$$
$$\left. + \int_0^1 \left\|\left(1 - h\left(t\right)\right)x + \left(h\left(t\right)\right)y\right\|^p dt\right]$$
$$\le \frac{\|x\|^p + \|y\|^p}{2}.$$

If we choose $h\left(t\right) = t$, we get the inequality obtained above,

$$\left\|\frac{x+y}{2}\right\|^p \le \int_0^1 \|tx + \left(1 - t\right)y\|^p dt \le \frac{\|x\|^p + \|y\|^p}{2},$$

for all $x, y \in X$.

(b) Let $f : C \subseteq X \to \mathbb{R}$ be a convex function on the convex set C of a linear space X, $t_i \in [0,1]$ $\left(i = \overline{1,n}\right)$. Then we have the inequality

$$f\left(\frac{x+y}{2}\right)$$

$$\leq \frac{1}{2}\left[f\left(\frac{1}{n}\sum_{i=1}^{n} t_i x + \frac{1}{n}\sum_{i=1}^{n}(1-t_i)\,y\right) + f\left(\frac{1}{n}\sum_{i=1}^{n}(1-t_i)\,x + \frac{1}{n}\sum_{i=1}^{n} t_i y\right)\right]$$

$$\leq \frac{1}{2n}\left[\sum_{i=1}^{n} f\left(t_i x + (1-t_i)\,y\right) + \sum_{i=1}^{n} f\left((1-t_i)\,x + t_i y\right)\right]$$

$$\leq \frac{f(x)+f(y)}{2}.$$

If we put in the above inequality $t_i = \sin^2 \alpha_i$, $\alpha_i \in \mathbb{R}$ $(i = \overline{1, n})$, then we have:

$$f\left(\frac{x+y}{2}\right) \leq \frac{1}{2}\left(f\left[\left(\frac{1}{n}\sum_{i=1}^{n} \sin^2 \alpha_i\right)x + \left(\frac{1}{n}\sum_{i=1}^{n}\cos^2 \alpha_i\right)y\right]\right.$$

$$\left. + f\left[\left(\frac{1}{n}\sum_{i=1}^{n}\cos^2 \alpha_i\right)x + \left(\frac{1}{n}\sum_{i=1}^{n}\sin^2 \alpha_i\right)y\right]\right)$$

$$\leq \frac{1}{2n}\sum_{i=1}^{n}\left(f\left[(\sin^2 \alpha_i)\,x + (\cos^2 \alpha_i)\,y\right]\right.$$

$$\left. + f\left[(\cos^2 \alpha_i)\,x + (\sin^2 \alpha_i)\,y\right]\right)$$

$$\leq \frac{f(x)+f(y)}{2}.$$

(c) For $x, y \geq 0$, let us consider the weighted means:

$$A_\alpha(x, y) := \alpha x + (1-\alpha)\,y$$

and

$$G_\alpha(x, y) := x^\alpha y^{1-\alpha}$$

where $\alpha \in [0, 1]$.

If $h : [0, 1] \to [0, 1]$ is an integrable mapping on $[0, 1]$, then, by (6.169) we have the inequality

$$A_{\int_0^1 h(t)dt}(x, y) \geq \exp\left[\int_0^1 \ln\left[A_{h(t)}(x, y)\right]dt\right] \geq G_{\int_0^1 h(t)dt}(x, y). \quad (6.173)$$

If $\int_0^1 h(t)\,dt = \frac{1}{2}$, we get

$$A(x, y) \geq \exp\left[\int_0^1 \ln\left[A_{h(t)}(x, y)\right]dt\right] \geq G(x, y), \quad (6.174)$$

which is a refinement of the classic arithmetic mean–geometric mean $(A.\text{–}G.)$ inequality.

In particular, if in this inequality we choose $h(t) = t$, $t \in [0,1]$, we recapture the well-known result for the identric mean:

$$A(x,y) \geq I(x,y) \geq G(x,y).$$

Now, if we use (6.172), we can state the following weighted refinement of the classical A.–G. inequality:

$$A(x,y) \geq G\left(A_{\int_0^1 h(t)dt}(x,y), A_{\int_0^1 h(t)dt}(x,y)\right) \qquad (6.175)$$

$$\geq \exp\left[\int_0^1 \ln\left[G\left(A_{h(t)}(x,y), A_{h(t)}(y,x)\right)\right] dt\right] \geq G(x,y).$$

If $\int_0^1 h(t)\,dt = \frac{1}{2}$, then, by (6.175) we get the following refinement of the A.–G. inequality:

$$A(x,y) \geq \exp\left[\int_0^1 \ln\left[G\left(A_{h(t)}(x,y), A_{h(t)}(y,x)\right)\right] dt\right] \geq G(x,y). \qquad (6.176)$$

If, in the above inequality we choose $h(t) = t$ for $t \in [0,1]$, then we get the inequality

$$A(x,y) \geq \exp\left[\int_0^1 \ln\left[G\left(A_t(x,y), A_t(y,x)\right)\right] dt\right] \geq G(x,y). \qquad (6.177)$$

(d) Some discrete refinements of arithmetic mean–geometric mean inequality can also be done.

If $\bar{x} = (x_1, \ldots, x_n) \in \mathbb{R}_+^n$, we can denote by $G_n(\bar{x})$ the geometric mean of $\bar{x}$, i.e., $G_n(\bar{x}) := \left(\prod_{i=1}^n x_i\right)^{\frac{1}{n}}$.

If $\bar{t} = (t_1, \ldots, t_n) \in [0,1]^n$, we can define the vector in $\mathbb{R}_+^n$ given by

$$\bar{A}_{\bar{t}}(x,y) := (A_{t_1}(x,y), \ldots, A_{t_n}(x,y))$$

where $x, y \geq 0$.

By applying (6.169) for the convex mapping $f(x) = -\ln x$ and the linear functional $A := \frac{1}{n}\sum_{i=1}^n t_i$, we get the inequality

$$A_{\tilde{t}}(x,y) \geq G_n\left(\bar{A}_{\bar{t}}(x,y)\right) \geq G_{\tilde{t}}(x,y) \qquad (6.178)$$

where $\tilde{t} := \frac{1}{n}\sum_{i=1}^n t_i \in [0,1]$ and $x, y \geq 0$.

If we choose t_i so that $\tilde{t} = \frac{1}{2}$, we get

$$A(x,y) \geq G_n\left(\bar{A}_{\bar{t}}(x,y)\right) \geq G(x,y) \qquad (6.179)$$

which is a discrete refinement of the classical arithmetic mean–geometric mean inequality.

In addition, if we use (6.172), we can state that

$$A(x,y) \geq G_n\left(A_{\bar{t}}(x,y), A_{\bar{t}}(y,x)\right) \qquad (6.180)$$

$$\geq G\left(G_n\left(\bar{A}_{\bar{t}}(x,y)\right), G_n\left(\bar{A}_{\bar{t}}(y,x)\right)\right) \geq G(x,y),$$

which is another refinement of the arithmetic mean–geometric mean inequality.

6.10 Generalisations of the Hermite-Hadamard Inequality for Isotonic Sublinear Functionals

Recall the isotonic linear functional A as defined in Section 6.8. The mapping A is said to be *normalised* if

(A3) $A(1) = 1$.

Isotonic, that is, order-preserving, linear functionals are natural objects in analysis which enjoy a number of convenient properties. Thus, they provide, for example, Jessen's inequality, which is a functional form of Jensen's inequality and a functional Hermite-Hadamard inequality.

In this section we show that these ideas carry over to a sublinear setting [104]. Let E be a nonempty set and K a class of real-valued functions $g : E \to \mathbb{R}$ having the properties

(K1) $1 \in K$;

(K2) $f, g \in K$ imply $f + g \in K$;

(K3) $f \in K$ implies $\alpha \cdot 1 + \beta \cdot f \in K$ for all $\alpha, \beta \in \mathbb{R}$.

We define the family of *isotonic sublinear functionals* $S : K \to \mathbb{R}$ by the properties

(S1) $S(f + g) \le S(f) + S(g)$ for all $f, g \in K$;

(S2) $S(\alpha f) = \alpha S(f)$ for all $\alpha \ge 0$ and $f \in K$;

(S3) If $f \ge g, f, g \in K$, then $S(f) \ge S(g)$.
　　An isotonic sublinear functional is said to be *normalised* if

(S4) $S(1) = 1$
　　and *totally normalised* if, in addition,

(S5) $S(-1) = -1$.

We note some immediate consequences. From (K2) and (K3), $f - g$ belongs to K whenever $f, g \in K$, so that from (S1)

$$S(f) = S((f - g) + g) \le S(f - g) + S(g)$$

and hence

(S6) $S(f - g) \ge S(f) - S(g)$ if $f, g \in K$.
　　Moreover, if S is a totally normalised isotonic sublinear functional, then we have

(S7) $S(\alpha \cdot \mathbf{1}) = \alpha$ for all $\alpha \in \mathbb{R}$

and

(S8) $S(f + \alpha \cdot \mathbf{1}) = S(f) + \alpha$ for all $\alpha \in \mathbb{R}$.

Equation (S7) is immediate from (S2) when $\alpha \geq 0$. When $\alpha < 0$ we have

$$S(\alpha \cdot \mathbf{1}) = S((-\alpha) \cdot (-\mathbf{1})) = (-\alpha) S(-\mathbf{1}) = (-\alpha)(-1) = \alpha.$$

Also, by (S6) and (S7), we have for $\alpha \in \mathbb{R}$

$$S(f - \alpha \cdot \mathbf{1}) \geq S(f) + S(-\alpha \cdot \mathbf{1}) = S(f) - \alpha,$$

which by (S1) and (S7)

$$S(f - \alpha \cdot \mathbf{1}) \leq S(f) + S(-\alpha \cdot \mathbf{1}) = S(f) - \alpha,$$

so that

$$S(f - \alpha \cdot \mathbf{1}) = S(f) - \alpha.$$

Since this holds for all $\alpha \in \mathbb{R}$, we have (S8).

It is clear that every normalised isotonic linear functional is a totally normalised isotonic sublinear functional.

In what follows, we shall present some simple examples of sublinear functionals that are not linear.

Let $A_1, \ldots, A_n : L \to \mathbb{R}$ be normalised isotonic linear functionals and $p_{i,j} \in \mathbb{R}$ $(i, j \in \{1, \ldots, n\})$ such that

$$p_{i,j} \geq 0 \text{ for all } i, j \in \{1, \ldots, n\} \text{ and } \sum_{i=1}^{n} p_{i,j} = 1 \text{ for all } j \in \{1, \ldots, n\}.$$

Define the mapping $S : L \to \mathbb{R}$ by

$$S(f) = \max_{1 \leq j \leq n} \left\{ \sum_{i=1}^{n} p_{i,j} A_i(f) \right\}.$$

Then S is a totally normalised isotonic sublinear functional on L. As particular cases of this functional, we have the mappings

$$S_0(f) := \max_{1 \leq j \leq n} \{ A_i(f) \}$$

and

$$S_Q(f) := \max_{1 \leq j \leq n} \left\{ \frac{1}{Q_j} \sum_{i=1}^{n} q_i A_i(f) \right\}$$

where $q_i \geq 0$ for all $i \in \{1, \ldots, n\}$ and $Q_j > 0$ for $j = 1, \ldots, n$. If we choose $q_i = 1$ for all $i \in \{1, \ldots, n\}$, we also have that

$$S_1(f) := \max_{1 \leq j \leq n} \left\{ \frac{1}{j} \sum_{i=1}^{n} A_i(f) \right\}$$

is a totally normalised isotonic sublinear functional on L.

If $A_1, \ldots, A_n$ are as above and $A : L \to \mathbb{R}$ is also a normalised isotonic linear functional, then the mapping

$$S_A (f) := \frac{1}{P_n} \sum_{i=1}^{n} p_i \max \{A (f), A_i (f)\},$$

where $p_i \geq 0$ $(1 \leq i \leq n)$ with $P_n = \sum_{i=1}^{n} p_i > 0$, is also a totally normalised isotonic sublinear functional.

The following provide concrete examples.

Suppose $x = (x_1, \ldots, x_n)$ and $y = (y_1, \ldots, y_n)$ are points in $\mathbb{R}^n$. Then the mappings

$$S (x) := \max_{1 \leq j \leq n} \left\{ \sum_{i=1}^{n} p_{i,j} x_i \right\},$$

where $p_i \geq 0$ and $\sum_{i=1}^{n} p_{i,j} = 1$ for $j \in \{1, \ldots, n\}$,

$$S_0 (x) := \max_{1 \leq i \leq n} \{x_i\}$$

and

$$S_Q (x) := \max_{1 \leq j \leq n} \left\{ \frac{1}{Q_j} \sum_{i=1}^{j} q_i A_x \right\}$$

where $q_i \geq 0$ and $Q_j > 0$ for all $i, j \in \{1, \ldots, n\}$, are totally normalised isotonic sublinear functionals on $\mathbb{R}^n$.

Suppose $i_0 \in \{1, .., n\}$ is fixed and $p_i \geq 0$ for all $i \in \{1, \ldots, n\}$, with $P_n > 0$. Then the mapping

$$S_{i0} (x) := \frac{1}{P_n} \sum_{i=1}^{n} p_i \max \{x_{i0}, x_i\}$$

is also totally normalised.

Denote by $R [a, b]$ the linear space of Riemann integrable functions on $[a, b]$. Suppose that $p \in R [a, b]$ with $p (t) > 0$ for all $t \in [a, b]$. Then the mappings

$$S_p (f) := \sup_{x \in (a,b]} \left[\frac{\int_a^x p (t) f (t) \, dt}{\int_a^x p (t) \, dt} \right]$$

and

$$s_1 (f) := \sup_{x \in (a,b]} \left[\frac{1}{x - a} \int_a^x f (t) \, dt \right]$$

are totally normalised isotonic sublinear functionals on $R [a, b]$.

If $C \in [a, b]$, then

$$S_{c,p} (f) := \frac{\int_a^b p (t) \max (f (c), f (t)) \, dt}{\int_a^b p (t) \, dt}$$

and

$$s_c(f) := \frac{1}{b-a} \int_a^b \max(f(c), f(t)) \, dt$$

are also totally normalised on $R[a, b]$.

We can give the following generalisation of the well-known Jensen's inequality due to Dragomir, Pearce, and Pečarić [104]:

Let $\phi : [\alpha, \beta] \subset \mathbb{R} \to \mathbb{R}$ be a continuous convex function and $f : E \to [\alpha, \beta]$ such that $f, \phi \circ f \in K$. Then, if S is a totally normalised isotonic sublinear functional on K, we have $S(f) \in [\alpha, \beta]$ and

$$S(\phi \circ f) \geq \phi(S(f)). \tag{6.181}$$

PROOF By (S3) and (S7), $\alpha \cdot \mathbf{1} \leq f \leq \beta \cdot \mathbf{1}$ implies

$$\alpha = S(\alpha \cdot \mathbf{1}) \leq S(f) \leq S(\beta \cdot \mathbf{1}) = \beta$$

so that $S(f) \in [\alpha, \beta]$.

Set $l_1(x) = x$ for all $x \in [\alpha, \beta]$. For an arbitrary but fixed $q > 0$, we have by convexity of ϕ that there exist real numbers $u, v \in \mathbb{R}$ such that

(i) $p \leq \phi$ and

(ii) $p(S(f)) \geq \phi(S(f)) - q$
  where
$$p(t) = u \cdot \mathbf{1} + v \cdot l_1(t).$$

If $\alpha < S(f) < \beta$ or ϕ has a finite derivative in $[\alpha, \beta]$, then we can replace (ii) by $p(S(f)) = \phi(S(f))$. Now (i) implies $p \circ f \leq \phi \circ f$. Hence, by (S3)

$$S(\phi \circ f) \geq S(p \circ f) = S(u \cdot \mathbf{1} + v \cdot f).$$

If $v \geq 0$ by (S8) and (S2), then we have

$$S(u \cdot \mathbf{1} + v \cdot f) = u + vS(f) = p(S(f)),$$

while if $v < 0$ by (S6), (S7), and (S2), then we have

$$S(u \cdot \mathbf{1} + v \cdot f) = S(u \cdot \mathbf{1} - |v| f) \geq u - S(|v| f)$$
$$= u - |v| S(f) = u + vS(f) = p(S(f)).$$

Therefore, we have in either case

$$S(\phi \circ f) \geq \phi(S(f)) - q.$$

Since q is arbitrary, the proof is complete.   ☐

If $S = A$, a normalised isotonic linear functional on L, then (6.108) becomes the well-known Jensen's inequality.

The following generalisations of Jensen's inequality for isotonic linear functionals also hold:

Let $A_1, \ldots, A_n : L \to \mathbb{R}$ be normalised isotonic linear functionals and $p_{i,j} \in \mathbb{R}$ be such that:

$$p_{i,j} \geq 0 \text{ and } \sum_{i=1}^{n} p_{i,j} = 1 \text{ for all } i, j \in \{1, \ldots, n\}.$$

If $\phi : [\alpha, \beta] \to \mathbb{R}$ is convex and $f : E \to [\alpha, \beta]$ is such that $f, \phi \circ f \in L$, then

$$\max_{1 \leq j \leq n} \left\{ \sum_{i=1}^{n} p_{i,j} A_i (\phi \circ f) \right\} \geq \phi \left(\max_{1 \leq j \leq n} \left\{ \sum_{i=1}^{n} p_{i,j} A_i (f) \right\} \right).$$

The proof follows by the above result applied for the following mapping

$$S(f) := \max_{1 \leq j \leq n} \left\{ \sum_{i=1}^{n} p_{i,j} A_i (f) \right\},$$

which is a totally normalised isotonic sublinear functional on L.

If $A_1, \ldots, A_n, \phi$ and f are as above, then

$$\max_{1 \leq j \leq n} \{ A_i (\phi \circ f) \} \geq \phi \left(\max_{1 \leq j \leq n} \{ A_i (f) \} \right)$$

and

$$\max_{1 \leq j \leq n} \left\{ \frac{1}{Q_j} \sum_{i=1}^{j} q_i A_i (\phi \circ f) \right\} \geq \phi \left(\max_{1 \leq j \leq n} \left\{ \frac{1}{Q_j} \sum_{i=1}^{j} q_i A_i (f) \right\} \right)$$

where $q_i \geq 0$ with $Q_j > 0$ for all $i, j \in \{1, \ldots, n\}$.

The following result may be stated as well:

If $A_1, \ldots, A_n, \phi$ and f are as shown, $p_i \geq 0, i \in \{1, \ldots, n\}, P_n > 0$ and $A : L \to \mathbb{R}$ are also normalised isotonic linear functionals, then we have the inequality

$$\frac{1}{P_n} \sum_{i=1}^{n} p_i \max \{ A (\phi \circ f), A_i (\phi \circ f) \} \geq \phi \left(\frac{1}{P_n} \sum_{i=1}^{n} p_i \max \{ A (f), A_i (f) \} \right).$$

The following reverse of Jensen's inequality for sublinear functionals was proved by Dragomir, Pearce, and Pečarić [104]:

Let $\phi : [\alpha, \beta] \subset \mathbb{R} \to \mathbb{R}$ be a convex function $(\alpha < \beta)$, $f : E \to [\alpha, \beta]$ such that $\phi \circ f, f \in K$ and $\lambda = \operatorname{sgn}(\phi(\beta) - \phi(\alpha))$. If S is a totally normalised isotonic sublinear functional on K, then

$$S(\phi \circ f) \le \frac{\beta\phi(\alpha) - \alpha\phi(\beta)}{\beta - \alpha} + \frac{|\phi(\beta) - \phi(\alpha)|}{\beta - \alpha} S(\lambda f). \tag{6.182}$$

PROOF   Since ϕ is convex on $[\alpha, \beta]$ we have

$$\phi(v) \le \frac{w - v}{w - u}\phi(u) + \frac{v - u}{w - u}\phi(w),$$

where $u \le v \le w$ and $u < w$.

Set $u = \alpha, v = f(t), w = \beta$. Then

$$\phi(f(t)) \le \frac{\beta - f(t)}{\beta - \alpha}\phi(\alpha) + \frac{f(t) - \alpha}{\beta - \alpha}\phi(\beta), t \in E,$$

or, alternatively,

$$\phi \circ f \le \frac{\beta\phi(\alpha) - \alpha\phi(\beta)}{\beta - \alpha} \cdot 1 + \frac{\phi(\beta) - \phi(\alpha)}{\beta - \alpha} \cdot f.$$

Applying the functional S and using its properties we have

$$
\begin{aligned}
S(\phi \circ f) &\le S\left(\frac{\beta\phi(\alpha) - \alpha\phi(\beta)}{\beta - \alpha} \cdot 1 + \frac{\phi(\beta) - \phi(\alpha)}{\beta - \alpha} \cdot f\right) \\
&= \frac{\beta\phi(\alpha) - \alpha\phi(\beta)}{\beta - \alpha} + S\left(\frac{\phi(\beta) - \phi(\alpha)}{\beta - \alpha} \cdot f\right) \\
&= \frac{\beta\phi(\alpha) - \alpha\phi(\beta)}{\beta - \alpha} + \frac{|\phi(\beta) - \phi(\alpha)|}{\beta - \alpha} S(\lambda f).
\end{aligned}
$$

Hence, the result is proved.   ▯

If $S = A$, and A is a normalised isotonic linear functional, then, by (6.182) we deduce the inequality

$$A(\phi(f)) \le \frac{\{(\beta - A(f))\phi(\alpha) + (A(f) - \alpha)\phi(\beta)\}}{(\beta - \alpha)}.$$

Note that this last inequality is a generalisation of the inequality

$$A(\phi) \le \frac{\{(b - A(l_1))\phi(a) + (A(l_1) - a)\phi(b)\}}{(b - a)}$$

due to Lupaş (see, for instance, Dragomir, Pearce, and Pečarić [104]). Here, $E = [a, b]$ $(-\infty < a < b < \infty)$, L satisfies (L1), (L2), $A : L \to \mathbb{R}$ satisfies

(A1), (A2), $A(1) = 1, \phi$ is convex on E and $\phi \in L, l_1 \in L$, where $l_1(x) = x, x \in [a, b]$.

By the use of Jensen's and Lupaş' inequalities for totally normalised sublinear functionals, we can state the following generalisation of the classical Hermite-Hadamard integral inequality due to Dragomir, Pearce, and Pečarić [104]:

Let $\phi : [\alpha, \beta] \to \mathbb{R}$ be a convex function and $e : E \to [\alpha, \beta]$ a mapping such that $\phi \circ e$ and e belong to K and let $\lambda := \operatorname{sgn}(\phi(\beta) - \phi(\alpha))$. If S is a totally normalised isotonic sublinear functional on K with

$$S(\lambda e) = \lambda \cdot \frac{\alpha + \beta}{2} \text{ and } S(e) = \frac{\alpha + \beta}{2},$$

then we have the inequality

$$\phi\left(\frac{\alpha + \beta}{2}\right) \le S(\phi \circ e) \le \frac{\phi(\alpha) + \phi(\beta)}{2}. \tag{6.183}$$

PROOF The first inequality in (6.183) follows by Jensen's inequality (6.181) applied to the mapping e.

By inequality (6.182), we have

$$S(\phi \circ e) \le \frac{\beta\phi(\alpha) - \alpha\phi(\beta)}{\beta - \alpha} + \frac{(\phi(\beta) - \phi(\alpha))(\beta + \alpha)}{2(\beta - \alpha)}$$

$$= \frac{\phi(\alpha) + \phi(\beta)}{2},$$

and the statement is proved. □

If $S = A, \phi$ is as above and $e : E \to [\alpha, \beta]$ is such that $\phi \circ e, e \in L$ and $A(e) = \frac{\alpha + \beta}{2}$, then the Hermite-Hadamard inequality

$$\phi\left(\frac{\alpha + \beta}{2}\right) \le A(\phi \circ e) \le \frac{\phi(\alpha) + \phi(\beta)}{2}$$

holds for normalised isotonic linear functionals (see also Pečarić and Dragomir [152] and Dragomir [49]).

The following fact may be stated as well [104]:

Let ϕ, f and S be defined as above with $\phi(\beta) \ge \phi(\alpha)$. Then

$$S(\phi(f)) \le \frac{\{(\beta - S(f))\phi(\alpha) + (S(f) - \alpha)\phi(\beta)\}}{\beta - \alpha}. \tag{6.184}$$

Finally, we have the following result [104]:

Let T be an interval which is such that $T \supset \phi([\alpha, \beta])$. If $F(u, v)$ is a real-valued function defined on $T \times T$ and increasing in u, then

$$F[S(\phi(f)), \phi(S(f))] \tag{6.185}$$

$$\leq \max_{x \in [a,b]} F\left[\frac{\beta - x}{\beta - a}\phi(\alpha) + \frac{x - \alpha}{\beta - \alpha}\phi(\beta), \phi(x)\right]$$

$$= \max_{\theta \in [0,1]} F[\theta\phi(\alpha) + (1 - \theta)\phi(\beta), \phi(\theta\alpha + (1 - \theta)\beta)].$$

PROOF By (6.184) and the increasing property of $F(\cdot, y)$ we have

$$F[S(\phi(f)), \phi(S(f))] \leq F\left[\frac{\beta - S(f)}{\beta - a}\phi(\alpha) + \frac{S(f) - \alpha}{\beta - \alpha}\phi(\beta, \phi(S(f)))\right]$$

$$\leq \max_{x \in [a,b]} F\left[\frac{\beta - x}{\beta - a}\phi(\alpha) + \frac{x - \alpha}{\beta - \alpha}\phi(\beta), \phi(x)\right].$$

Equality in (6.185) follows immediately from the change of variable $\theta = \frac{\beta - x}{\beta - a}$, so that $x = \theta\alpha + (1 - \theta)\beta$ with $0 \leq \theta \leq 1$. ∎

Comments

(a) Suppose that $e \in K$, $p \geq 1$, $e^p \in K$, and S is as above. We define the mean

$$L_p(s, e) := [S(e^p)]^{\frac{1}{p}}.$$

By the use of (6.183) we have the inequality

$$A(\alpha, \beta) \leq L_p(s, e) \leq [A(\alpha^p, \beta^p)]^{\frac{1}{p}},$$

provided that

$$S(e) = \frac{\alpha + \beta}{2}.$$

A particular case which generates in its turn the classical L_p-mean is where $S = A$, where A is a linear isotonic functional defined on K.

(b) Now, if $e \in K$ is such that $e^{-1} \in K$, then we define the mean as

$$L(s, e) := [S(e^{-1})]^{-1}.$$

If we assume that $S(-e) = -\frac{\alpha + \beta}{2}$ and $S(e) = \frac{\alpha + \beta}{2}$, then, by (6.183) we have the inequality:

$$H(\alpha, \beta) \leq L(S, e) \leq A(\alpha, \beta).$$

A particular case which generalises in its turn the classical logarithmic mean is where $S = A$, where A is as above.

(c) Finally, if we suppose that $e \in K$ is such that $\ln e \in K$, we can also define the mean

$$I(S, e) := \exp[-S(-\ln e)].$$

Now, if we assume that $S(-e) = -\frac{\alpha+\beta}{2}$ and $S(e) = \frac{\alpha+\beta}{2}$, then, by (6.183) we get the inequality:

$$G(\alpha, \beta) \leq I(S, e) \leq A(\alpha, \beta),$$

which generalises the corresponding inequality for the identric mean.

References

[1] R.P. Agarwal and S.S. Dragomir, An application of Hayashi's inequality for differentiable functions, *Computers Math. Appl.*, 32(6) (1996), 95–99.

[2] D. Andrica and C. Badea, Grüss inequality for positive linear functionals, *Periodica Math. Hungar.*, 19(2) (1988), 155–167.

[3] K.E. Atkinson, *An Introduction to Numerical Analysis,* Wiley and Sons, Second Edition, 1989.

[4] N.S. Barnett, P. Cerone, and S.S. Dragomir, A sharp bound for the error in the corrected trapezoid rule and applications, *Tamkang J. Math.*, 33(3) (2002), 233–253.

[5] N.S. Barnett, P. Cerone, S.S. Dragomir, and A.M. Fink, Comparing two integral means for absolutely continuous mappings whose derivatives are in $L_\infty[a,b]$ and applications, *Comput. Math. Appli.*, 44 (2002), 241–251.

[6] E. Beckenbach and R. Bellman, *An Introduction to Inequalities,* Random House, The L.W. Singer Company, 1961.

[7] R. Bellman, Almost orthogonal series, *Bull. Amer. Math. Soc.*, 50 (1944), 517–519.

[8] E. Berkson, Some types of Banach spaces, Hermitian systems and Bade functionals, *Trans. Amer. Math. Soc.*, 106 (1965), 376–385.

[9] M. Biernacki, Sur une inégalité entre les intégrales due à Tchebyscheff, *Ann. Univ. Mariae Curie-Sklodowska*, A5 (1951), 23–29.

[10] C. Blatter, Zur Riemannschen Geometrie im Grossen auf dem Möbiusband, (German) *Compositio Math.*, 15 (1961), 88–107.

[11] R.P. Boas, A general moment problem, *Amer. J. Math.*, 63 (1941), 361–370.

[12] E. Bombieri, A note on the large sieve, *Acta Arith.*, 18 (1971), 401–404.

[13] N.G. de Bruijn, Problem 12, *Wisk. Opgaven*, 21 (1960), 12–14.

[14] M.L. Buzano, Generalizzazione della diseguaglianza di Cauchy-Schwarz, (Italian), *Rend. Sem. Mat. Univ. e Politech. Torino*, 31 (1971/73), 405–409.

[15] P.L. Čebyšev, O približennyh vyraženijah odnih integralov čerez drugie, *Soobščenija i protokoly zasedaniĭ Matemmatičeskogo občestva pri Imperatorskom Har'kovskom Universitete*, (2) (1882), 93–98; *Polnoe sobranie sočineniĭ P. L. Čebyševa*, Moskva–Leningrad, 1948a, 128–131.

[16] P.L. Čebyšev, Ob odnom rjade, dostavljajušćem predel'nye veličiny integralov pri razloženii podintegral'noĭ funkcii na množeteli, *Priloženi k 57 tomu Zapisok Imp. Akad. Nauk*, (4) (1883); *Polnoe sobranie sočineniĭ P. L. Čebyševa*, Moskva–Leningrad, 1948b, 157–169.

[17] P. Cerone, Multidimensional integration via dimension reduction, *Proceedings of the International Conference on Numerical Analysis and Applied Mathematics*, T. Simos (Ed.), 936 (2007), 682–685.

[18] P. Cerone, Generalised trapezoidal rules with error involving bounds of the nth derivative, *Math. Ineq. Applics.*, 5(3) (2002), 451–462.

[19] P. Cerone, On perturbed trapezoidal and midpoint rules, *Korean J. Comput. Appl. Math.*, 2 (2002), 423–435.

[20] P. Cerone, Approximate multidimensional integration through dimension reduction via the Ostrowski functional, *Nonlinear Func. Anal. Applics.*, 8 (2003), 313–333.

[21] P. Cerone, Bounding the Gini mean difference, *International Series of Numerical Mathematics*, Springer, C. Bandle, L. Losonczi, A. Gilányi, Z. Páles, and M. Plum (Eds.), 157 (2008), 77–89.

[22] P. Cerone, Multidimensional integration via trapezoidal and three point generators, *J. Korean Math. Soc.*, 40(2) (2000), 251–272.

[23] P. Cerone, On an identity for the Chebyshev functional and some ramifications, *J. Inequal. Pure Appl. Math.*, 3(1) (2002), Art. 4. [ONLINE: http://www.emis.de/journals/JIPAM/article157.html?sid=157].

[24] P. Cerone, On odd zeta and other special function bounds, *Inequality Theory and Applications. Vol. 5*, Nova Science Publishers, New York, 2007, 13–31.

[25] P. Cerone, On relationships between Ostrowski, trapezoidal and Chebyshev identities and inequalities, *Soochow J. Math.*, 28(3) (2002), 311–328.

[26] P. Cerone, On some generalisations of Steffensen's inequality and related results, *J. Ineq. Pure Appl. Math.*, 2(3) (2001), Art. 28. [ONLINE: http://www.emis.de/journals/JIPAM/article144.html?sid=144].

[27] P. Cerone, Special functions: Approximations and bounds, *Appl. Anal. Discrete Math.*, 1(1) (2007), 72–91.

[28] P. Cerone, A new Ostrowski type inequality involving integral means over end intervals, *Tamkang J. Math.*, 33(2) (2002), 109–118.

[29] P. Cerone and S.S. Dragomir, A refinement of the Grüss inequality and applications, *Tamkang J. Math.*, 38(1) (2007), 37–49.

[30] P. Cerone and S.S. Dragomir, Lobatto type quadrature rules for functions with bounded derivative, *Math. Ineq. Appl.*, 3(2) (2000), 197–209.

[31] P. Cerone and S.S. Dragomir, On some inequalities for the expectation and variance, *Korean J. Comp. Appl. Math.*, 8(2) (2000), 357–380.

[32] P. Cerone and S.S. Dragomir, Differences between means with bounds from a Riemann-Stieltjes integral, *Comp. Math. Applics.*, 46(2–3) (2003), 445–453.

[33] P. Cerone and S.S. Dragomir, Generalisations of the Grüss, Chebyshev and Lupaş inequalities for integrals over different intervals, *Inter. J. Appl. Math.*, 6(2) (2001), 117–128.

[34] P. Cerone and S.S. Dragomir, Three point quadrature rules involving, at most, a first derivative, *RGMIA Res. Rep. Coll.*, 2(4) (1999), Art. 8. [ONLINE: http://rgmia.org/v2n4.asp].

[35] P. Cerone and S.S. Dragomir, Midpoint type rules from an inequalities point of view, *Handbook of Analytic-Computational Methods in Applied Mathematics*, G.A. Anastassiou (Ed.), CRC Press, New York (2000), 135–200.

[36] P. Cerone and S.S. Dragomir, New upper and lower bounds for the Cebysev functional, *J. Inequal. Pure Appl. Math.*, 3(5) (2002), Art. 77. [ONLINE: http://www.emis.de/journals/JIPAM /article229.html?sid=229].

[37] P. Cerone and S.S. Dragomir, On some inequalities arising from Montgomery's identity, *J. Comput. Anal. Appli.*, 5(4) (2003), 341–367.

[38] P. Cerone and S.S. Dragomir, Trapezoidal-type rules from an inequalities point of view, *Handbook of Analytic-Computational Methods in Applied Mathematics*, G. Anastassiou (Ed.), CRC Press, New York (2000), 65–134.

[39] P. Cerone, S.S. Dragomir, and C.E.M. Pearce, A generalised trapezoid inequality for functions of bounded variation, *Turkish J. Math.*, 24(2) (2000), 147–163.

[40] P. Cerone, S.S. Dragomir, and J. Roumeliotis, Some Ostrowski type inequalities for n-time differentiable mappings and applications, *Demonstratio Mathematica*, 32(4) (1999), 133–138.

[41] P. Cerone, S.S. Dragomir, J. Roumeliotis, and J. Šunde, A new generalisation of the trapezoid formula for n-time differentiable mappings and applications, *Demonstratio Math.*, 33(4) (2000), 719–736.

[42] P. Cerone, J. Roumeliotis, and G. Hanna, On weighted three point quadrature rules, *ANZIAM J.*, 42(E) (2000), C340–361.

[43] P.L. Chebyshev, Sue les expressions approximatives des intégrales définies par les autres prises entre les même limites, *Proc. Math. Soc. Charkov*, 2 (1982), 93–98.

[44] D.E. Daykin, C.J. Eliezer, and C. Carlitz, Problem 5563, *Amer. Math. Monthly*, 75 (1968), p. 198 and 76 (1969), 98–100.

[45] J.B. Diaz and F.T. Metcalf, A complementary triangle inequality in Hilbert and Banach spaces, *Proc. Amer. Math. Soc.*, 17(1) (1966), 88–97.

[46] S.S. Dragomir, A counterpart of Bessel's inequality in inner product spaces and some Grüss type related results, *Bull. Korean Math. Soc.*, 43(1) (2006), 27–41.

[47] S.S. Dragomir, A generalisation of Grüss' inequality in inner product spaces and applications, *J. Math. Anal. Appl.*, 237 (1999), 74–82.

[48] S.S. Dragomir, A generalisation of Kurepa's inequality, *Filomat*, 19 (2005), 7–17.

[49] S.S. Dragomir, A refinement of Hadamard's inequality for isotonic linear functionals, *Tamkang J. Math.* (Taiwan), 24 (1993), 101–106.

[50] S.S. Dragomir, A refinement of Ostrowski's inequality for the Čebyšev functional and applications, *Analysis* (Germany), 23 (2003), 287–297.

[51] S.S. Dragomir, A reverse of the generalised triangle inequality in normed spaces and applications, *RGMIA Res. Rep. Coll.*, 7(2004), Supplement, Art. 15. [ONLINE: `http://rgmia.org/v7(E).asp`].

[52] S.S. Dragomir, Additive reverses of the continuous triangle inequality for Bochner integral of vector valued functions in Hilbert spaces, *Analysis* (Munich), 24(4) (2004), 287–304.

[53] S.S. Dragomir, *Advances in Inequalities of the Schwarz, Grüss and Bessel Type in Inner Product Spaces*, Nova Science Publishers, New York, 2005.

[54] S.S. Dragomir, Better bounds in some Ostrowski-Grüss type inequalities, *Tamkang J. Math.*, 32(3) (2001), 211–216.

[55] S.S. Dragomir, *Discrete Inequalities of the Cauchy - Bunyakovsky - Schwarz Type*, Nova Science Publishers, Inc., Hauppauge, New York, 2004. x+225 pp. ISBN: 1-59454-049-7.

[56] S.S. Dragomir, Generalizations of Precupanu's inequality for orthornormal families of vectors in inner product spaces, *Riv. Mat. Univ. Parma*, (7) 3 (2004), 49–60.

[57] S.S. Dragomir, Grüss type integral inequality for mappings of r-Hölder type and applications for trapezoid formula, *Tamkang J. Math.*, 31(1) (2000), 43–47.

[58] S.S. Dragomir, Inequalities for orthornormal families of vectors in inner product spaces related to Buzano's, Richard's and Kurepa's results, *Tamkang J. Math.*, 37(3) (2006), 227–235.

[59] S.S. Dragomir, Inequalities of Grüss type for the Stieltjes integral and applications, *Kragujevac J. Math.*, 26 (2004), 89–122.

[60] S.S. Dragomir, New estimates of the Čebyšev functional for Stieltjes integrals and applications, *J. Korean Math. Soc.*, 41(2) (2004), 249–264.

[61] S.S. Dragomir, New reverses of Schwarz, triangle and Bessel inequalities in inner product spaces, *Austral. J. Math. Anal. Appl.*, 1(2004), No. 1, Art. 1, [ONLINE: http://ajmaa.org/cgi-bin/paper.pl?string=nrstbiips.tex].

[62] S.S. Dragomir, On Hadamard's inequality for convex functions, *Mat. Balkanica*, 6 (1992), 219–222.

[63] S.S. Dragomir, On some improvements of Čebyšev's inequality for sequences and integrals, *Studia Univ. Babeş-Bolyai, Mathematica*, XXXV (4), (1990), 35–40.

[64] S.S. Dragomir, On the Boas-Bellman inequality in inner product spaces, *Bull. Austral. Math. Soc.*, 69(2) (2004), 217–225.

[65] S.S. Dragomir, On the Bombieri inequality in inner product spaces, *Libertas Math.*, 25 (2005), 13–26.

[66] S.S. Dragomir, On the Cauchy-Buniakowski-Schwartz inequality for real numbers (Romanian), *Caiete Metodico-Stiinţifice*, No. 57, Timisoara University, 1989.

[67] S.S. Dragomir, On the Čebyšev's inequality for weighted means, *Acta Math. Hungar.*, 104(4) (2004), 345–355.

[68] S.S. Dragomir, On the Ostrowski's inequality for Riemann-Stieltjes integral and applications, *Korean J. Comput. Appl. Math.*, 7(3) (2000), 611–627.

[69] S.S. Dragomir, On the Ostrowski inequality for Riemann-Stieltjes integral where f is of Hölder type and u is of bounded variation and applications, *J. Korean Soc. Indust. Appl. Math.*, 5(1) (2001), 35–45.

[70] S.S. Dragomir, On the Ostrowski's integral inequality for mappings of bounded variation and applications, *Math. Ineq. Appl.*, 4(1) (2001), 59–66.

[71] S.S. Dragomir, Refinements of Buzano's and Kurepa's inequalities in inner product spaces, *Facta Univ. (Niš), Ser. Math. Inform.*, 20 (2005), 65–73.

[72] S. S. Dragomir, Refinements of the Hermite-Hadamard integral inequality for convex functions, *Tamsui J. Math. Sci.*, 17(2) (2001), 97–111.

[73] S.S. Dragomir, Refinements of the Schwarz and Heisenberg inequalities in Hilbert spaces, *J. Inequal. Pure Appl. Math.*, 5(3) (2004), Art. 60. [ONLINE: http://www.emis.de/journals/JIPAM /article446.html?sid=446].

[74] S.S. Dragomir, Reverses of Schwarz, triangle and Bessel inequalities in inner product spaces, *J. Inequal. Pure Appl. Math.*, 5(3) (2004), Art. 76. [ONLINE: http://www.emis.de/journals/JIPAM /article432.html?sid=432].

[75] S.S. Dragomir, Reverses of the continuous triangle inequality for Bochner integral of vector valued function in Hilbert spaces, *J. Inequal. Pure Appl. Math.*, 6(2) (2005), Art. 46. [ONLINE: http://www.emis.de/journals/JIPAM/article515.html?sid=515].

[76] S.S. Dragomir, Reverses of the triangle inequality in Banach spaces, *J. Inequal. Pure Appl. Math.*, 6(5) (2005), Art. 129 [ONLINE: http://www.emis.de/journals/JIPAM/article603.html?sid=603].

[77] S.S. Dragomir, *Semi-Inner Products and Applications*, Nova Science Publishers Inc., New York, 2004.

[78] S.S. Dragomir, Sharp bounds of Čebyšev functional for Stieltjes integrals and applications, *Bull. Austral. Math. Soc.*, 67(2) (2003), 257–266.

[79] S.S. Dragomir, Some Grüss type inequalities in inner product spaces, *J. Inequal. Pure Appl. Math.*, 4(2) (2003), Art. 42. [ONLINE: http://www.emis.de/journals/JIPAM/article280.html?sid=280].

[80] S.S. Dragomir, Some inequalities for (m, M)-convex mappings and applications for the Csiszár Φ-divergence in information theory, *Math. J. Ibaraki Univ.*, 33 (2001), 35–50.

[81] S.S. Dragomir, Some inequalities for Riemann-Stieltjes integral and applications, *Optimisation and Related Topics*, Editor: A. Rubinov, Dordrecht/Boston/London, Kluwer Academic Publishers, (2000), 197–235.

[82] S.S. Dragomir, Some inequalities of midpoint and trapezoid type for the Riemann-Stieltjes integral, *Nonlinear Anal.*, 47(4) (2001), 2333–2340

[83] S.S. Dragomir, Some integral inequality of Grüss type, *Indian J. Pure Appl. Math.*, 31(4) (2000), 397–415.

[84] S.S. Dragomir, Some properties of quasilinearity and monotonicity of Hölder and Minkowski's inequality, *Tamkang J. Math.*, 26(5) (1999), 21–24.

[85] S. S. Dragomir, Some remarks on Hadamard's inequality for convex functions, *Extracta Math.* (Spain), 9(2) (1994), 88–94.

[86] S.S. Dragomir, Some reverses of the generalised triangle inequality in complex inner product spaces, *Linear Algebra Appl.*, 402 (2005), 245–254.

[87] S.S. Dragomir, The Ostrowski's integral inequality for Lipschitzian mappings and applications, *Comput. Math. Appli.*, 38 (1999), 33–37.

[88] S.S. Dragomir, P. Cerone, and A. Sofo, Some remarks on the midpoint rule in numerical integration, *Studia Univ. Babeş-Bolyai, Math.*, XLV(1) (2000), 63–74.

[89] S.S. Dragomir, P. Cerone, and A. Sofo, Some remarks on the trapezoid rule in numerical integration, *Indian J. Pure Appl. Math.*, 31(5) (2000), 475–494.

[90] S.S. Dragomir and I. Fedotov, An inequality of Grüss type for Riemann-Stieltjes integral and applications for special means, *Tamkang J. Math.*, 29(4) (1998), 287–292.

[91] S.S. Dragomir and I. Fedotov, A Grüss type inequality for mappings of bounded variation and applications to numerical analysis, *Non. Funct. Anal. Appl.*, 6(3) (2001), 425–437.

[92] S.S. Dragomir and S. Fitzpatrick, The Hadamard's inequality for s-convex functions in the first sense, *Demonstratio Math.*, 31(3) (1998), 633–642.

[93] S.S. Dragomir and S. Fitzpatrick, The Hadamard inequalities for s-convex functions in the second sense, *Demonstratio Math.*, 32(4) (1999), 687–696.

[94] S.S. Dragomir and C.J. Goh, A counterpart of Jensen's discrete inequality for differentiable convex mappings and applications in information theory, *Math. Comput. Modelling*, 24(2) (1996), 1–11.

[95] S.S. Dragomir and N.M. Ionescu, Some refinements of Cauchy-Buniakowski-Schwartz inequality for sequences, *Symp. Math. Appl.*, Timişoara, 3–4 Nov. 1989, pp. 79–82.

[96] S.S. Dragomir and N.M. Ionescu, Some converse of Jensen's inequality and applications, *Anal. Numer. Theor. Approx.*, 23 (1994), 71–78.

[97] S.S. Dragomir and N.M. Ionescu, Some integral inequalities for differentiable convex functions, *Coll. Pap. of the Fac. of Sci. Krogujevac* (Yugoslavia), 13 (1992), 11–16.

[98] S.S. Dragomir and B. Mond, On the superadditivity and monotonicity of Schwartz's inequality in inner product spaces, *Contributions, Macedonian Acad. Sci. Arts,* 15(2) (1994), 5–22.

[99] S.S. Dragomir and B. Mond, Integral inequalities of Hadamard type for log-convex functions, *Demonstratio Math.,* 31(2) (1998), 354–364.

[100] S.S. Dragomir and B. Mond, On the Boas-Bellman generalisation of Bessel's inequality in inner product spaces, *Italian J. Pure Appl. Math.,* 3 (1998), 29–35.

[101] S.S. Dragomir, B. Mond, and J.E. Pečarić, Some remarks on Bessel's inequality in inner product spaces, *Studia Univ. Babeş-Bolyai, Mathematica,* 37(4) (1992), 77–86.

[102] S.S. Dragomir and C.E.M Pearce, Quasi-convex functions and Hadamard's inequality, *Bull. Australian Math. Soc.,* 57(1998), 377–385.

[103] S.S. Dragomir and C.E.M. Pearce, *Selected Topics on Hermite-Hadamard Inequalities and Applications,* RGMIA Monographs, Victoria University, Melbourne, Victoria, Australia, 2000. [ONLINE: `http://rgmia.org/monographs/hermite_hadamard.html`].

[104] S.S. Dragomir, C.E.M. Pearce, and J.E. Pečarić, On Jensen's and related inequalities for isotonic sublinear functionals, *Acta. Sci. Math.* (Szeged), 61 (1995), 373–382.

[105] S.S. Dragomir, C.E.M. Pearce, and J. Šunde, Abel-type inequalities for complex numbers and Gauss-Polya type integral inequalities, *Math. Comm.,* 3 (1998), 99–101.

[106] S.S. Dragomir and E. Pearce, A refinement of the second part of the Hadamard inequality, *The 6th Symp. of Math. and Its Appl.,* Technical University of Timisoara, November 1995.

[107] S.S. Dragomir and J.E. Pečarić, Refinements of some inequalities for isotonic functionals, *Anal. Num. Theor. Approx.,* 18(1) (1989), 61–65.

[108] S.S. Dragomir, J.E. Pečarić, and L.E. Persson, Some inequalities of Hadamard type, *Soochow J. Math.* (Taiwan), 21 (1995), 335–341.

[109] S.S. Dragomir and T.M. Rassias (Eds.), *Ostrowski Type Inequalities and Applications in Numerical Integration,* Kluwer Academic Publishers, Dordrecht, 2002.

[110] S.S. Dragomir and J. Sándor, On Bessel's and Grüss' inequality in pre-hilbertian spaces, *Periodica Math. Hungar.,* 29(3) (1994), 197–205.

[111] S.S. Dragomir and S. Wang, Applications of Ostrowski's inequality to the estimation of error bounds for some special means and for some numerical quadrature rules, *Appl. Math. Lett.,* 11 (1998), 105–109.

[112] S.S. Dragomir and S. Wang, An inequality of Ostrowski-Grüss type and its applications to the estimation of error bounds for some special means and for some numerical quadrature rules, *Comput. Math. Applic.*, 33 (1997), 15–22.

[113] S.S. Dragomir and S. Wang, A new inequality of Ostrowski's type in L_p-norm, *Indian J. Math.*, 40(3) (1998), 299–304.

[114] S.S. Dragomir and S. Wang, A new inequality of Ostrowski's type in L_1-norm and applications to some special means and to some numerical quadrature rules, *Tamkang J. Math.*, 28(3) (1997), 239–244.

[115] S.S. Dragomir and S. Wang, Applications of Iyengar's type inequalities to the estimation of error bounds for the trapezoidal quadrature rule, *Tamkang J. Math.*, 29(1) (1998), 55–58.

[116] C.F. Dunkl and K.S. Williams, A simple inequality, *Amer. Math. Monthly*, 71 (164), 53–54.

[117] M. Fujii and F. Kubo, Buzano's inequality and bounds for roots of algebraic equations, *Proc. Amer. Math. Soc.*, 117(2) (1993), 359–361.

[118] H. Gauchman, A Steffensen type inequality, *J. Ineq. Pure and Appl. Math.*, 1(1) (2000), Art. 3. [ONLINE: http://www.emis.de/journals /JIPAM/article96.html?sid=96].

[119] G. Grüss, Über das Maximum des absoluten Betrages von $\frac{1}{b-a}\int_a^b f(x)g(x)\,dx - \frac{1}{(b-a)^2}\int_a^b f(x)\,dx \int_a^b g(x)\,dx$, *Math. Z.*, 39 (1934), 215–226.

[120] S. Gudder and D. Strawther, Strictly convex normed linear spaces, *Proc. Amer. Math. Soc.*, 59(2) (1976), 263–267.

[121] G.H. Hardy, J.E. Littlewood, and G. Pólya, *Inequalities*, 1st Ed. and 2nd Ed. Cambridge University Press, Cambridge, England, (1934, 1952).

[122] H. Heilbronn, On the averages of some arithmetical functions of two variables, *Mathematica*, 5(1958), 1–7.

[123] H. Hudzik and L. Maligranda, Some remarks on s-convex functions, *Preprint, LULEÅ University of Technology*, 1992, Sweden.

[124] K.S.K. Iyengar, Note on an inequality, *Math. Student*, 6 (1938), 75–76.

[125] J. Karamata, Inégalités relatives aux quotients et à la difference de $\int fg$ et $\int f \int g$, *Acad. Serbe Sci. Publ. Inst. Math.*, 2 (1948), 131–145.

[126] J. Karamata, *Teorija i Praksa Stieltjesova Integrala* (Serbo-Croatian) (Stieltjes Integral, Theory and Practice), SANU, Posebna izdanja, 154, Beograd, 1949.

[127] H. Kenyon, Note on convex functions, *Amer. Math. Monthly,* 63 (1956), 107.

[128] V.L. Klee, Solution of a problem of E.M. Wright on convex functions, *Amer. Math. Monthly,* 63 (1956), 106–107.

[129] A.N. Korkine, Sur une théorème de M. Tchebychef, *C.R. Acad. Sci. Paris,* 96 (1883), 326–327.

[130] S. Kurepa, On the Buniakowsky-Cauchy-Schwarz inequality, *Glasnick Mathematički,* 1(21)(2) (1966), 147–158.

[131] Z. Liu, Note on a theorem of P. Cerone, *Soochow J. Math.,* 33(3) (2006), 323–326.

[132] G. Lumer, Semi-inner product spaces, *Trans. Amer. Math. Soc.,* 100 (1961), 29–43.

[133] A. Lupaş, A remark on the Schweitzer and Kantorovich inequalities, *Univ. Beograd Publ. Elektrotehn. Fak. Ser. Mat. Fiz.,* No. 381–409 (1972), 13–15.

[134] A. Lupaş, On two inequalities of Karamata, *Univ. Beograd Publ. Elektrotehn. Fak. Ser. Mat. Fiz.,* No. 602–633 (1978), 119–123.

[135] M. Matić, J.E. Pečarić, and N. Ujević, On new estimation of the remainder in generalised Taylor's formula, *Math. Ineq. Appl.,* 2(3) (1999), 343–361.

[136] W. Matuszewska and W. Orlicz, A note of the theory of s-normed spaces of ψ-integrable functions, *Studia Math.,* 21 (1961), 107–115.

[137] G.V. Milovanović and J.E. Pečarić, Some considerations of Iyengar's inequality and some related applications, *Univ. Beograd. Publ. Elektrotehn Fak. Ser. Mat. Fiz.,* No. 544–576 (1976), 166–170.

[138] D.S. Mitrinović, *Analytic Inequalities,* Springer Verlag, Berlin, 1970.

[139] D.S. Mitrinović, The Steffensen inequality, *Univ. Beograd. Publ. Elektrotehn. Fak. Ser. Mat. Fiz.,* No. 247–273 (1969), 1–14.

[140] D.S. Mitrinović and J.E. Pečarić, On an identity of D.Z. Djoković, *Prilozi Mak. Akad. Nauk. Umj. (Skopje),* 12(1) (1991), 21–22.

[141] D.S. Mitrinović, J.E. Pečarić, and A.M. Fink, *Classical and New Inequalities in Analysis,* Kluwer Academic Publishers, Dordrecht/Boston/London.

[142] D.S. Mitrinović, J.E. Pečarić, and A.M. Fink, *Inequalities for Functions and Their Integrals and Derivatives,* Kluwer Academic Publishers, Dordrecht/Boston/London, 1994.

[143] M.H. Moore, An inner product inequality, *SIAM J. Math. Anal.*, 4(3) (1973), 514–518.

[144] J. Musielak, Orlicz Spaces and Modular Spaces, *Lecture Notes in Mathematics*, 1034, Springer-Verlag, New York/Berlin, 1983.

[145] W. Orlicz, A note on modular spaces I, *Bull. Acad. Polon Sci. Ser. Math. Astronom. Phys.*, 9 (1961), 157–162.

[146] A.M. Ostrowski, On an integral inequality, *Aequat. Math.*, 4 (1970), 358–373.

[147] A. Ostrowski, Uber die Absolutabweichung einer differentienbaren Funktionen von ihren Integralmittelwert, *Comment. Math. Hel*, 10 (1938), 226–227.

[148] T.C. Peachey, A. McAndrew, and S.S. Dragomir, The best constant in an inequality of Ostrowski type, *Tamkang J. Math.*, 30(3) (1999), 219–222.

[149] J.E. Pečarić, On some classical inequalities in unitary spaces, *Mat. Bilten* (Scopje), 16(1992), 63–72.

[150] J.E. Pečarić and S.S. Dragomir, Some remarks on Čebyšev's inequality, *L'Anal. Num. Théor de L'Approx.*, 19(1) (1990), 58–65.

[151] J.E. Pečarić, F. Proschan, and Y.L. Tong, *Convex Functions, Partial Orderings and Statistical Applications*, Academic Press, Inc., Boston/San Diego/New York, 1991.

[152] J.E. Pečarić and S.S. Dragomir, A generalization of Hadamard's integral inequality for isotonic linear functionals, *Rudovi Mat.* (Sarajevo), 7(1991), 103–107.

[153] G. Pólya and G. Szegö, *Problems and Theorems in Analysis*, Volume 1: Series, Integral Calculus, Theory of Functions (in English), translated from German by D. Aeppli, corrected printing of the revised translation of the fourth German edition, Springer Verlag, New York, 1972.

[154] T. Precupanu, On a generalisation of Cauchy-Buniakowski-Schwarz inequality, *Anal. St. Univ. "Al. I. Cuza" Iaşi*, 22(2) (1976), 173–175.

[155] F. Qi, Further generalisations of inequalities for an integral, *Univ. Beograd. Publ. Elektrotehn Fak. Ser. Mat.*, 8 (1997), 79–83.

[156] U. Richard, Sur des inégalités du type Wirtinger et leurs application aux équations différentielles ordinaires, Collquium of Analysis held in Rio de Janeiro, August, 1972, pp. 233–244.

[157] J. Roumeliotis, P. Cerone, and S.S. Dragomir, An Ostrowski type inequality for weighted mappings with bounded second derivative, *J. Korean Soc. Indust. Appl. Math.*, 3(2) (1999), 107–118.

[158] J. Sándor, Some integral inequalities, *El. Math.*, 43 (1988), 177–180.

[159] P. Schweitzer, An inequality about the arithmetic mean (Hungarian), *Math. Phys. Lapok* (Budapest), 23 (1914), 257–261.

[160] M.S. Slater, A companion inequality to Jensen's inequality, *J. Approx. Theory*, 32 (1981), 160–166.

[161] N. Ja. Sonin, O nekotoryh neravenstvah, ostnosjaščihsja k oprede-lennym integralam, *Zap. Imp. Akad. Nauk po Fiziko-Matem. Otd.*, 6 (1898), 1–54.

[162] J.F. Steffensen, On certain inequalities between mean values, and their applications to actuarial problems, *Strand. Aktuarietidskrift*, 82–97.

[163] N. Ujević, A generalisation of Grüss inequality in prehilbertian spaces, *Math. Ineq. Appl.*, 6(4) (2003), 617–623.

[164] P.M. Vasić and J.E. Pečarić, Note on the Steffensen inequality, *Univ. Beograd. Publ. Elektrotechn. Fak. Ser. Mat. Fiz.*, No. 716–734, 80–82.

[165] G.S. Watson, G. Hlpargu, and G.P.H. Styan, Some comments on six inequalities associated with the inefficiency of ordinary least squares with one regressor, *Linear Algebra Appl.*, 264 (1997), 13–54.

[166] A. Witkowski, On Young's inequality, *J. Inequal. Pure Appl. Math.*, 7(5) (2006), Art. 164. [ONLINE: http://www.emis.de/journals/JIPAM/article782.html?sid=782].

[167] E.M. Wright, An inequality for convex functions, *Amer. Math. Monthly*, 61 (1954), 620–622.

Index

9 780367 383275